Avril Robarts LRC

Liverpool John Moores University

A SHORT COURSE IN
GEOTECHNICAL SITE INVESTIGATION

NOEL SIMONS, BRUCE MENZIES and MARCUS MATTHEWS

First published by Thomas Telford Publishing, Thomas Telford Ltd, 1 Heron Quay, London E14 4JD.
URL: http://www.thomastelford.com

Distributors for Thomas Telford books are
USA: ASCE Press, 1801 Alexander Bell Drive, Reston, VA 20191-4400, USA
Japan: Maruzen Co. Ltd, Book Department, 3–10 Nihonbashi 2-chome, Chuo-ku, Tokyo 103
Australia: DA Books and Journals, 648 Whitehorse Road, Mitcham 3132, Victoria

Cover photograph: rock fall at Charmouth, Dorset, UK, by kind permission of Charmouth Heritage Coast Centre.
Photographer: Adrian Adams, © Charmouth Heritage Coast Centre

A catalogue record for this book is available from the British Library

ISBN: 07277 2948 9

© The Authors, and Thomas Telford Limited 2002.

All rights, including translation, reserved. Except as permitted by the Copyright, Designs and Patents Act 1988, no part of this publication may be reproduced, stored in a retrieval system or transmitted in any form or by any means, electronic, mechanical, photocopying or otherwise, without the prior written permission of the Publishing Director, Thomas Telford Publishing, Thomas Telford Ltd, 1 Heron Quay, London E14 4JD.

This book is published on the understanding that the authors are solely responsible for the statements made and opinions expressed in it and that its publication does not necessarily imply that such statements and/or opinions are or reflect the views or opinions of the publishers. While every effort has been made to ensure that the statements made and the opinions expressed in this publication provide a safe and accurate guide, no liability or responsibility can be accepted in this respect by the authors or publishers.

Typeset by Academic + Technical, Bristol
Printed and bound in Great Britain by MPG Books, Bodmin

Preface

A Short Course in Geotechnical Site Investigation is based on University of Surrey short courses. These courses were designed not only to familiarize students with the practicalities of geotechnical engineering but also to refresh the knowledge of practising engineering geologists and civil and structural engineers.

To provide the necessary focus required by this short course (and therefore short) book, we have tried to minimize detail so that the fundamental principles of geotechnical site investigation will stand out. Sometimes, however, the subject considered was of such importance (e.g. desk study) or complexity (e.g. parameter determination) that we have not been as short as we had hoped. For this we apologize to the reader! Because of space considerations, geoenvironmental site investigation is not considered here, and we have also restricted our remit to the *planning* of the geotechnical site investigation rather than its execution.

To quickly familiarise the reader with the principles and realities of geotechnical site investigation, we begin in Chapter 1 with how to plan a site investigation, develop the geological model and carry out conceptual design. This is followed in Chapter 2 by the desk study and the walk-over survey. The nature of geotechnical hazards and the management of geotechnical risk, including the role of the risk register, are explained in Chapter 3. We then consider modern and classic means of parameter determination in soils in Chapter 4, again with a somewhat restricted remit. Some very brief 'hand written' short course notes are included in Chapter 4. We conclude with an appendix that presents a reproduction of *Writing reports* by the late David Palmer – a timeless monograph published by Soil Mechanics Ltd in 1957. We also append to the inside back cover of this book copies of *Identification and Description of Soils* and *Identification and Description of Rocks* supplied by Environmental Services Group Limited. This card is designed to be taken into the field during the walk-over survey.

Frequently we refer the reader to *Site Investigation*, 2nd Edition, by Clayton, Matthews and Simons (1995) for particular detail on execution, methods, equipment and interpretation. For considerations of soil mechanics, rock mechanics, foundation engineering and slope stability problems, we also refer the reader to the companion volumes in the Short Course Series: *A short course in foundation engineering*, 2nd Edition, by Simons and Menzies (2000) and *A short course in soil and rock slope engineering* by Simons, Menzies and Matthews (2001). For instrumentation, we refer the reader to *Geotechnical instrumentation for monitoring field performance* by Dunnicliff with Green (1988).

For more general topics, we mention BS5930; 1999 *Code of practice for site investigations*, *Inadequate Site Investigation* (ICE, 1991), *Without site investigation ground is a hazard* by Littlejohn, Cole and Mellors (1994), and *Eurocode 7: A Commentary* by Simpson and Driscoll (1998).

We make reference to these other sources not only because of the impossibility of covering these important aspects of site investigation in sufficient detail in a book of this size to be useful, but also because of our remit to concentrate on those topics essential to the *planning* of a site investigation.

Above all, it is the prime aim of this book, as with the others in our Short Course Series, to provide geotechnical engineers and engineering geologists with the means to check that ground properties (inputs) and design predictions (outputs) collectively pass the 'sanity test'!

Noel Simons, Bruce Menzies, Marcus Matthews
University of Surrey
Guildford 2002

Acknowledgements

We particularly thank the following.

- Chris Clayton for his help and advice and for generously giving permission to make verbatim extracts from his published papers and unpublished notes.
- David Hight for generously giving permission to make wide-ranging verbatim extracts from his published papers.
- Andrew Bowden for his guidance from the perspective of a site investigation practitioner.
- Robert Whittle who kindly contributed the section on pressuremeter testing.
- Sylvia Palmer, Peter Eldred, Soil Mechanics Ltd and Environmental Services Group Limited, for permission to reproduce *Writing reports* by D.J. Palmer.
- Environmental Services Group Limited for copies of their card included with this book: *Identification and Description of Soils* and *Identification and Description of Rocks.*

We warmly thank our colleagues at the University of Surrey for all their support over many years: Mike Huxley, Ab Tarzi, Mike Gunn, Vicki Hope, Rick Woods, Stephan Jefferis and Emma Hellawell. We are most grateful for the help and advice of Volker Berhorst, Paul Mayne, Martin Fahey, Serge Leroueil, Christos Vrettos, Eddie Bromhead, Andrew Malone, Leong Eng Choon, Clive Dalton, Tim Spink, Tony Bracegirdle, Keith Gabriel, Mike Newton, John Chantler, Colin Warren, Martin Culshaw, Tony Cooper, Peter Jackson, Raymond Coe, Matthew Duthy, Andy Powell, Sylvia Palmer, David Churcher, Fin Jardine, Alex Woodcraft, Jim Hall, Patrick Godfrey, Ian Cruickshank, Geoff Davis, Alan Clark, Jerry Sutton, Richard Tinsley, Alan Moxhay, Gordon Wilson and Chris Pamplin. Of course, the comments expressed in this book are those of the authors and do not necessarily reflect the views of any of the above. We also gratefully acknowledge the important contributions to this book of past research students: Ayad Madhloom, Peter Wilkes, Malcolm Roy, Howard Sutton,

Caesar Merrifield, Ray Telling, John Gumbel, Joachim Rodriques, Jarbas Milititsky, Kyriacos Kyrou and the late Nick Kalteziotis. We are also most grateful to Margaret Harris who drew many of the original figures.

We acknowledge permissions from the following.

- Prof. Chris Clayton and Thomas Telford Publishing Ltd, to make verbatim extracts from
  - Clayton, C.R.I., Siddique, A. and Hopper, R.J. (1998). Effects of sampler design on tube sampling disturbance – numerical and analytical investigations. *Géotechnique*, 48, No. 6, 847–867.
  - Clayton, C.R.I. and Siddique, A. (1999). Tube sampling disturbance – forgotten truths and new perspectives. *Proc. Instn Civ. Eng. Geotech. Eng.*, 137, July, 127–135.
  - Clayton, C.R.I. (2001). Managing geotechnical risk: time for a change? *Proc. ICE, Geotechnical Engineering*. Paper 149. Pages 3–11.
- Dr David Hight, to make verbatim extracts from
  - Hight, D.W. (1996). Moderators' report on Session 3: drilling, boring, sampling and description. *Advances in Site Investigation Practice*. Thomas Telford, London.
  - Hight, D.W. (2000). Sampling methods: evaluation of disturbance and new practical techniques for high quality sampling in soils. Keynote Lecture, *Proc. 7th Nat. Cong. of the Portuguese Geotech. Soc.*, Porto.
  - Hight, D.W., Hamza, M.M. and El Sayed, A.S. (2000). Engineering characterisation of the Nile Delta clays. *Proc. Conf. IS Yokohama 2000. Coastal Geotechnical Engineering in Practice*. To be published in Volume 2.
- Dr Jim Hall, Ian Cruickshank, Patrick Godfrey and Thomas Telford Publishing Ltd, to make extracts from
  - Hall, J.W., Cruickshank, I.C. and Godfrey, P.S. (2001). Software-supported risk management for the construction industry. *Proc. ICE, Civil Engineering*. Paper 12272. Pages 42–48.
- Prof. Martin Fahey, to reproduce text and figures from his presentation 'Measuring soil stiffness for settlement prediction' to Session 1.2 of the 15th International Conference on Soil Mechanics and Geotechnical Engineering, Istanbul, August 2001.
- Prof. Paul Mayne, to reproduce text and figures from his presentation 'Soil property characterisation by insitu tests' to Session 1.2 of the 15th International Conference on Soil Mechanics and Geotechnical Engineering, Istanbul, August 2001.

- Norwest Holst Soil Engineering Ltd, to reproduce photographs of site investigation equipment.
- GeoMil Equipment B.V., to reproduce photographs and text and drawings on penetration testing equipment.
- Cambridge Insitu Ltd, to reproduce photographs and text and drawings of pressuremeter testing equipment.
- GDS Instruments Ltd, to reproduce photographs of triaxial test and continuous surface wave equipment and accessories.
- CIRIA, to make extracts from 'RiskCom: Software tool for managing and communicating risks'.
- Charmouth Heritage Coast Centre, to reproduce photos of a rock fall at Charmouth, Dorset, England. Copyright: Charmouth Heritage Coast Centre. Photographer Adrian Adams.
- The Chief Engineer for the Somerset sub-unit of the South-Western Road Construction Unit for data on the Brent Knoll trial embankment.
- Gordon Cantlay and Oscar Faber & Partners, to use data from the investigation for Necom House, Lagos.
- Prof. A. Viana da Fonceca and Prof Luís de Sousa, editors of the Proceedings of the 7th National Congress of the Portuguese Geotechnical Society, Porto, (VII Congresso Nacional de Geotecnia 2000) to make extracts from Hight, D.W. (2000). Sampling methods: evaluation of disturbance and new practical techniques for high quality sampling in soils. Keynote Lecture, *Proc. 7th Nat. Cong. of the Portuguese Geotech. Soc.*, Porto.
- Swets & Zeitlinger Publishers and A.A. Balkema Publishers to make extracts from Hight, D.W., Hamza, M.M. and El Sayed, A.S. (2000). Engineering characterisation of the Nile Delta clays. *Proc. Conf. IS Yokohama 2000. Coastal Geotechnical Engineering in Practice*. To be published in Volume 2.
- Taylor & Francis Books Ltd and Professor E.N. Bromhead to make use of text and figures from approximately two pages of text and two figures from *The Stability of Slopes*, 2nd edition, 1992, by E.N. Bromhead, Blackie Academic.
- Sealand Aerial Photography, to reproduce the aerial photograph of Littleheath Road, Fontwell, showing sinkholes.

Contents

	List of case studies	xi
Chapter 1	**Planning and conceptual design**	**1**
	Overview	1
	Introduction	3
	Key attributes of successful geotechnical engineering practice	6
	Key elements in planning the investigation	8
Chapter 2	**The desk study and walk-over survey**	**68**
	What is the desk study?	68
	Why do desk studies?	68
	Overview of how a desk study is done	73
	What to look for in a desk study	75
	Aerial photography	105
	Overview of the walk-over survey	117
	Equipment needed for a walk-over survey	117
	Feature identification	118
	Local enquiries	120
Chapter 3	**Geotechnical hazards and risk management**	**122**
	Overview	122
	Hazards from natural and man-made materials	123
	Geotechnical hazards	127
	Managing geotechnical risk	139
	Why ground conditions are a risk	146
	Problems with the current approach	148
	Eliminating uncertainty – new methods of working	151
Chapter 4	**Parameter determination: classic and modern methods**	**165**
	Overview: key terminology, parameters and test types	165
	Milestones in research: the past 30 years	167

Introduction to key laboratory and field methods 169
Laboratory tests 170
Field tests 220

Appendix 1 Recommended list of units 292

Appendix 2 *Writing reports* by David Palmer 295

References and bibliography 321

Index 341

List of case studies

1. *Underground car park at the House of Commons, London* 20
Demonstrates the limitations of a classic site investigation. Design parameters adopted were not those resulting from the site investigation but were based on field measurements of the deformations of the ground around a similar structure also in London Clay.
2. *140 m high office block, Necom House, Lagos, Nigeria* 24
Demonstrates use of simple classification tests that indicated overconsolidated underlying clays allowing a shallow raft foundation design following densification of upper loose sands using compaction piles.
3. *Runway at Fornebu Airport, Oslo* 29
Demonstrates the scope and content of a good site investigation. Feasibility of sand drains investigated to speed up settlements of embankment on soft clay.
4. *New quay wall at East Port Said, Egypt* 35
Demonstrates the scope and content of a good modern site investigation. Use of piezocone for soil profiling and high quality sampling and sample assessment to demonstrate Nile delta clays are lightly overconsolidated when poor samples data excluded thus removing a prevailing view that the clays were underconsolidated.
5. *Short term failure of a cutting in London Clay, Bradwell* 41
Demonstrates dangers of using shear strengths exceeding 50% of the values measured in conventional undrained triaxial tests on 38 mm diameter samples.
6. *Short term failure of a cut-off trench in London Clay, Wraysbury* 42
Demonstrates effects of strength anisotropy and size of test specimen on measurements of undrained strength.
7. *Failure of a natural slope in soft intact clay, Drammen, Norway* 50
Demonstrates the scope and content of a good site investigation. Good agreement between field vane and lab

measurements of undrained strength. Piezometer observations unusually show pore pressures were hydrostatic.

8. *Ground improvement by stone columns, Hoeidah, Yemen Arab Republic* 61
Demonstrates decision to use stone columns more cost effective than piling even when the number of columns had to be doubled because of unexpected silt layers.

9. *Lewisham extension to the Docklands Light Railway, London* 69
Demonstrates the type and range of information that can be available to the desk study.

10. *Redevelopment of a site, Milton Keynes, UK* 85
Demonstrates the use of topographical and geological maps in the desk study.

11. *Redevelopment of a site, Bristol, UK* 87
Demonstrates the use of geological maps in the desk study.

12. *Biddulph Moor, UK* 92
Demonstrates the use of historical maps in determining the site history.

13. *Basildon, UK* 95
Demonstrates the use of aerial photography to study site history.

14. *Paddington Station, UK* 95
Demonstrates the type and range of information that can be useful for studying site history.

15. *Stag Hill, Guildford, UK* 101
Demonstrates the use of aerial photography for the identification of slope instability.

16. *Sevenoaks by-pass, UK* 102
Demonstrates the use of aerial photography for the identification and mapping the extent of slope instability.

17. *Subsidence from sinkholes, Fontwell, West Sussex, UK* 127
Demonstrates the role of aerial photography for identifying collapse subsidence ('sinkholes') caused by groundwater.

18. *Bang Bo test excavation, Thailand* 191
Demonstrates deterioration of soil strength by disturbance of heavy equipment.

19. *Compensation grouting in Singapore marine clay* 191
Demonstrates destructuration by grouting leading to either no heave or settlement.

20. *Brent Knoll trial embankment, UK* 226
Demonstrates prediction of field failure by using shear vane data that was empirically corrected as well as adjusted for strength anisotropy.

21. *Kilburn, London* 260
Demonstrates determination of ground parameters of London Clay from pressuremeter testing.
22. *Quarry rehabilitation, Swanscombe, UK* 288
Demonstrates ground improvement by stone columns and with time from installation of the stone columns by using continuous surface wave stiffness profiling.
23. *Industrial development, Basildon, UK* 288
Demonstrates ground improvement by stone columns by using continuous surface wave stiffness profiling.

CHAPTER ONE

Planning and conceptual design

Overview

Aims and content of this book

When planning this book, we sought advice from site investigation practitioners. In particular, we asked them to tell us what young civil engineers and engineering geologists would want from our book; the following reply from Andrew Bowden was adopted as our theme.

Comment (Andrew Bowden, Mouchel Consulting Ltd, UK)

What all young engineering geologists ask me is: how do you plan a site investigation? How many holes, where, how deep, what samples and why? What to test them for and how many tests should be done? How do you know if the tests are correct? My reply to them is that you first have to learn some site investigation. Then learn some design and then some more site investigation until the picture emerges. You cannot plan a site investigation until you understand what is needed for design and you cannot understand what is needed for design until you know about soil properties which you understand by doing site investigation and getting dirty. .

Your book should explain how the site investigation elements interact and connect together. Start with desk studies to formulate a ground model. After that do outline design in order to find out what will matter from the site investigation e.g. what matters for spread foundations (the 2B rule), piles, and slopes? Cover effective stress, groundwater, excavations and earthworks. Refer to your book A Short Course in Foundation Engineering *for the detail. Then plan the site investigation to test the ground model – geological control holes then filling in the information between them. Follow this with analysis of the geology – this is critically important.*

Next classification of the soils – include weathering scales and principles of description and classification. After that: planning the analysis of the soil properties, using cross checks and rules of thumb to verify one test with another (give the reader a lot of index parameters and graphs of correlations). Then report writing –

> *structure and what matters, followed by verification of SI during construction – making sure the assumptions and deductions (e.g. from the desk study) on which the SI was based are correct by monitoring of instruments, checking the soil encountered, etc. Include plenty of handy hints and reference tables but not too much detail – give references to easily available up-to-date books and codes (not obscure or exotic ones please).*
>
> *Leave out contamination – that would be the subject for a book itself. The whole book should be no bigger than* A Short Course in Foundation Engineering *and also be affordable.*

Accordingly, we have written this book concentrating on *how to plan a site investigation*. We saw this restriction of content as necessary to our remit of writing a short course (and therefore short) book.

To quickly familiarize the reader with the principles and realities of geotechnical site investigation, we begin in Chapter 1 with how to plan a site investigation, develop the geological model, and carry out conceptual design. This is followed in Chapter 2 by descriptions of the desk study and the walk-over survey. The nature of geotechnical hazards and the management of geotechnical risk and the role of the risk register are explained in Chapter 3. We then consider modern and classic means of parameter determination in soils in Chapter 4, again with a somewhat restricted remit. Some very brief 'hand written' short course notes are included in Chapter 4. We conclude with a reproduction of *Writing reports* by the late David Palmer – a timeless monograph published by Soil Mechanics Ltd in 1957. We also include with this volume a copy of *Identification and Description of Soils* and *Identification and Description of Rocks* supplied by Environmental Services Group Limited. This card is designed to be taken into the field during the walk-over survey.

Frequently, we refer the reader to *Site Investigation*, Second Edition, by Clayton *et al.* (1995b) for particular detail on execution, methods, equipment and interpretation. For considerations of soil mechanics, rock mechanics, foundation engineering and slope stability problems, we also refer the reader to the companion volumes in the Short Course Series: *A Short Course in Foundation Engineering*, Second Edition, by Simons and Menzies (2000) and *A Short Course in Soil and Rock Slope Engineering* by Simons *et al.* (2001). For instrumentation, we refer the reader to *Geotechnical Instrumentation for Monitoring Field Performance* by Dunnicliff and Green (1988).

For more general topics, we mention BS 5930: 1999 Code of Practice for Site Investigations (British Standards Institution, 1999); *Inadequate Site Investigation* (ICE, 1991); *Without Site Investigation Ground is a Hazard* by Littlejohn *et al.* (1994); and *Eurocode 7: A Commentary* by Simpson and Driscoll (1998). We make reference to these other sources not only because of the impossibility of covering these important aspects of site investigation in sufficient detail in a book of this size to be useful, but also because of our remit to concentrate on those topics essential to the *planning* of a site investigation.

Introduction

What is geotechnical site investigation?

Site investigation is the process whereby all relevant information concerning the site of a proposed civil engineering or building development and its surrounding area is gathered. Ground investigation is a narrower process, involving the acquisition of information on the ground conditions in and around a site. The two terms are, however, often used to mean the same thing.

What are the quality indicators of a good investigation?

When Keith Gabriel was elected chairman of the Association of Geotechnical and Geoenvironmental Specialists (AGS) in March 2001, a survey of the members was carried out to establish views on the Association's activities and future priorities. Top of the list was site investigation quality. Gabriel expressed his views in a contribution to the journal *Ground Engineering* (Gabriel, 2001). He pointed out that:

> *Rates remain extremely competitive, restricting investment in new equipment and techniques. We must continue to encourage clients to consider best value rather than lowest cost.*

The authors totally concur with this view which is the underlying theme of this book.

Gabriel (2001) proposed that an assessment of a site investigation should be based on 'critical success factors' and seven 'key performance indicators' which in turn are based on the AGS site investigation code of conduct.

Critical success factors include:

- identification of ground hazards
- provision for better management of ground risk
- provision of better value for clients and users
- efficient processes which continuously improve
- provision of relevant, reliable information and effective supply chain management.

The key performance indicators are:

- preparation – desk study and walk-over survey
- design

- procurement
- management – project, risk and quality
- supervision
- reporting – factual, interpretative and ground model
- outcome – client satisfaction, project review and user feedback.

What are the investigation objectives?

The objectives of carrying out a site investigation are:

- to assess the general suitability of the site and its environs
- to enable an adequate and economic design to be prepared, including the design of temporary works, ground improvement techniques and groundwater control schemes
- to plan the best method of construction, and to foresee difficulties and delays which may arise for whatever reason
- the design of remedial works if any failures have occurred
- to explore sources of indigenous materials for use in construction
- to select sites for the disposal of waste or surplus materials
- to carry out safety checks on existing slopes, dams or structures
- to determine the changes which may arise in the ground and environmental conditions, either naturally or as a result of the works, on adjacent works and on the environment in general.

How do you plan a site investigation?

At the outset it is important to bear in mind the valuable advice given by Glossop (1968):

> *If you do not know what you should be looking for in a site investigation, you are not likely to find much of value.*

In other words, site investigations should be planned by experienced personnel.

A site investigation will normally be carried out in stages, as follows:

- evaluate the client's requirements
- carry out the desk study (see Chapter 2)
- carry out the walk-over survey (see Chapter 2)
- consider possible conceptual designs (see Chapter 1)
- evaluate hazards and start the risk register (see Chapter 3)
- set up the geological model (see Chapter 1)
- carry out borings, test pits, in situ tests and evaluate pore water pressure distribution
- perform laboratory and field tests and determine design parameters – see Chapter 4 (chemical testing of ground and groundwater will be

required to determine whether the conditions will be aggressive to steel or concrete placed in the ground or if geotechnical procedures such as grouting are contemplated)

- write the report (see Appendix 2: *Writing reports*)
- review conceptual design and revise if necessary
- review hazards and update risk register
- carry out further inspections, testing, etc., at the groundwork stage of construction to verify assumptions.

A flow chart summarizing these steps is given in Fig. 1.1.

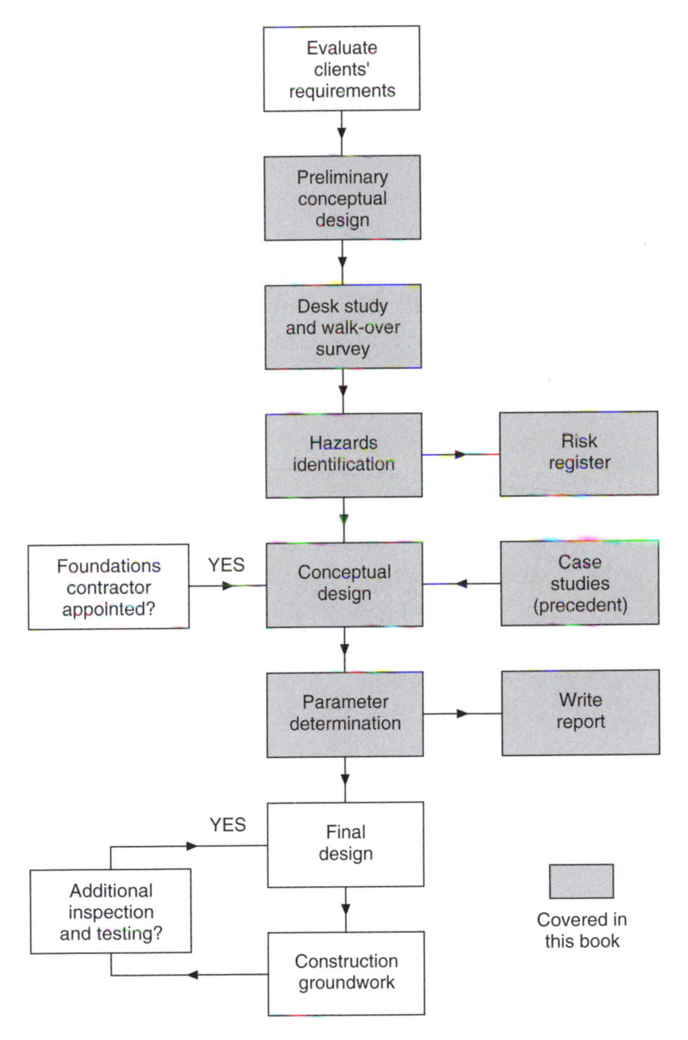

Fig. 1.1 Information and activity flow chart for geotechnical site investigation

It should be noted that 'conceptual design' is the identification of acceptable design solutions by qualitatively assessing the advantages and disadvantages of a number of possible design schemes without, however, carrying out a detailed analysis. It requires the input of an experienced geotechnical engineer or engineering geologist.

When planning a site investigation, it is crucial to be aware that geotechnical information is required to cover three main areas:

- the development of the geological model
- the identification and evaluation of hazards
- the geotechnical design of the project.

Conceptual designs must be carried out at the earliest possible stage so that all the relevant, and only the relevant, parameters are obtained. It is therefore necessary to assess the possible engineering solutions which may be adopted to deal with the specific project under consideration so that the required geotechnical parameters are obtained. Expenditure should not be incurred to obtain information which is not required and will not be used in the design process.

Key attributes of successful geotechnical engineering practice

It is important to bear in mind the requirements for the successful practice of geotechnical engineering which are given below, after Peck (1962).

- *Knowledge of precedents.* It is important to build up a database of case records so that one is aware of the existing state of knowledge relating to the problem under consideration. For example, if a prediction of settlement is made in a particular manner, is the agreement with the observed settlement of a structure founded on a comparable soil accurate enough for practical purposes? If not, a different approach is indicated which may dictate that an alternative form of site investigation is required. Another example is the investigation of slope stability. Should the problem be analysed in terms of total stress or is an effective stress approach appropriate? If the latter, can a design be based on peak parameters or residual values or, as is sometimes the case, on parameters intermediate between peak and residual, e.g. 'fully softened' or critical state values? Such decisions can only be safely made on the basis of reliable and appropriate case records of field failures and a site investigation should be planned to provide the correct values required. The input of an experienced geotechnical engineer is essential.
- *Familiarity with soil mechanics procedures.* This requires knowledge of the computational methods available to a geotechnical engineer

to help solve engineering problems, including calculations of earth pressure, bearing capacity, settlement, slope stability, seepage, etc.

- *Understanding of geology*. It is vital that a sound appreciation of the site geology is built up. What is the origin of the materials encountered on site? How did they get there? Have the properties been altered since deposition? Are the soils residual or transported? What is the distribution of pore water pressure across the site? Will the engineering operations to be carried out alter the geotechnical properties?
- *Search for all possible failure mechanisms*. In addition to the 'standard' problems to be considered (such as bearing capacity, settlement, earth pressure, slope stability and seepage), could there be any potential hazards lurking in the background which could trap the unwary? A list of potential hazards is given in Chapter 3. It is crucial that at every stage of a site investigation the possibility of a potential hazard developing is always borne in mind. If a potential hazard exists and has been overlooked then a major disaster could result. The collapse of an existing retaining wall due to abnormal rainfall, or the failure of a nearby slope, or ground deformations due to the presence of an underground undetected void could have serious consequences. The search for potential hazards should extend to adjacent sites. *Try to anticipate the unexpected!*
- *Develop a cynical pessimistic approach*. Never forget Murphy's Law: 'If something can go wrong, it will!' (this law has other names!). Check and cross check everything. Take nothing for granted.
- *Similitude*. When tackling a geotechnical problem, the best we can do is to model as well as possible the field conditions in our testing procedures and in our design calculations. The match can never be perfect and precise similitude is not attainable between the test model, analytical model and field prototype. There will always be differences in field and predicted behaviour. It is the duty of a geotechnical engineer to adopt the highest possible standards to keep these differences as small as possible.

To emphasize this point, we define for our students in a half-joking way the difference between 'bad' and 'good' geotechnical engineering design as follows:

> *'Bad' geotechnical design is where you put the wrong parameters into the wrong analysis and get the wrong answer. 'Good' geotechnical design is where you put the wrong parameters into the wrong analysis and get the right answer! This is because in good design we assess how wrong our parameter determinations are and how wrong our analyses are and make a compensating correction based on precedent.*

In addition, potential hazards should be considered so that the site investigation will identify such problems at an early stage. *Starting a risk register at the outset is the key to best practice.*

The site investigation should be specifically designed for each individual job. The days of a standard site investigation of two boreholes to a depth of 10 m, with U100 samples at depth intervals of 1·5 m in clays and Standard Penetration Tests (SPTs) at depth intervals of 1·5 m in granular soils are long over!

Key elements in planning the investigation

So that the reader can quickly get a feel for site investigation practice, in this chapter we consider the key elements involved in planning the investigation as follows:

- borehole layout and spacing
- procurement
- development of the geological model (Fookes' model)
- conceptual design and case studies (i.e. precedent).

Further key elements considered separately in subsequent chapters are:

- the desk study and walk-over survey (Chapter 2)
- geotechnical hazards and risk management (Chapter 3)
- parameter determination: classic and modern methods (Chapter 4).

Borehole layout and spacing

The number, type, location and depth of investigations depend on the nature and size of the project, on the variation of ground conditions across the site, and on the cost of the investigations, which is sometimes related to the cost of the project. Ideally, of course, the extent of the investigation should be governed by technical considerations rather than cost. It is difficult to offer specific advice on the required number of boreholes which need to be sunk on a particular site as many factors, such as cost, the time available for the investigation and, on occasions, the availability of equipment and personnel, need to be considered. Some guidance to borehole depth, layout and spacing may be obtained from some of the case records given later.

A typical site investigation for a motorway in the UK may consist of one borehole on the centreline of the motorway at a spacing of say 150 m taken down to a depth of about 5 m. Under a high embankment, a greater depth of investigation may be necessary as well as locating boreholes off the centreline. Deep cuttings will pose special problems and several boreholes may be required with relevant laboratory testing and analysis. The possibility of the bottom of a cutting failing because of water pressure

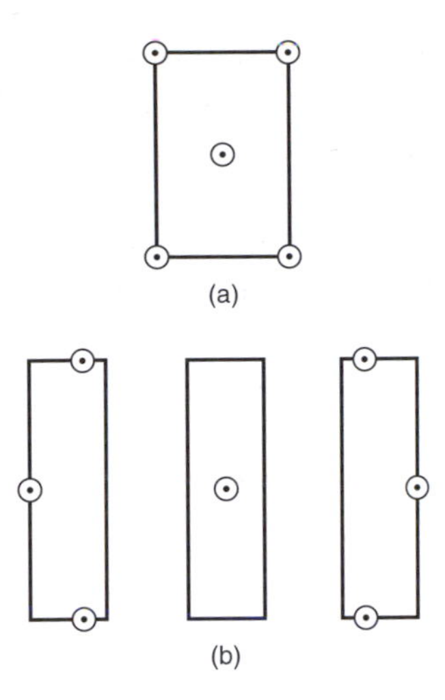

Fig. 1.2 Typical borehole layouts for (a) a large warehouse and (b) a multi-storey block of flats

in an underlying permeable stratum should be investigated. The position of each bridge structure should be carefully investigated by up to four boreholes 25 m to 30 m deep.

The spacing of boreholes for foundations for structures often lies in the range of 20 m to 40 m and some typical borehole layouts are shown in Fig. 1.2. Boreholes should be sunk as close to the proposed foundations as possible, and this is particularly important when the depth to the bearing stratum is irregular. If the layout of the structures has not been decided at the time of carrying out the site investigation, an evenly spaced grid of boreholes may be adopted, possibly with infilling by dynamic or static probings or by seismic methods (see Chapter 4). If trial pits are to be used they should not be located close to the positions of the intended foundations because of the disturbance of the ground caused by the relatively large and deep excavations. Borehole layout and frequency are partly controlled by the complexity of the geological conditions. If the ground conditions are relatively uniform, a wide spacing of boreholes may be satisfactory but if the ground conditions are complex a closer spacing of boreholes will be required.

The depth to which boreholes should be sunk is governed by the depth to which the soil is significantly affected by the foundation loading. If a piled foundation is a possible solution then the boreholes should extend to such a depth below the elevation of the pile tips where the stress increase due to the foundation loading will not have an adverse effect on the performance of the structure.

It is common for boreholes to be taken down to a depth of $1{\cdot}5B$ or $2B$ below the elevation of a foundation of breadth B and it is assumed that stress increases below this level will not have an adverse effect on the behaviour of a structure. Some authorities take a stress increase of 10% of the applied foundation pressure at any depth as an acceptable limit. It should be noted that for a square or circular foundation the increase in vertical stress below an applied loading is about 15% of the applied loading at a depth of 1·5*B* and about 10% at a depth of 2*B*. For an infinitely long foundation the corresponding figures are 40% and 30% respectively and in such cases it may be prudent to investigate the ground to a depth below foundation level of more than 2*B* to be on the safe side. Figure 1.3 gives suggested borehole depths for various foundation conditions.

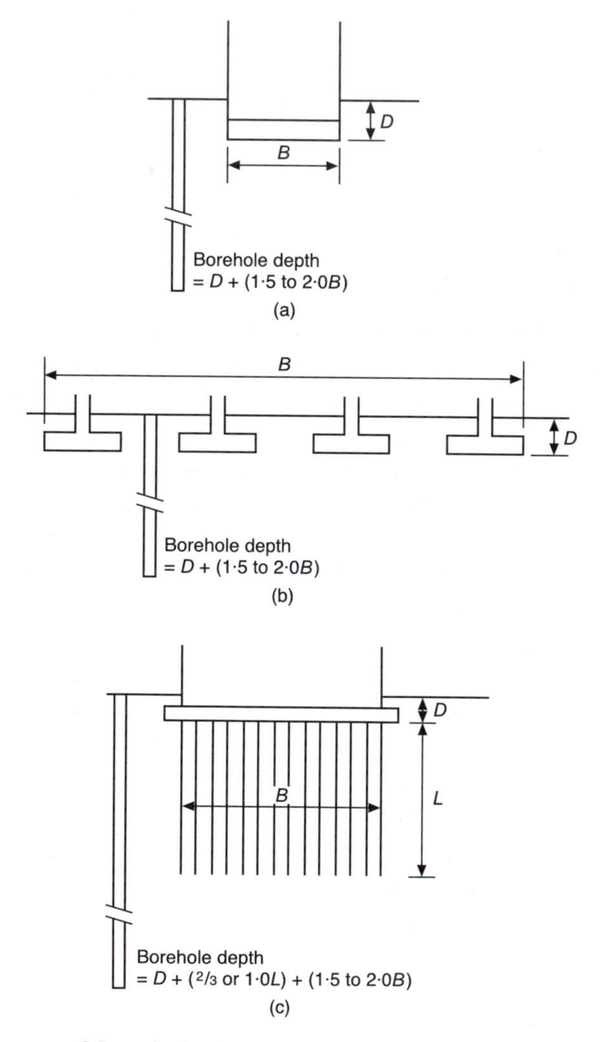

Fig. 1.3 Suggested borehole depths for various foundation conditions. (a) Individual footing or raft. (b) Closely spaced footings. (c) Piled raft.

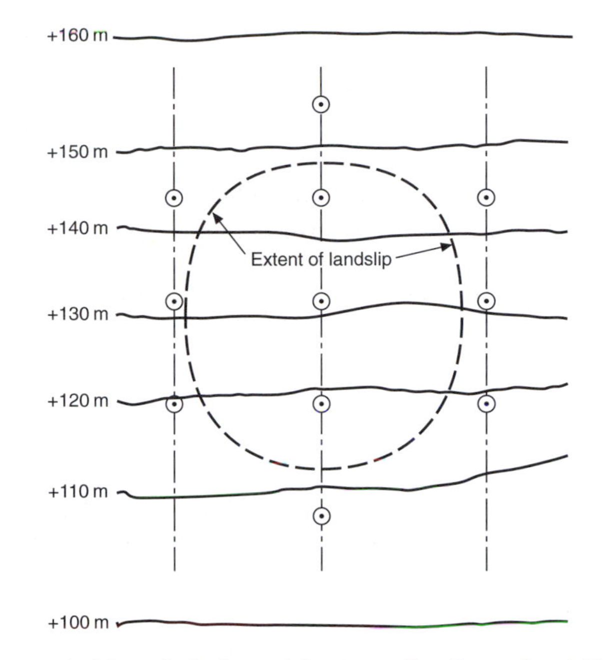

Fig. 1.4 Suggested borehole layout for investigating a landslip

When investigating the stability of a slope it is usually convenient to place three to five boreholes in a straight line up the slope and if the area under consideration is extensive then more than one line may be necessary. Figure 1.4 gives a suggested layout for investigating a landslip. Boreholes should be terminated only when a competent stratum has been reached. The distribution of water pressure throughout the slope must be carefully investigated and this will require piezometers to be placed at various depths and read at regular time intervals in an attempt to observe the most adverse water pressure distribution. The location of any slip surfaces needs to be thoroughly investigated and this may require continuous samples or test pits – great care must be taken to ensure the safety of any personnel working in the test pit.

It is advisable on any site to sink at least one borehole to a sufficient depth to establish the solid geology and to determine the depth of weathering of the bedrock. If a structure is to be founded on rock then it must be established that actual bedrock has been reached, not just an isolated boulder. It is good practice to extend some boreholes 3 m to 6 m into rock to check the depth of weathering and the possible existence of a boulder.

Procurement

Much concern has been expressed about delays and escalating costs of construction projects both in the UK and abroad. The delays are frequently attributed to inadequate site and ground investigations. The primary causes of shortcomings in ground investigation include:

- unfair or unsuitable methods of competition
- inappropriate conditions of contract
- insufficient and inadequate supervision
- inadequate and unenforceable specification of work
- lack of client awareness
- inadequate finance
- insufficient time to carry out a proper investigation
- lack of geotechnical expertise.

Clients should be aware of the inherent risks associated with site investigation. It is vital that financial decision-makers appreciate that *you pay for a site investigation whether you have one or not*!

While there are many ways in which ground investigation could be adequately procured, the following two broad systems of procurement have evolved in the UK (Uff and Clayton, 1986).

- *System 1*. Design of the investigation and supervision by a Consulting Engineer or other design professional employed by the Client; physical work, testing and reporting carried out as required under a separate contract by a Contractor, chosen by selective tendering. System 1 has the advantage of using well known forms of contract, and it can demonstrate cost accountability through the tendering process. It also allows the design engineer to participate in the investigation process. Its disadvantages include the difficulty of ensuring that adequate expertise and supervision are provided by the Engineer.
- *System 2*. Design and direction, together with all physical work, testing and reporting as required, by a single contracting party in a 'package' arrangement made with the Employer. System 2 is capable of overcoming the problems of System 1, and it allows full use to be made of the expertise of the specialist contractor. It also avoids any division or confusion of responsibilities, but there is less available experience in the use of System 2, particularly in large contracts (see also Chapter 3).

Development of the geological model (Fookes' Model)

Sources

This section is based on the following sources which we gratefully acknowledge.

- Fookes, P. G. (1997). Geology for engineers: the geological model, prediction and performance. The First Glossop Lecture. *Q. J. Eng. Geol.*, **30**, 293–424.

- Fookes, P. G., Baynes, F. and Hutchinson, J. N. (2000). Total geological history: a model approach to the anticipation, observation and understanding of site conditions. *Proc. Int. Conf. on Geotech. and Geol. Eng., GeoEng 2000*. Vol. 1, Technomic, Pennsylvania, 370–421.
- Fookes, P. G., Baynes, F. and Hutchinson, J. N. (2001). Total geological history: a model approach to understanding site conditions. *Ground Eng.*, **34**(3), 22–23.

Overview

One of the major problems in geotechnical engineering is the risk of encountering unexpected geological conditions. Failure to anticipate such conditions is generally due to an inadequate geological understanding of the site. Fookes (1997) and Fookes *et al.* (2001) describe an approach of systematically building up a geological model of a site as more information becomes available during a project so that finally a sound understanding of the geology and the geomorphology is obtained. This section draws heavily on the work of Fookes and his colleagues.

Ground conditions at any site are a product of its total geological and geomorphological history, which includes the stratigraphy, the structure and the past and present geomorphological processes and climatic conditions. There should be a specific and determined endeavour to understand the engineering geology and geomorphology environment of a site and to incorporate that understanding into the project design. Each site evaluation, however small, should have at least one engineering geologist involved in the work who should have a sound knowledge of geomorphology or should work together with an experienced geomorphologist.

A site investigation generally proceeds in stages:

- desk study
- walk-over survey, together with geological mapping and trial pitting
- main ground investigation
- supplementary investigations
- observations made during construction.

After the desk study has been carried out an initial geological model is set up and this model is then improved as more and more information becomes available. If the work is carried out by dedicated and competent personnel then the risk of encountering unexpected geological conditions on site will have been reduced to the absolute minimum.

The systematic build-up of the geological model as an investigation proceeds is illustrated in Figs 1.5(a)–(e), taken directly from Fookes (1997). Here, the build-up of knowledge is shown visually. Detailed comment may be found in Fookes (1997) and Fookes *et al.* (2001).

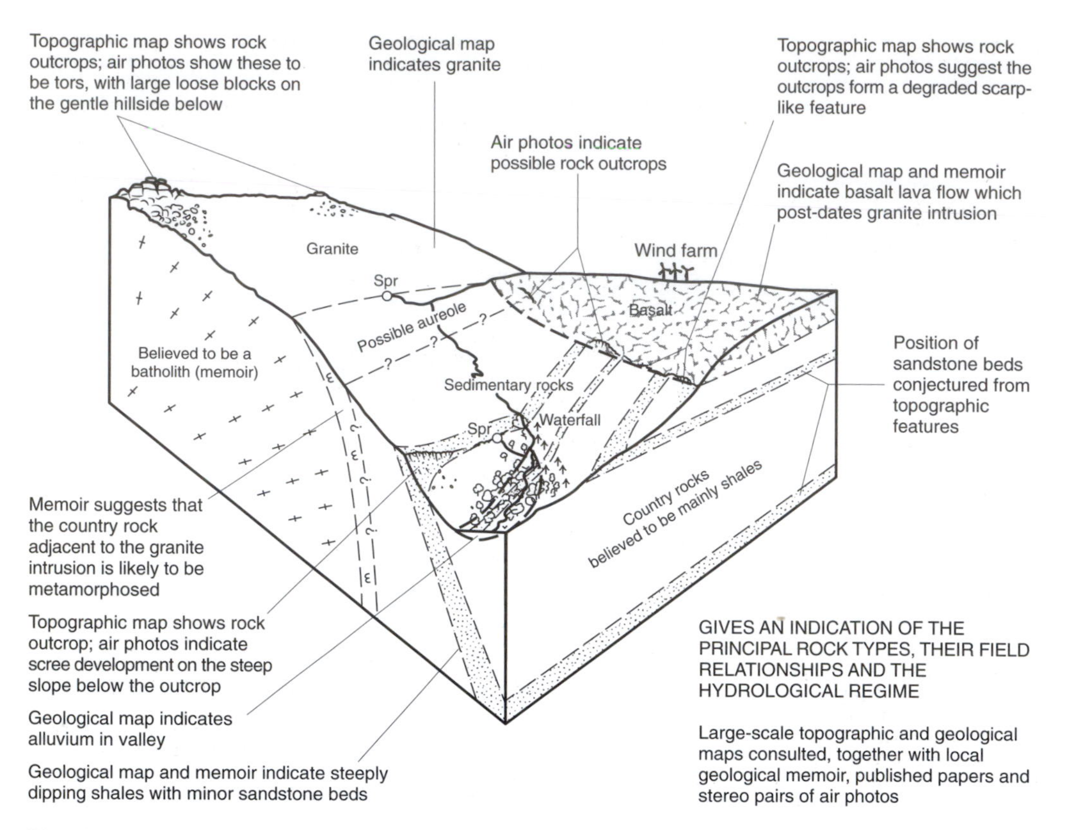

Fig. 1.5 (a) Progressive increase in information building up the geological model during the course of a site investigation, after Fookes (1997) – desk study (superimposed on basic geology)

Conceptual design and case studies

Overview

When planning a site investigation it is necessary to assess the possible engineering solutions that may be adopted to deal with the specific project under consideration so that the required geotechnical parameters are obtained. Conceptual design is thus a key activity in the investigation process. The following points should be addressed. Is it a slope stability problem? Is it a long term situation, for example, the stability of a natural slope? Is the clay intact or fissured? Clearly the undrained shear strength is not relevant to a long term condition, and what is required are the shear strength parameters with respect to effective stress, but are the peak parameters the controlling values or should we be considering lower parameters, for example critical state, or 'fully softened', or residual? Have there been large trees on the site that have been cut down? Do we know the most critical distribution of pore water pressure?

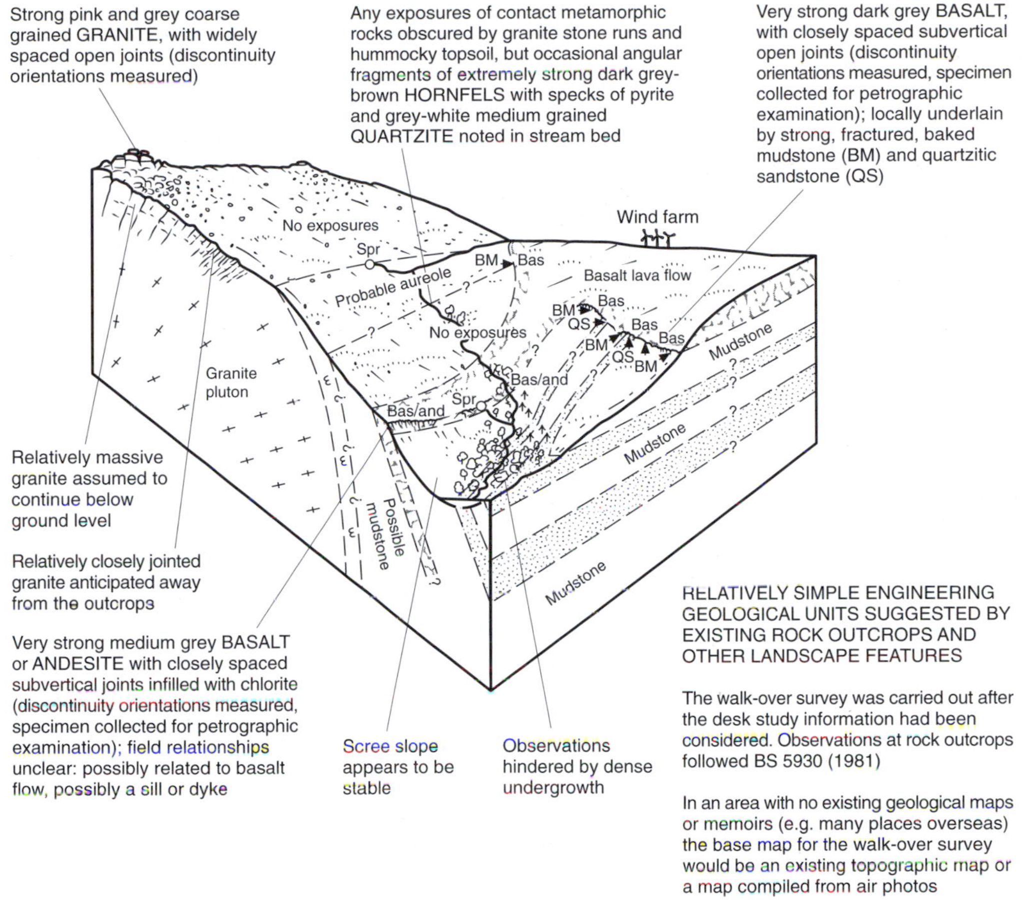

Strong pink and grey coarse grained GRANITE, with widely spaced open joints (discontinuity orientations measured)
Any exposures of contact metamorphic rocks obscured by granite stone runs and hummocky topsoil, but occasional angular fragments of extremely strong dark grey-brown HORNFELS with specks of pyrite and grey-white medium grained QUARTZITE noted in stream bed
Very strong dark grey BASALT, with closely spaced subvertical open joints (discontinuity orientations measured, specimen collected for petrographic examination); locally underlain by strong, fractured, baked mudstone (BM) and quartzitic sandstone (QS)
Wind farm
No exposures
Spr
BM
Bas
Basalt lava flow
Probable aureole
No exposures
QS
Mudstone
Bas/and
Granite pluton
Possible mudstone
Mudstone
Mudstone
Relatively massive granite assumed to continue below ground level
Relatively closely jointed granite anticipated away from the outcrops
Very strong medium grey BASALT or ANDESITE with closely spaced subvertical joints infilled with chlorite (discontinuity orientations measured, specimen collected for petrographic examination); field relationships unclear: possibly related to basalt flow, possibly a sill or dyke
Scree slope appears to be stable
Observations hindered by dense undergrowth
RELATIVELY SIMPLE ENGINEERING GEOLOGICAL UNITS SUGGESTED BY EXISTING ROCK OUTCROPS AND OTHER LANDSCAPE FEATURES
The walk-over survey was carried out after the desk study information had been considered. Observations at rock outcrops followed BS 5930 (1981)
In an area with no existing geological maps or memoirs (e.g. many places overseas) the base map for the walk-over survey would be an existing topographic map or a map compiled from air photos

Fig. 1.5 (b) Progressive increase in information building up the geological model during the course of a site investigation, after Fookes (1997) – walk-over survey (superimposed on basic geology)

If the problem is short term can we use the total stress analysis? If so, what is the most reliable method of measuring the undrained shear of the clay in question? Should we use the in situ vane test, or plate bearing tests taken to failure, or undrained shear box tests in the laboratory or in the field, or can we rely on strength values inferred from in situ cone tests or Standard Penetration Tests? Should we carry out undrained triaxial tests, in compression or extension, what size sample should be tested, and what type of sampler will give the most suitable results?

If we are dealing with a foundation problem is it a question of deformation or of strength or could either be the controlling factor? Could shallow foundations suffice or is there the possibility that piles might be the most effective solution? The site investigation should be designed with both

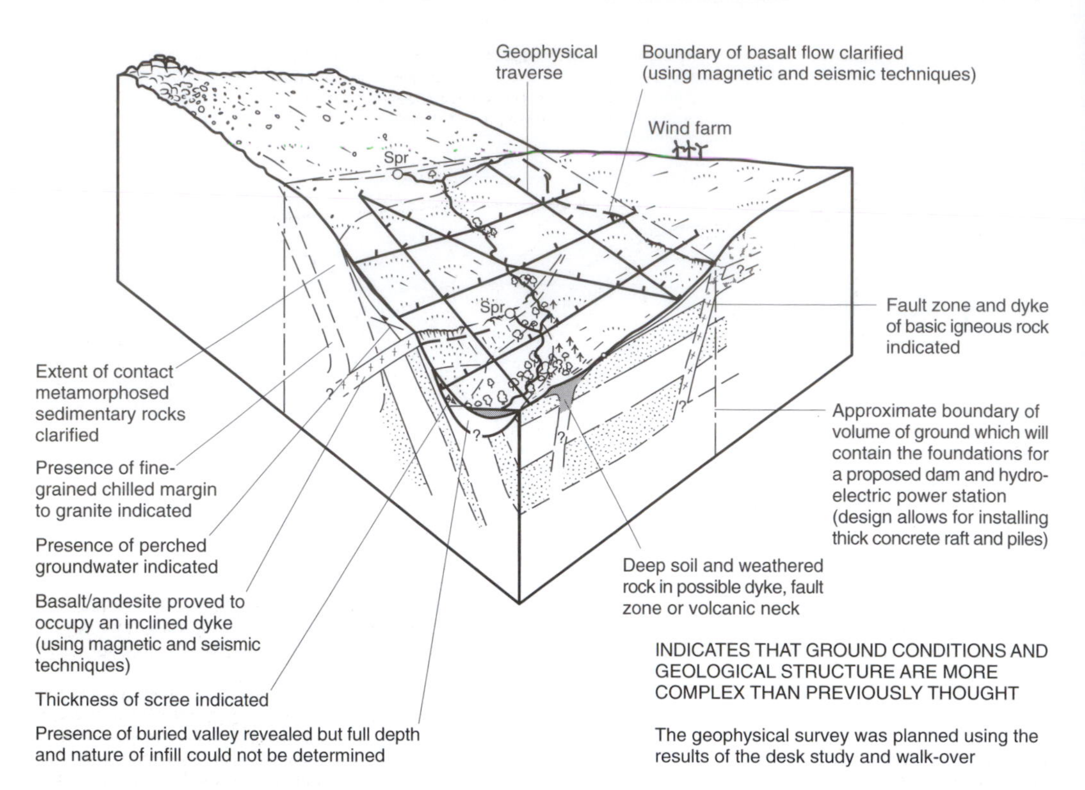

Fig. 1.5 (c) Progressive increase in information building up the geological model during the course of a site investigation, after Fookes (1997) – geophysical survey (superimposed on basic geology)

possibilities in mind. If a deep excavation is required will de-watering be involved and how will the sides of the excavation be supported?

The list of factors to be considered when planning a site investigation is thus immense and the input of an experienced geotechnical engineer is essential. In the following sections a number of conceptual designs and case studies are considered and discussed. We believe these analyses will be helpful to those responsible for planning geotechnical site investigations.

Geotechnical investigations on clays

Conceptual design: ten-storey office building on stiff fissured clay

There are at least three possible founding alternatives:

(a) a raft at shallow depth, say 1·5 m
(b) a deep raft incorporating basements at a founding depth of, say, 6 m, or
(c) a shallow raft with bored piles transferring the loading to stronger and less compressible material at greater depth.

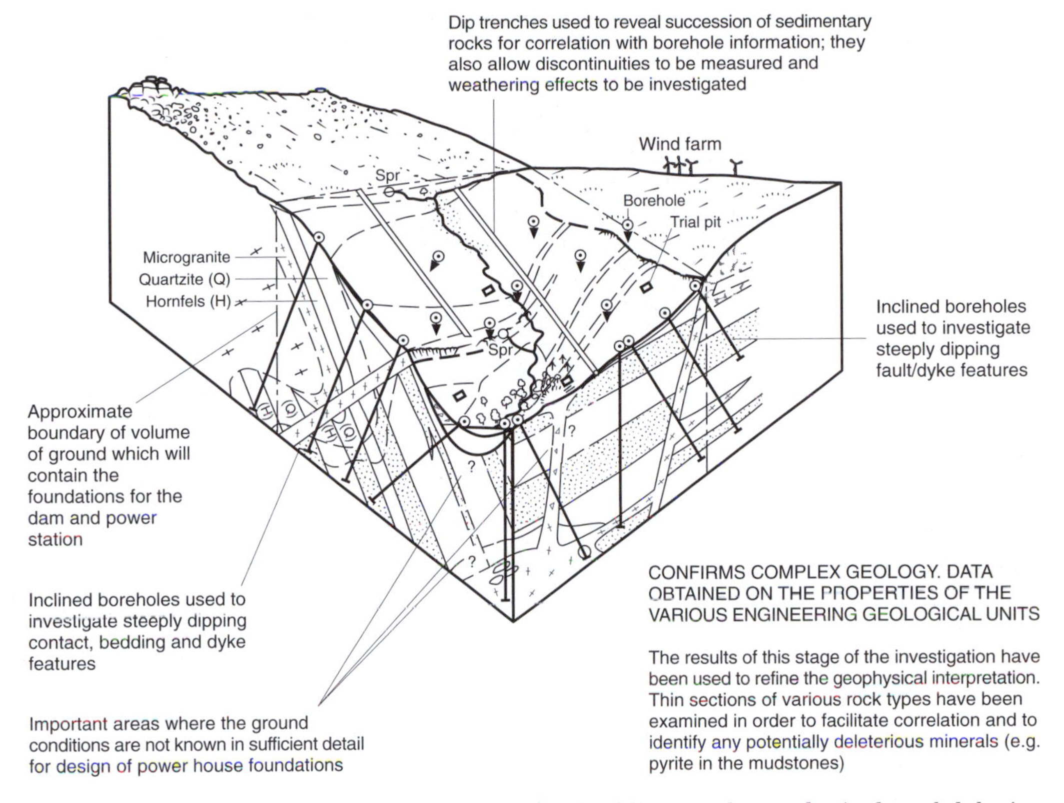

Fig. 1.5 (d) Progressive increase in information building up the geological model during the course of a site investigation, after Fookes (1997) – main ground investigation (superimposed on basic geology)

These possibilities are shown in Fig. 1.6. The approach for each alternative would be as follows.

(a) For case (a) it will be necessary to check firstly that an acceptable factor of safety against a bearing capacity failure can be obtained and, if so, then to estimate what the settlement is likely to be. This will require a knowledge of the undrained shear strength, the undrained stiffness modulus and the compressibility coefficients m_v and c_v over a depth interval below the founding level of 1·5 m down to a depth of say 1·5 to 2 times the least width of the structure. Investigating the soil profile over this depth interval should ensure that the increase in vertical stress due to the foundation loading below this level is likely to be insignificant.

(b) Similar parameters will be required for case (b) but this time the depth range is from 6 m to 6 m plus 1·5 to 2 times the least width of the structure.

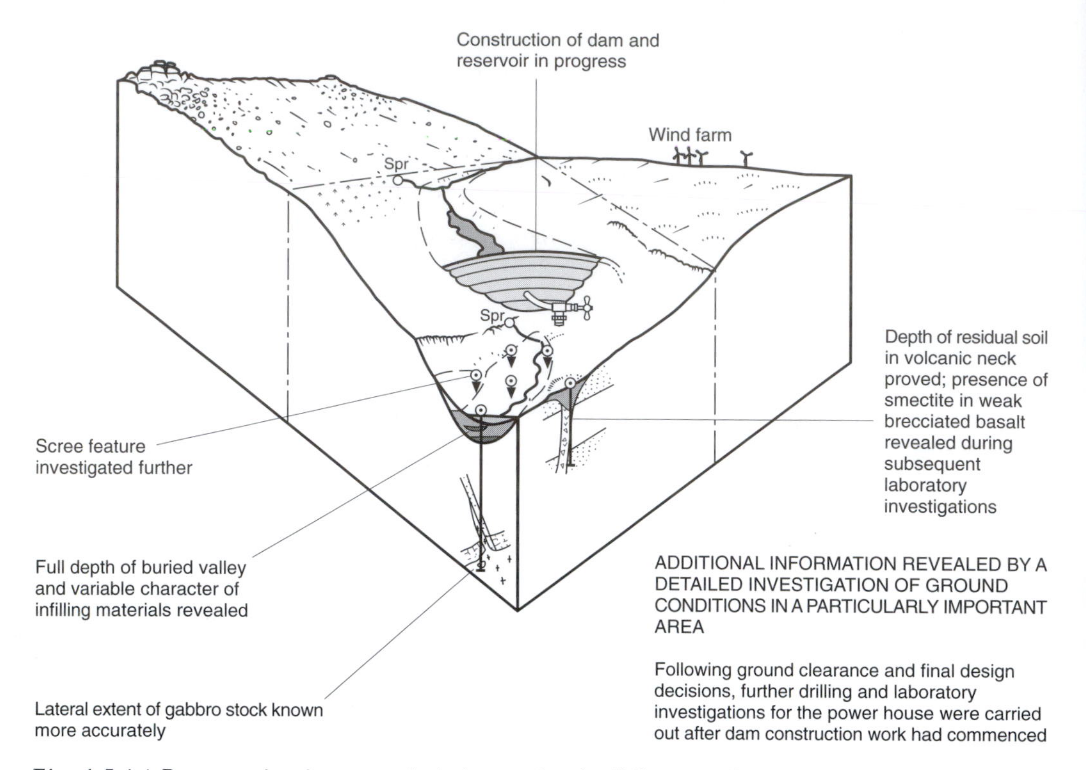

Fig. 1.5 (e) Progressive increase in information building up the geological model during the course of a site investigation, after Fookes (1997) – supplementary ground investigation (superimposed on basic geology)

(c) For case (c), the same parameters will be needed over a depth interval ranging from a depth of say two-thirds the proposed pile length to this depth plus 1·5 to 2 times the least width of the structure. The two-thirds depth rule is chosen because it is sometimes assumed that the piles transfer the loading to this level. If a different assumption is made the depth of boreholes should be adjusted accordingly. It may be sensible to choose a load transfer depth of two-thirds the pile length if the piles are driven and the full pile length if the piles are bored. This is because the disturbance to the ground and the excess pore water pressures resulting from driving piles may lead to greater settlements than would be the case with bored piles. Calculating settlements from a higher elevation would give greater settlements than for a lower assumed elevation.

While the same geotechnical parameters are required for the three possible founding alternatives, the required depth ranges are quite different. It is clearly essential that all possible foundation solutions should be carefully considered before embarking on a site investigation.

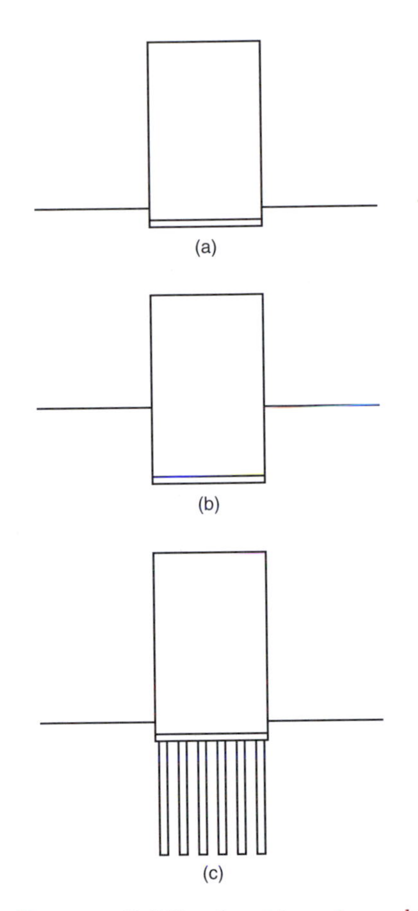

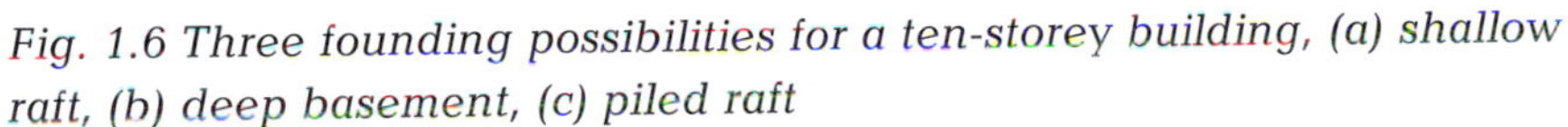

Fig. 1.6 Three founding possibilities for a ten-storey building, (a) shallow raft, (b) deep basement, (c) piled raft

The design of bored piles in stiff fissured clays is usually based on the undrained shear strength. If it is intended that the pile design will be carried out using effective stress methods, however, then effective stress triaxial tests will have to be carried out to obtain the relevant parameters.

As always, knowledge of the distribution of the pore water pressures in the ground is essential. Unfortunately, on many occasions insufficient attention is paid to this important aspect of site investigation.

In passing, it is of interest to note that it is often the case that while stiff clay is generally strong enough to carry the load safely at shallow depth with an adequate factor of safety, settlement considerations may dictate that another foundation solution will be required. In this case the deep raft may result in smaller settlements than a bored pile solution. This is because for the deep raft the net increase in foundation pressure at foundation level, which governs consolidation settlement, has been reduced by the weight of excavated soil removed. If it is possible to arrange that the net increase in foundation loading is zero, then the consolidation settlement would be zero even for the most highly compressible soils. This is known as a floating foundation.

If the job is important enough it may be advisable to check the design by carrying out pile loading tests, either on a test pile loaded to failure or on a working pile under an applied load of not greater than 1·5 times the design load. Consideration should be given to carrying out large-diameter plate loading tests at various depths to provide additional information on the stress–strain behaviour in end bearing of large-diameter piles. Loading tests on plugs of concrete cast in bored piles with a void below may result in significant economies in the design of the shaft friction component of pile capacity.

It can be seen that all possible founding solutions must be considered when planning a site investigation so that when the choice is finally made the relevant parameters are available and the ground has been explored to a sufficient depth below each possible founding level.

Case study: underground car park at the House of Commons, London

This important investigation has been described by Burland and Hancock (1977). The site finally chosen for the car park, New Palace Yard, presented a major engineering problem as it involved the construction of a large excavation 18 m deep in close proximity to the Clock Tower

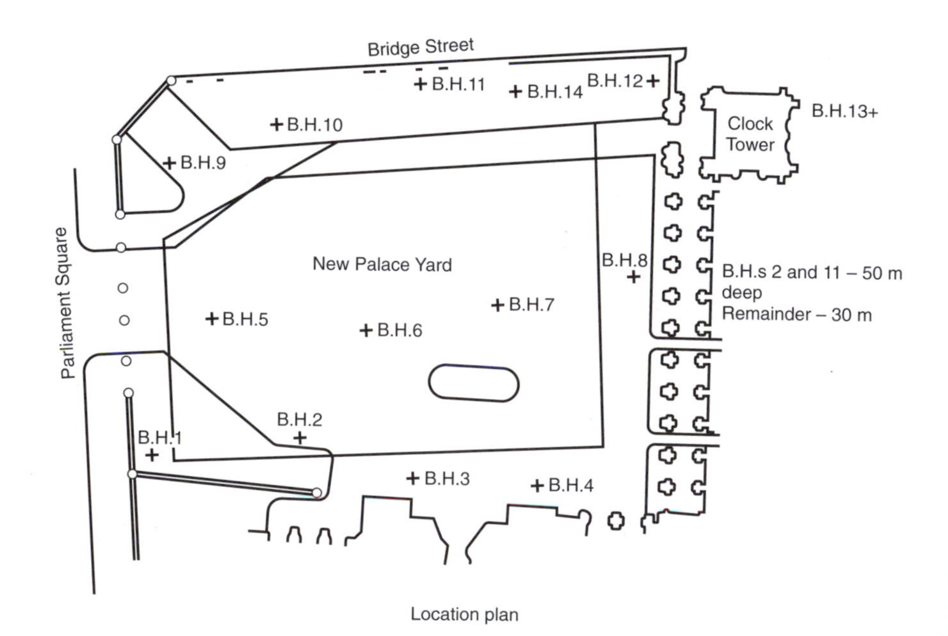

Fig. 1.7 Location plan for boreholes, House of Commons car park, from Burland and Hancock (1977)

(Big Ben), Parliament building and the 14th century Westminster Hall. A very thorough site investigation was carried out and the soil profile and groundwater conditions revealed had a major influence on the design of the retaining walls and foundations.

Fourteen boreholes were sunk on the site to explore the soil and to investigate the groundwater conditions. Their locations are shown in Fig. 1.7 and a generalized borehole log is given in Fig. 1.8. A detailed visual examination was carried out on a number of open drive samples and a good correlation of the various strata was found between the boreholes. The visual examination revealed that immediately below the lowest level of the proposed car park there existed a layer of London Clay containing partings of fine sand and silt up to 10 mm thick and at

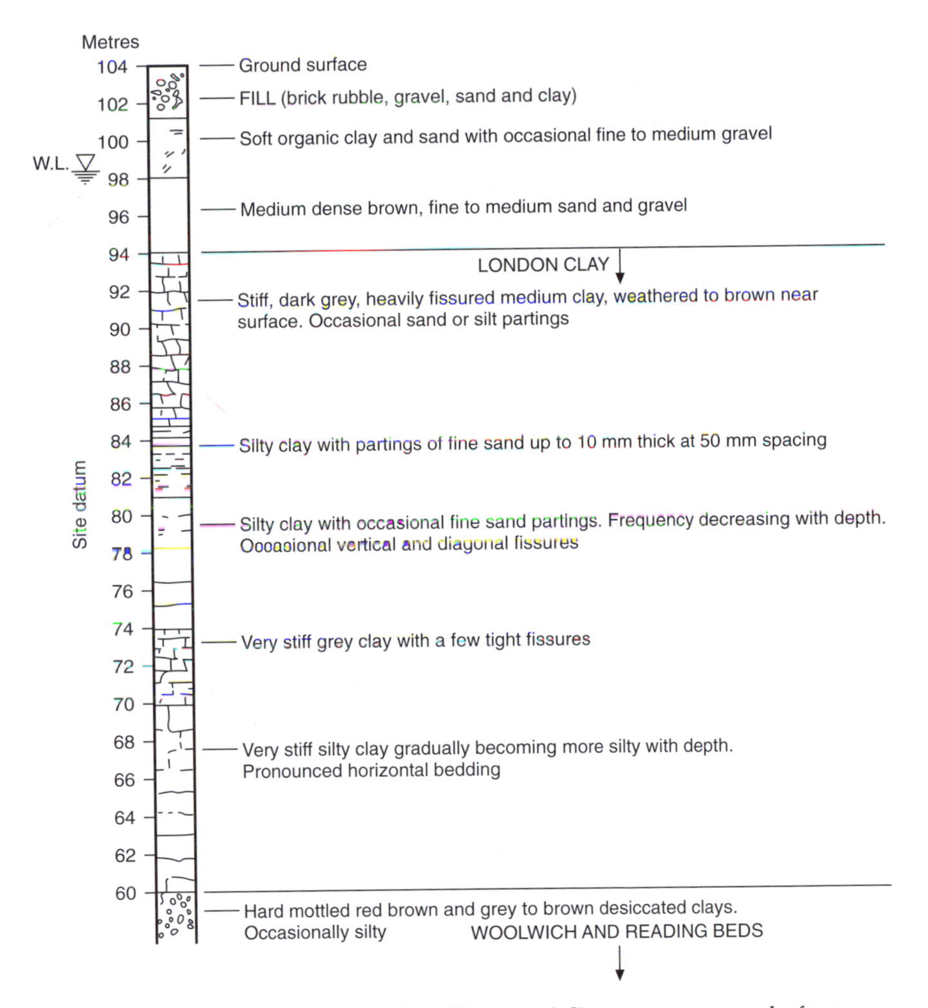

Fig. 1.8 Generalized borehole log, House of Commons car park, from Burland and Hancock (1977)

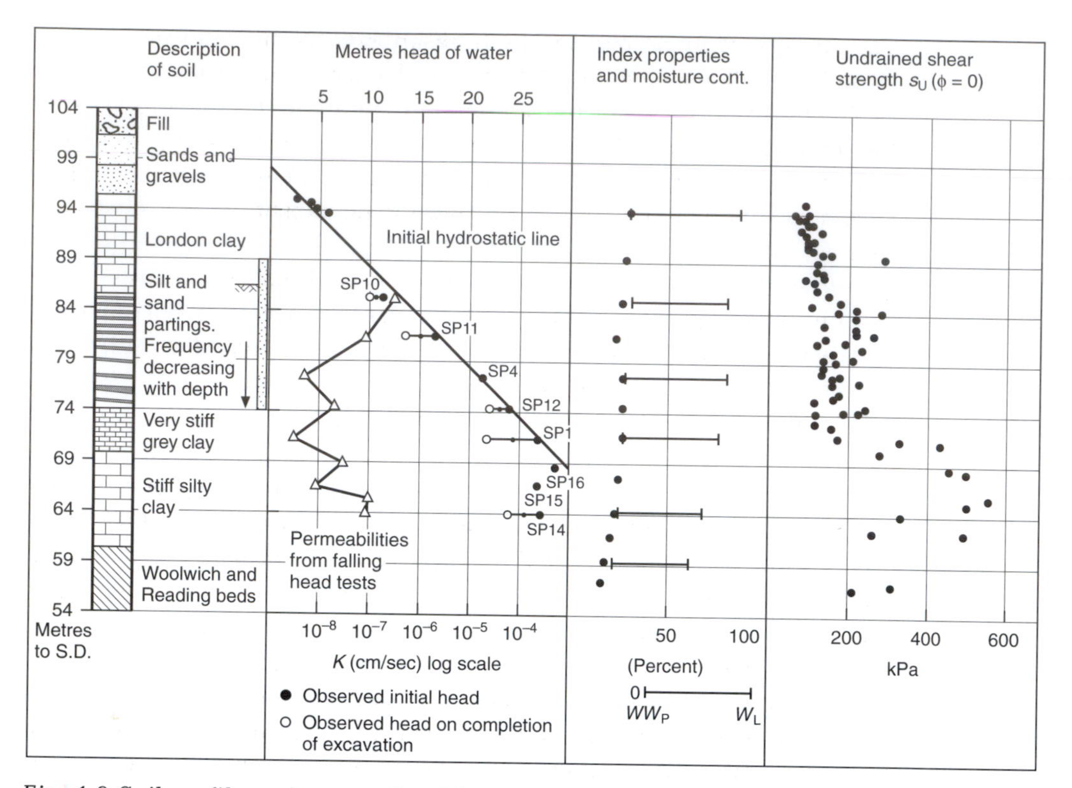

Fig. 1.9 Soil profile and properties, House of Commons car park, from Burland and Hancock (1977)

50 mm spacing near the top of the layer. The finding of this layer had very important implications on the design of the foundations and retaining walls since the relatively high horizontal permeability could give rise to high water pressures beneath the excavation leading to hydraulic uplift and possible base failure. Furthermore, the presence of the sand layers could lead to difficulties in the construction of the piles, particularly under-reams due to water seepage.

Casagrande standpipes were installed in most of the boreholes. In addition to the standpipes, three rapid response pneumatic piezometers were installed beneath the centre of the excavation. The in situ permeability of the London Clay at various depths was obtained by carrying out falling head tests on all the standpipes. Unconsolidated undrained triaxial tests were carried out on samples from some of the boreholes and the results are given on Fig. 1.9 showing the usual scatter associated with stiff fissured clay. Special oedometer tests were carried out on selected samples to measure the compressibility. A finite element analysis was carried out to predict ground deformations. This required an assessment of the variation of Young's Modulus with depth and this was based on the values that were obtained by back-analysis of the measurements of retaining wall movements at Britannic House in the City of London (Cole and Burland, 1972). It is important to note that these were three to five times larger than the values derived from the laboratory oedometer tests. Of course, this was before the discovery of the dependence of stiffness on strain level (see Fig. 4.1). There is no substitute for field-based data!

In view of depth of excavation, the close proximity of such priceless historic buildings and the sensitivity of the job, a very comprehensive programme of monitoring was undertaken during all phases of excavation and construction and is described in Burland and Hancock (1977). Briefly, the monitoring may be considered under three broad headings: surface surveying, ground movements at depth and pore water pressures. Approximately 60 movement points were established by grouting BRS levelling stations into the masonry of the surrounding buildings. Changes in the verticality of the Clock Tower were measured using a Hilger and Watts 'Autoplumb' and readings were taken every day during the construction period. The horizontal deflections of the diaphragm walls were measured by inclinometers using the Soil Instruments Ltd Mk II version. The heave at various depths was measured using two magnet extensometers. Pore water pressures were observed using a number of standpipes and pneumatic piezometers chosen for their rapid response time. Unfortunately the pneumatic piezometers ceased to function at an early stage during excavation because dirt entered the air lines. The location of the instrumentation is shown in Fig. 1.10.

The thorough site investigation, the finite element analysis and the comprehensive monitoring programme all contributed to the successful completion of this complex and sensitive civil engineering project. An important point to note is that the design parameters adopted were *not* those resulting from the site investigation but were based on field measurements of the deformations of the ground around a similar wall also in the London Clay.

The parameters derived from the site investigation were based on the sampling procedures and laboratory testing which were in use at that time, i.e. during the late 1970s. To obtain reliable values of, for example, the maximum shear modulus, G_{max}, bender element testing (see the section 'Seismic methods' in Chapter 4) of high-quality samples should be used, together with in situ geophysical surveys using surface waves. Tou *et al.* (2001) published values of G_{max} for the Singapore Jurong formation residual soils determined by these two techniques and found reasonable agreement. It would be expected that such values would be close to the in situ parameters.

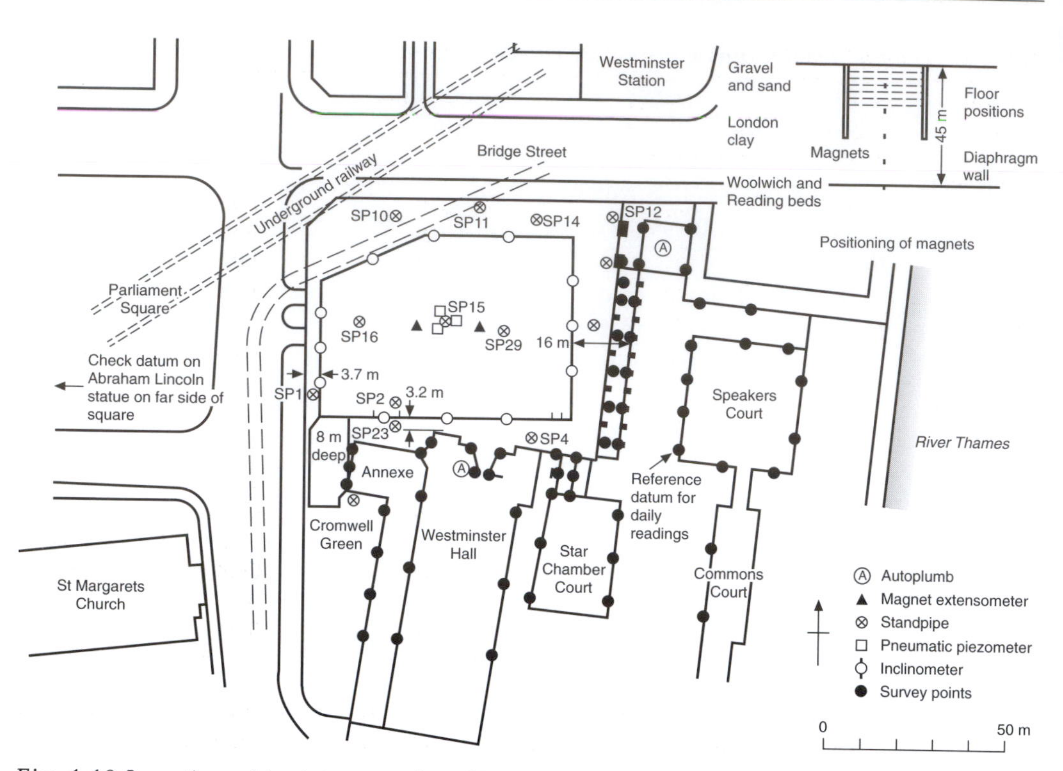

Fig. 1.10 Location of instrumentation, House of Commons car park, from Burland and Hancock (1977)

Case study: 140 m high office block, Necom House, Lagos, Nigeria

The structure with an approximate plan area of 45 m × 23 m, had a reinforced concrete raft about 2 m thick, applying a ground pressure of 230 kPa. The ground conditions were as follows.

- Upper sands, 18 to 21 m thick, and being very loose to loose from ground level to a depth of 7·5 m with a Dutch Cone point resistance (CR) varying from 1–5 MPa, and medium dense below a depth of 7·5 m with a CR of 9–15 MPa. The SPT N value was very variable.
- Upper clays, 3 to 7 m thick, firm to stiff sandy silty clays, with an undrained shear strength (s_u) varying from 40 to 70 kPa.
- Lower sands, 15 to 17 m thick, dense to very dense, CR about 20 MPa, $N = 45$.
- Lower clays, about 10 m thick, stiff to very stiff silty clays, $s_u = 65$ to 170 kPa.

The groundwater table was found at a depth of 2 m. The upper sands had a D_{10} grain size varying from 0·09 to 0·19 mm with an average of 0·14 mm, while for the lower sands D_{10} was in the range of 0·07 to 0·20 mm with an average of 0·11 mm. Both sands had zero silt content and a gravel content

Table 1.1 Geotechnical properties, Necom House, Lagos

Liquid limit	Plastic limit	Water content: %	Liquidity index	C_c (calc.)	C_c (oed.)
Upper clays					
78	18	41	0·38	0·61	0·18
85	19	36	0·26	0·68	0·16
111	26	67	0·48	0·91	0·55
74	22	48	0·50	0·58	0·43
38	12	29	0·65	0·25	0·15
137	31	53	0·21	1·14	0·49
Lower clays					
66	21	37	0·36	0·51	0·15
95	21	35	0·19	0·77	0·36
89	33	34	0·02	0·71	0·19
101	30	47	0·24	0·82	0·57
81	20	32	0·20	0·64	0·28
70	20	17	−0·06	0·54	0·10
55	18	15	0·46	0·41	0·21
66	22	19	−0·07	0·51	0·26

of less than 5%. Below the lower clays, alternating layers of very stiff clays and dense sands were penetrated by boreholes for a further 16 m.

There were two founding possibilities:

- a raft at shallow depth, in which case the compressibility of the underlying soils would govern whether or not this was an acceptable solution
- a piled raft with the piles terminating near the bottom of the upper sands or in lower sands. In either case the loading would be taken down close to an underlying clay layer which could possibly lead to unacceptable settlements. In any event the cost of piling would be considerable.

The shallow raft solution was therefore an attractive possibility, provided that the settlements were acceptable. Unfortunately, the oedometer tests were only loaded up to 800 kPa and it was not possible to assess with any degree of confidence whether the clays were over-consolidated or not.

The liquid limit, plastic limit and natural moisture content test results are given in Table 1.1. It can be seen that the liquidity index ranged from 0·21 to 0·65 for the upper clays and from −0·07 to 0·46 for the lower clays. These values suggest that both clays could be over-consolidated.

The compression index was calculated from the expression $C_c = 0.009(W_l - 100)$ which is valid for normally consolidated clays and the results are given in Table 1.1 in the column C_c (calc.). Values of C_c were taken from the oedometer tests curves covering the stress interval just greater than the effective vertical stress in the field and these values are

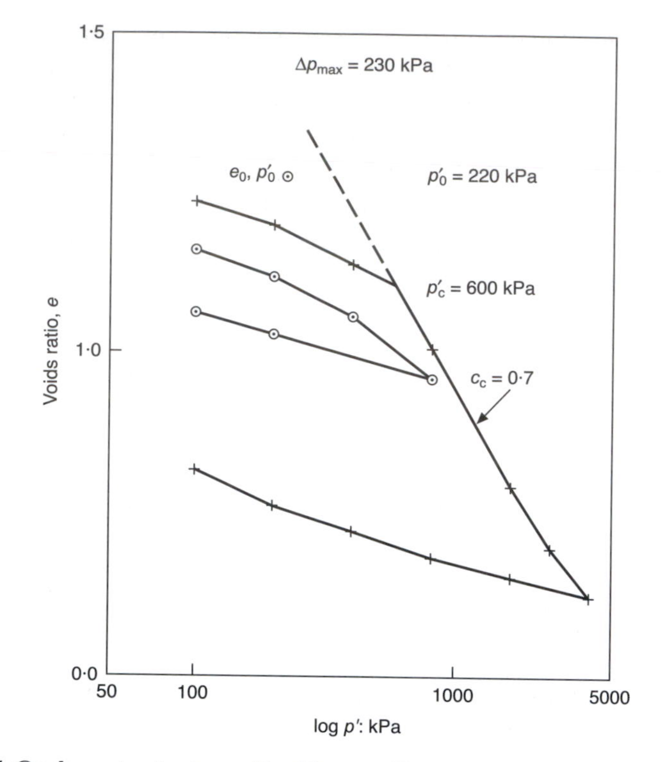

Fig. 1.11 Oedometer test results, Necom House

also given in Table 1.1 in the column C_c (oed.). It can be seen that in all cases C_c (oed.) is much less than C_c (calc.), suggesting that at stress levels just greater than the effective vertical stress in the field the clay is over-consolidated.

It was then decided to commission a second site investigation to obtain further data concerning the compressibility characteristics of the clays. The results of an oedometer test taken to a higher stress level than had been used previously are shown in Fig. 1.11. The pre-consolidation pressure was estimated to be about 600 kPa, the present effective vertical stress in the field was 220 kPa and the clay is seen to be over-consolidated by 380 kPa. As the applied foundation loading pressure was 230 kPa, the clay would remain in the over-consolidated state even after the foundation loading was applied.

A trial compaction of the very loose 7·5 m thick upper sand layer was carried out using 200 mm square reinforced piles purely as compaction piles, 7·6 m long on a triangular grid with a spacing of 0·91 m. Before compaction, the CR varied from 2 to 3 MPa, and after the trial compaction the CR lay in the range 5 to 6 MPa. It was finally decided to use 230 mm square reinforced concrete compaction piles 7·6 m long on a triangular grid with a spacing of 0·85 m and to found the structure on a 2 m thick reinforced concrete raft at shallow depth without deep piles. By placing the raft at as high an elevation as was practical the intensity of applied loading on the surface of the upper clay was kept as low as possible.

The compaction was controlled using the Dutch Cone apparatus and it was specified that the compaction would be considered acceptable when the average CR over a depth range of 1·5 m to 7·5 m was 8 MPa, with readings being taken every 250 mm at the centroid of a pile group. No single reading should be less than 5 MPa and any reading over 20 MPa would be taken as 20 MPa when calculating averages. This specification was achieved.

Settlement calculations were carried out as follows.

- Using the results of the original site investigation with no correction factors being applied, the calculated settlement for the shallow raft was in the range 250 to 375 mm, depending on the assumptions made.
- Using all the results from both site investigations and taking into account that the clays were over-consolidated, the calculated settlement for the shallow raft with compaction of the upper very loose sands was 90 mm, and for a raft piled into the lower sands was 80 mm.

It was clear that, from the settlement point of view, little was to be gained by using the far more expensive piled solution and it was decided to adopt the shallow raft. It should be noted that the settlement calculations were made prior to construction. Settlement observations were taken at six points on the raft grouped in three sets each of two points, and the results are shown in Fig. 1.12. The end of construction

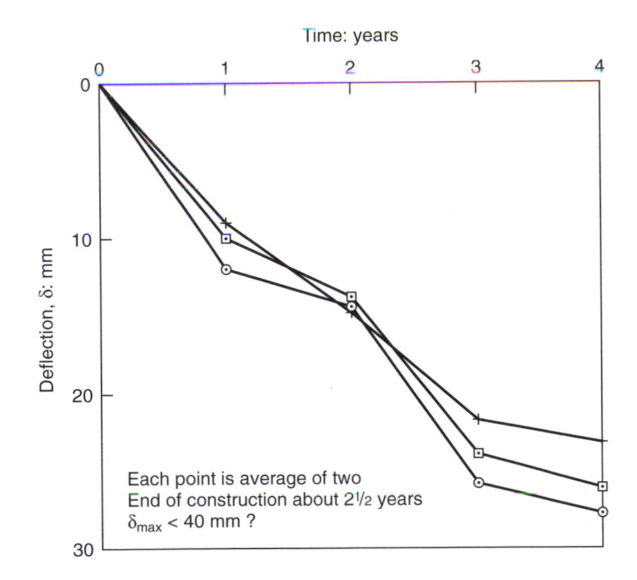

Fig. 1.12 Settlement observations, Necom House

took place after about 2·5 years and the observations were discontinued after 4 years. The maximum observed settlement was less than 30 mm and from the shape of the time settlement curves it would seem that the total final settlement will not exceed 40 mm.

This is an extremely important case record showing that the use of simple classification tests can point the way to a safe and economical foundation solution for a large structure located over a soil profile which initially could be interpreted as being unsatisfactory.

Oscar Faber and Partners were the Consulting Engineers for the project and the work was carried out under the supervision of Gordon Cantlay. The authors gratefully acknowledge their permission to make use of the valuable data obtained during the investigation.

Construction on soft clay

Conceptual design: embankments on soft clay

The geotechnical problems which need to be addressed are, firstly, the stability of the embankments and, secondly, the magnitude and rate of settlement. The assessment of stability will require reliable measurements of the undrained shear strength of the soft clay. This can be achieved using the in situ vane test at a number of locations and with close depth intervals so that a sound coverage of the whole clay deposit is obtained. High-quality sampling tubes should be used to obtain the highest quality samples possible. These samples should then be tested in undrained triaxial compression tests to obtain the undrained shear strength. Other factors to be taken into account include strength anisotropy, time to undrained failure, sampling disturbance, and size of test specimen. It may be prudent to apply a correction factor to the measured results, after Bjerrum (1973) as shown in the section 'Case study: the Brent Knoll trial embankment' in Chapter 4.

A potential failure surface could pass through both the soft clay and the embankment fill which will be compacted. These materials will have quite different stress–strain relationships and the possibility of progressive failure must be considered. Large lateral movements in the soft clay are likely to develop which could lead to the formation of vertical cracks in the fill wider at the bottom than at the top, long before the strength of the clay has been fully developed. In a stability analysis it would be prudent to ensure that a factor of safety greater than one is obtained assuming vertical tension cracks in the fill of zero strength.

Oedometer tests on the quality clay samples would enable an estimate to be made of the amount of settlement to be expected. Should it be important to estimate the rate of settlement, either to determine the post-construction settlement or because stage construction may be necessary, then the oedometer tests on small specimens are unlikely to yield satisfactory predictions. The undisturbed samples should be closely examined for fabric, for example laminations of silt or fine sand, which would suggest that the permeability of the mass in situ may well be much greater than that reflected by the small oedometer tests. Consideration should then be given to obtaining large-diameter high-quality samples, for example 250 mm in diameter, which could then be tested in large hydraulic oedometers to obtain more realistic measurements of c_v, the coefficient of consolidation.

In situ permeability tests could be carried out to obtain more reliable values of the coefficient of permeability but it should be noted that only a limited stress range is possible. No indication would be given of a possible significant reduction in permeability with increasing stress level as the height of the embankment increases.

The use of the piezocone should be considered. Considerable success has been achieved in South Africa using piezocones to predict the rate of settlement of embankments (Jones, 1974; Jones, 1975; Jones and Rust, 1981; Jones and Rust, 1982; Rust and Jones, 1990). Limitations of the piezocone include the small size of the filter area and the fact that the observed rate of dissipation of the excess pore water pressure set up by penetration of the cone into the ground may be different from the dissipation of excess pore water in the field under the *constant* load of an embankment. The dissipation of pore water pressure around a cone may be influenced by the reduction in total stresses in the ground resulting from volume change caused by flow of water around the piezocone.

It may be worthwhile to construct a trial embankment to check the predictions made. If so, every attempt should be made to ensure that the trial embankment is the same height as the prototype.

Case study: runway at Fornebu Airport, Oslo

A 500 m length of the new East–West runway for Fornebu Airport, Oslo, was located above a deposit of very soft to soft highly compressible grey silty clay with depths varying from 0 to 20 m, occasionally underlain by thin layers of sand and gravel, resting on bedrock. Over this section, the runway was required to run at about 8 to 10 m above ground level in order to tie in with the existing North–South runway. Several possible solutions to the problem were considered, for example excavation or displacement of the clay, or using piles to support the runway, but it was finally decided to construct the embankment on the clay and to accelerate the rate of consolidation by using vertical sand drains. From chainage 1200 m to chainage 1330 m the original ground surface was covered by miscellaneous fill material, the top of which varied from elevation +2·0 m to +4·5 m. From chainage 1330 m to chainage 1700 m

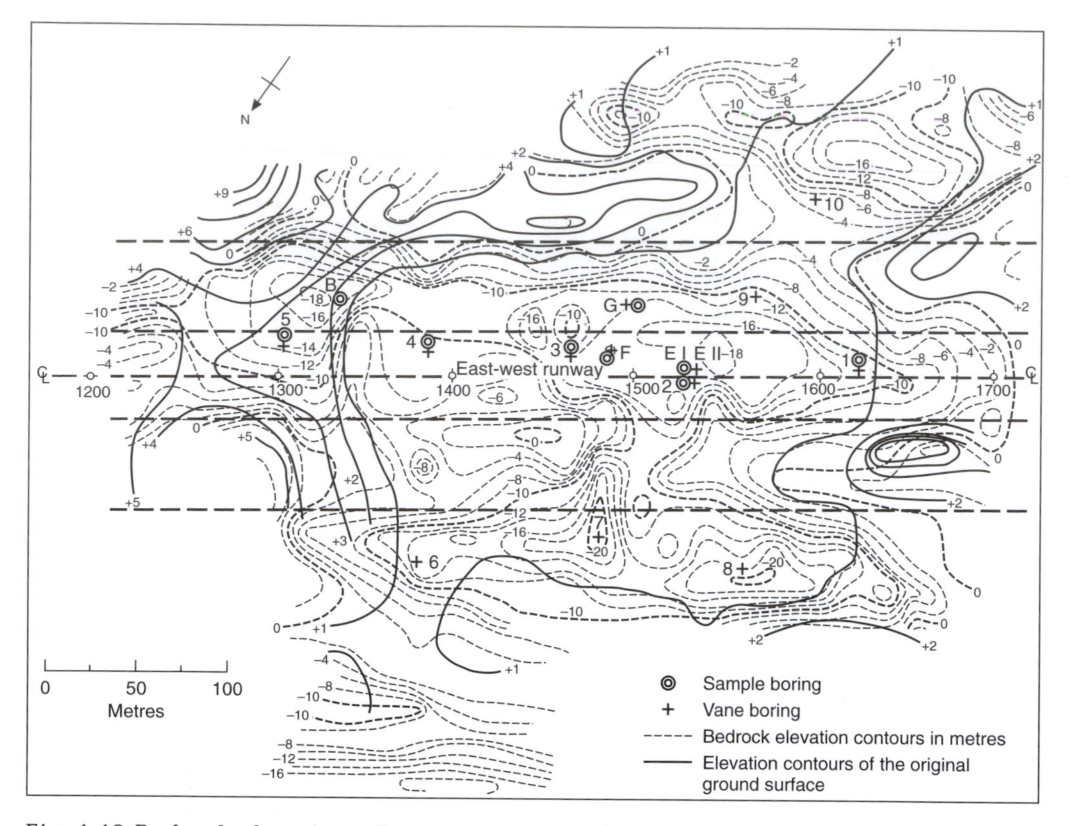

Fig. 1.13 Bedrock elevations, Fornebu airport, Oslo, from Rolfsen and Simons (1971)

the original ground surface was free of any superimposed loading and the surface elevation varied from +0·3 m to +0·7 m.

The contours of bedrock elevation are shown on Fig. 1.13. These were determined by the Swedish rotary sounding apparatus on a 20 m square grid. This was a rapid and economical operation.

The Fornebu clay was deposited in the marine environment of the Oslo Fjord late in the Pleistocene epoch during and after the retreat of the glaciers from southern Norway. The clay was analysed by micro-paleontological methods and divided into zones according to the microfossils which are indicated on a typical borehole log (Fig. 1.14).

Seven undisturbed sample borings with adjacent vane borings numbered 1 to 5, B and F, and a further five vane borings numbered 6 to 10, were carried out prior to construction, at locations shown on Fig. 1.13. The samples were taken using the Norwegian Geotechnical Institute (NGI) 54 mm diameter fixed piston sampler and the vane borings made with the Swedish Geotechnical Institute (SGI) penetration vane equipment.

It should be noted that particular care was taken to investigate whether any signs of laminations or varves could be found. A thin strip of clay was

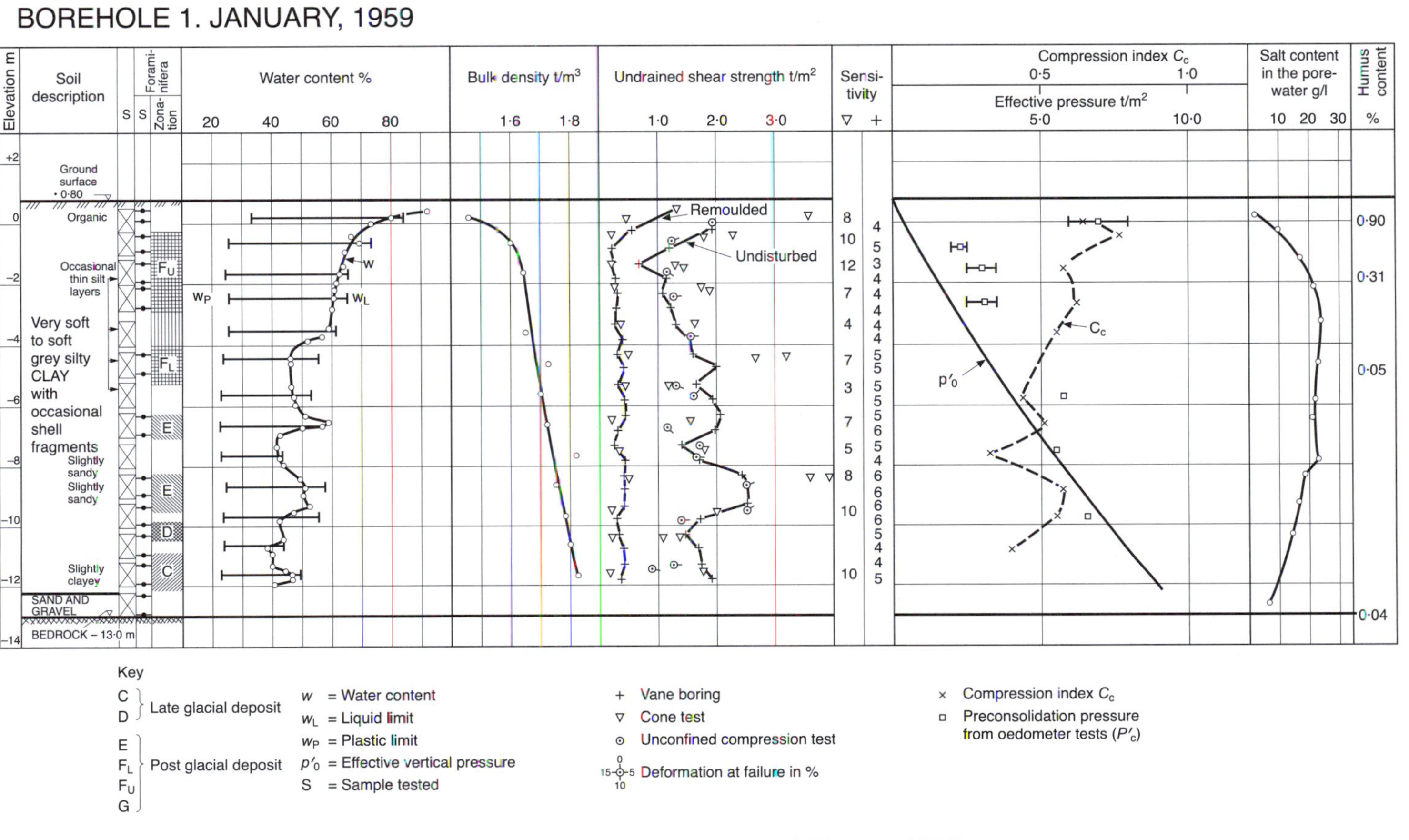

Fig. 1.14 Typical borehole log, Fornebu airport, Oslo, from Rolfsen and Simons (1971)

cut down the whole length of every sample recovered and allowed to dry slowly. It was then examined visually at various times to investigate its structure. Apart from four samples in borehole 1, two samples in borehole B and two samples in borehole 3 which showed occasional very thin silt layers, no structural features were apparent.

Field pumping tests. Many of the soundings carried out by the Swedish rotary sounding apparatus indicated the possible existence of a sand or gravel layer immediately overlying the bedrock and in three boreholes sand or gravel was encountered above the bedrock. It was therefore decided to carry out field pumping tests to determine whether, by pumping from deep holes, a lowering of the water pressure under the clay deposit could be achieved. Such a reduction, particularly if sand drains could be installed to penetrate into an underlying permeable layer, would lead to significant increases in the magnitude and rate of consolidation settlement of the clay layer. By terminating such pumping immediately before the runway was constructed, it was hoped that subsequent settlements would be reduced. The observed reductions in water pressure were small, less than 1 m head of water, and the trial pumping was accordingly terminated.

The following parameters were determined using laboratory tests:

- liquid and plastic limit and natural moisture content
- undrained shear strength by undrained triaxial test and the fall cone test, which provides a convenient way of determining sensitivity
- bulk unit weight
- salt content of the pore water, which is a useful indicator when dealing with marine clays
- clay and organic content.

Standard oedometer tests were performed and, in addition, special oedometer tests were carried out to:

- investigate the effects of temperature on the rate of consolidation (the laboratory temperature being significantly higher than the average temperature in the field)
- provide tests on samples with the axis horizontal to determine horizontal permeability
- determine K_0 by comparing pre-consolidation pressures obtained from horizontal and vertical samples
- provide tests with different durations of the loading increment
- provide tests with model sand drains
- simulate surcharge tests, where different loading intensities were applied for different periods of time and then unloaded.

Field instrumentation. Prior to any filling operations, instruments were installed to record:

- settlement of the ground surface
- settlement at various depths in the clay
- lateral movements at the ground surface near the toe of the embankment, to provide an early warning of any impending ground movement which could possibly cut off the sand drains
- lateral movements within the clay mass
- pore water pressures.

The locations of these installations can be seen in Fig. 1.15 and details of their construction in Fig. 1.16. A full description of the instrumentation and the observations made over a period of time are given by Rolfsen

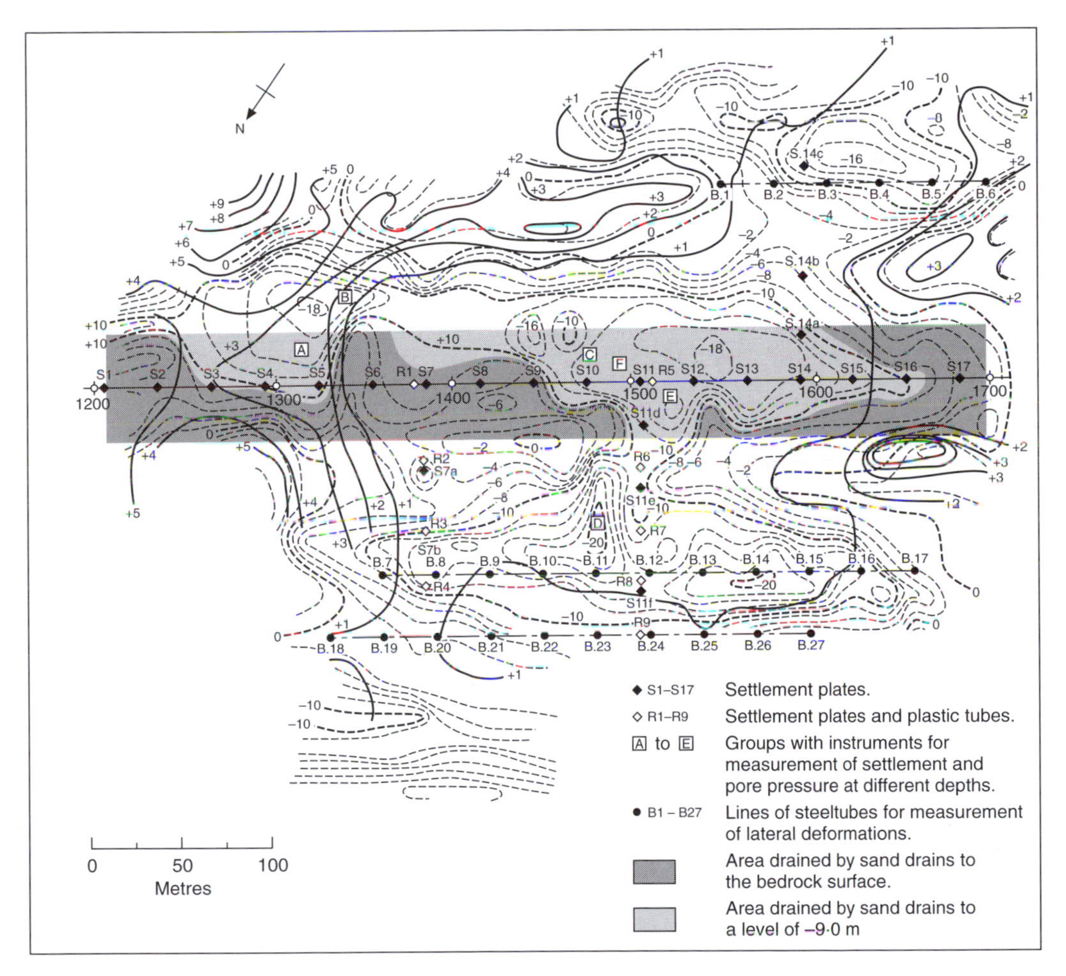

Fig. 1.15 Location plan of field instrumentation, Fornebu airport, Oslo, from Rolfsen and Simons (1971)

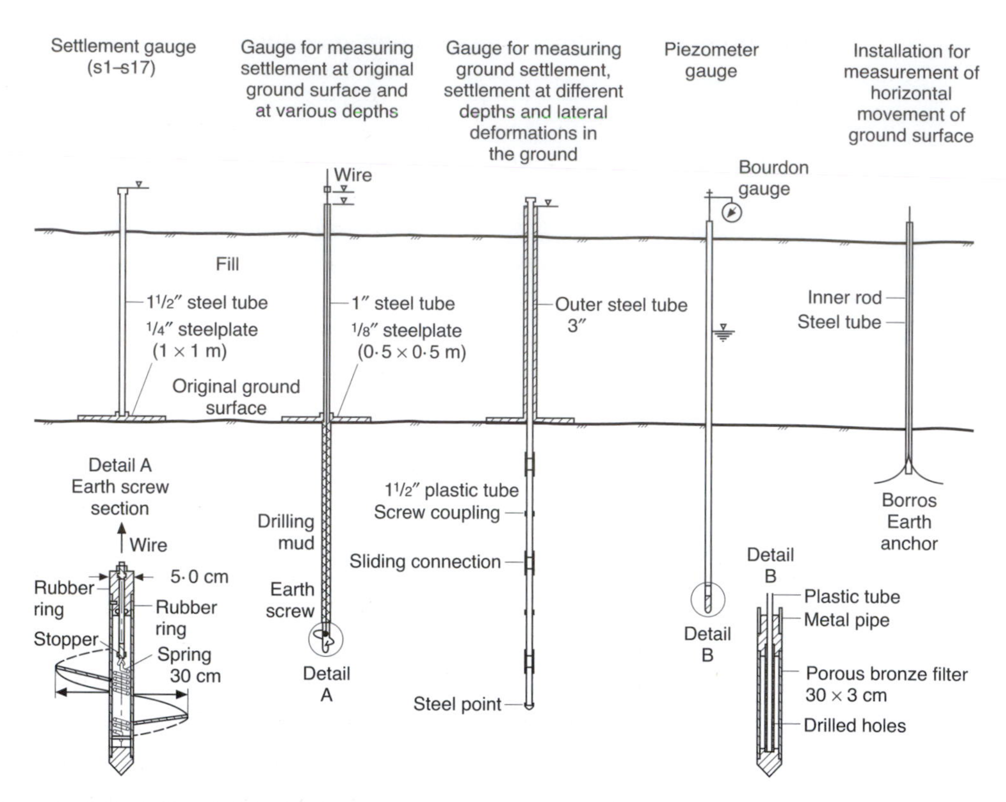

Fig. 1.16 Details of field instrumentation, Fornebu airport, Oslo, from Rolfsen and Simons (1971)

and Simons (1971). The observations show that the sand drains were effective in increasing the rate of consolidation and although appreciable horizontal ground movements developed the efficiency of the sand drain installation was not impaired.

Observations were made of the pore water pressures set by driving the mandrels necessary for the construction of the sand drains. The pore water pressures developed approached the total vertical stress in the ground so that the effective vertical stress was very small immediately after sand drain installation. There was concern for the stability of the side slopes so the undrained shear strength of the soft clay in the close proximity of the drains was checked using in situ vane tests. Only a very small reduction in undrained shear strength was recorded and it followed that the high pore water pressures observed because of the installation of the drains was primarily due to increases in the total lateral stresses in the ground as the clay was displaced laterally by driving the mandrels. When considering effective stresses in the ground it is important to bear in mind not only the vertical stresses but also the horizontal stresses.

The main points to note are as follows.

- The good agreement between the results of in situ vane tests and undrained triaxial compression tests on undisturbed samples taken by the NGI 54 mm diameter thin walled piston sampler. At depths greater than 10 m there is an indication that the laboratory test results are somewhat less than those recorded by the in situ vane; the greater the depth of the sample, the greater is the possibility of sample disturbance.
- The measured undrained shear strengths allowed the side slopes of the embankment to be designed using low factors of safety. No failures occurred although at one location in particular signs of overstress were evident with significant horizontal movements taking place in the soft clay. There is no evidence to suggest that the horizontal movements were of sufficient magnitude to sever the vertical sand drains which would have adversely affected the efficiency of the drains.
- The vertical sand drains were successful in reducing the post-construction settlement to such a level that maintenance of the runway did not pose a major problem. The previously constructed north–south runway, without sand drains, had to be re-levelled on a number of occasions, which naturally affected the operation of the airport.

The installation of the displacement mandrels used to form the sand drains disturbed the soft clay and set up very high excess pore water pressures, which approached the total vertical stresses, thereby reducing the vertical effective stress to very small values. However, in situ vane tests carried out immediately after sand drain installation showed very little reduction in undrained shear strength, indicating that the observed high excess pore water pressures were due to increases in horizontal total stresses resulting from the displacement of the clay during sand drain installation There were no signs that the stability of the side slopes was impaired. When considering effective stresses in the ground it is therefore necessary to take into account both vertical and horizontal total stresses, and not only vertical stresses, which is generally the case.

The Norwegian Geotechnical Institute acted as the geotechnical consulting engineers to the project and the work was carried out under the general direction of Ove Eide.

Case study: new quay wall at East Port Said, Egypt

This investigation is described by Hight *et al.* (2001) and the summary which follows is taken from this most instructive paper which sets the standard for the planning and execution of a ground investigation.

Extensive deposits of the Nile Delta Clays are present at the site and extend to depths of 60 m below ground level. The investigation focussed on the engineering properties of a thick deposit of highly plastic clay which extended from about 20 m to 60 m below ground level. This deposit was subdivided into two units denoted D and E. Previous investigations had indicated the possibility of the deposits being under-consolidated.

The quay wall has been designed as a novel retaining structure, using deep rectangular barrettes and involving T-panel diaphragm walls.

The aims of the ground investigation were to provide:

- details of the stratigraphy along the line of the wall
- the best estimate of engineering properties from which suitable model parameters could be selected for 3D analysis
- conventional design parameters for check calculations to be made
- sufficient information to assess construction risks.

The philosophy that was adopted to achieve these aims was:

- to determine lateral and vertical variability using the piezocone (CPTu) as a profiling tool
- to take the highest quality samples and assess the quality of individual samples
- to determine engineering properties in the minimum number of high-quality laboratory tests
- to record the detailed fabric of the soils, including photographs, in continuously sampled boreholes
- to establish site-specific correlations between CPTu parameters, fabric and engineering properties.

The layout of the ground investigation is shown in Fig. 1.17 and comprised:

- four boreholes (BH1, 6, 6A and 10), in which fixed piston, pushed open tube and Mazier samplings were carried out
- three borings (SPT1, SPT6, SPT10), alongside BH1, 6 and 10, in which Standard Penetration Tests (SPTs) and in situ vane tests were carried out
- ten cone penetration tests with measurement of penetration pore pressures (CPTu 1 to 10).

In order to obtain high-quality tube samples, 700 mm long stainless steel Shelby tubes were used, on which the cutting edge was sharpened to approximately 5° (see Chapter 4: Sampling disturbance). The tubes had no inside clearance, outside diameters of 105 mm or 86 mm, and wall thicknesses of 3 mm. The sampling tubes were used either in conjunction with hydraulic Osterberg piston samplers down to 25 m below ground level, and as pushed open tubes at depths below 25 m below ground

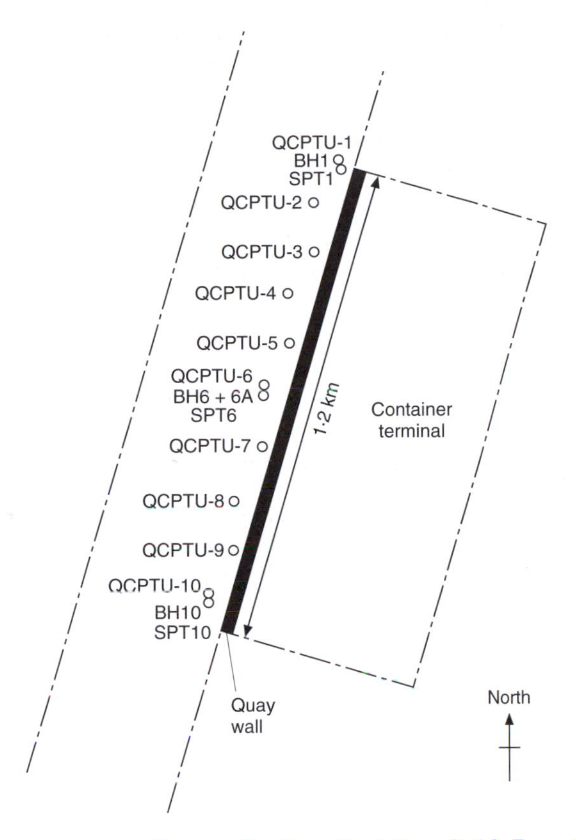

Fig. 1.17 Layout of ground investigation, East Port Said, Egypt, after Hight et al. *(2001)*

level, where the piston sampler would not function satisfactorily. 105 mm and 86 mm OD pushed open tubes were used for the full depth in BH6A. The Mazier rotary coring system utilized a triple tube core barrel the inner tube of which was rigid plastic. Cores of 101 mm and 86 mm were obtained. Profiles of index properties are shown in Fig. 1.18 and undrained strength profiles are shown in Fig. 1.19.

The particular sample tube geometry should have minimized the strains imposed during the penetration of the tube. Because of the clay's high plasticity, the soil should have been tolerant of these levels of strain, and high-quality samples should have been obtained.

Sample quality has been evaluated on the basis of

- the initial effective stress, p_i', measured in UU triaxial compression tests
- values of G_{max} measured on samples reconsolidated to in situ stresses in the simple shear, DSS, apparatus and compared to in situ values

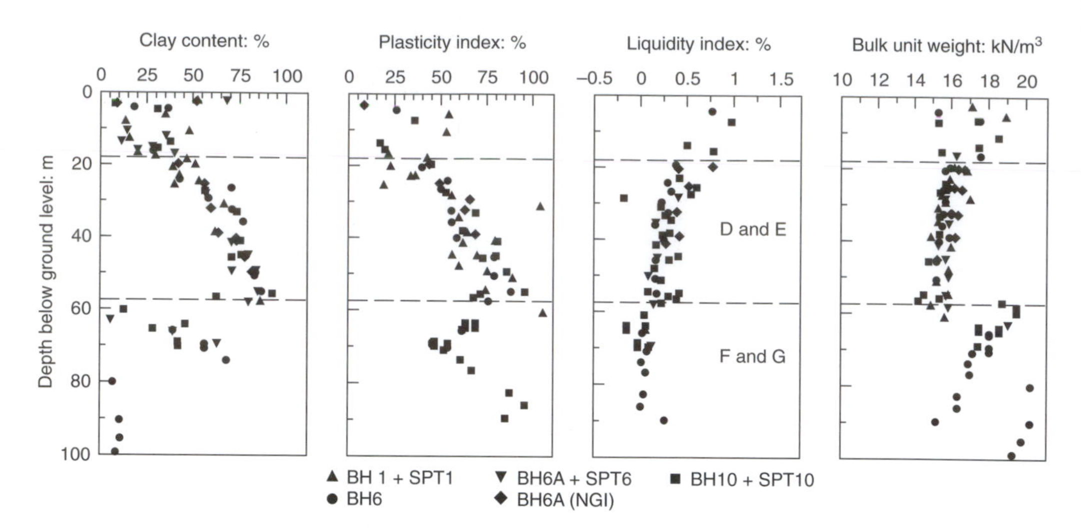

Fig. 1.18 Index properties, East Port Said, Egypt, after Hight et al. *(2001)*

- the magnitude of the volumetric strains during reconsolidation to in situ effective stresses in the oedometer and consolidated-anisotropically and consolidated-isotropically undrained triaxial tests, expressed as the ratio de/e_0 where de is the change in voids ratio and e_0 is the original voids ratio (Lunne *et al.* 1997a – see Chapter 4: Sampling disturbance).

After eliminating the results of tests on samples affected by gas exsolution and by other forms of disturbance, estimated values of overconsolidation ratio, OCR, were obtained and are plotted in Fig. 1.20.

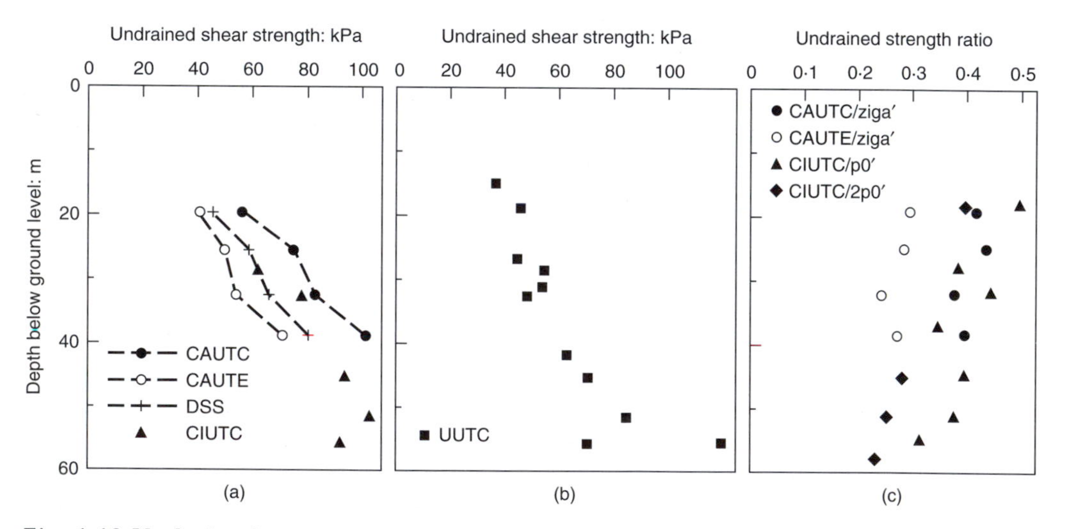

Fig. 1.19 Undrained strength profiles, East Port Said, Egypt, after Hight et al. *(2001)*

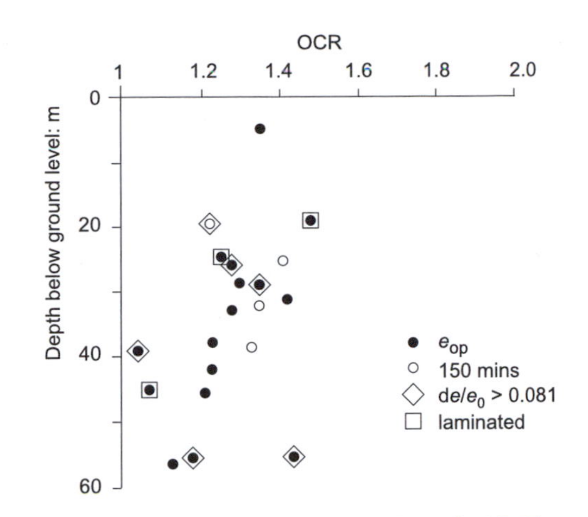

Fig. 1.20 Assessment of stress history, East Port Said, Egypt, after Hight et al. *(2001)*

Interpretation of these data requires that account is taken of the quality of each sample and the degree of saturation, S_r, in each. Eliminating results from poor-quality samples, having $de/e_0 > 0.081$ (modified boundary between good to fair and poor samples) and $S_r < 0.9$, and from laminated samples, a clear picture emerges of OCR reducing with depth through layers D and E, from approximately 1.4 to 1.2 – contrary to the prevailing view that the Nile delta clays are underconsolidated! The precise value of OCR depends on the method of interpretation and the test duration. The OCR has been determined from the NGI incremental load oedometer tests using both the Casagrande construction (see Chapter 4) and a method due to Janbu (1963). The values determined using the Janbu method have been plotted in Fig. 1.20.

Referring to Fig. 1.18, the liquidity index varied from about 0.5 at a depth of 19 m to about 0.2 m at a depth of 58 m (units D and E). These values indicate that the material is lightly overconsolidated and do not support the original view emanating from previous investigations that the deposits could be underconsolidated. It is the view of the authors that in general insufficient attention is given to liquidity index values. On a number of occasions the liquidity index has provided useful indications of the behaviour of a clay deposit, see the section on Necom House in this chapter. We suggest that the liquidity index should always be calculated and listed and plotted in the same way as natural moisture content, liquid limit, plastic limit and plasticity index.

The paper concludes that the investigation has demonstrated the benefits of taking high-quality samples and of taking into account sample quality in a systematic way in the interpretation of data. The benefits of profiling with the piezocone, using this to assess natural variability, and of relying on a limited amount of non-routine laboratory testing have also been demonstrated.

Contrary to the prevailing view, it has been shown that the deltaic clays forming units D and E are lightly over-consolidated, with over-consolidation ratio (OCR) reducing with depth. They have anisotropic strengths. The clays have unusually high plasticity indices at depth and a significant organic content which is probably responsible for the exsolution of gas on stress release. Methods to prevent gas exsolution and to allow deep piston sampling need to be introduced.

Much can be learnt from the paper of Hight *et al.* In particular, it is crucial that the aims of an investigation are clearly considered at the outset and that a philosophy is established to achieve these aims. Following Gilbert (1879), it can be said that the paper is 'the very model of a modern major general' geotechnical investigation!

Cuttings in stiff fissured clays

This is a common problem as many deep cuttings in stiff fissured clays are excavated every year. If the cutting will only be open for a short period of time, for example, to construct a cut-off for the clay core of an earth dam, an end of construction (undrained) situation exists and the stability analysis can be carried out in terms of total stress. This will require reliable measurements of the undrained shear strength.

If the cutting is required to remain operational for a long period of time, for example for a motorway, it is important to realize that after a cutting has been made the stability will decrease with time as the pore water pressures in the slopes, resulting from the unloading due to the excavation process, increase to come into equilibrium with the prevailing boundary conditions. This process may take many decades, depending on the permeability of the clay. The problem cannot be analysed in terms of total stress as the clay is swelling with time so measurements of the undrained shear strength are irrelevant. An effective stress stability analysis must be carried out and this will require the determination of the relevant shear strength parameters with respect to effective stress, as well as an assessment of the long term pore pressure distribution in the slopes.

Short term stability

When cuttings are only required to be open for a short time, it is usual for reasons of economy to operate with comparatively low factors of safety, for example about 1·4, which may be compared with the much higher factors, in the range of 2·5 to 3·0, generally adopted for the design of foundations. There is, therefore, much less margin for error and the measurement of the undrained strength must be carefully considered. Two case studies are relevant and these are now summarized below.

Case study: Short term failure of a cutting in London Clay, Bradwell
This classic case of a short term failure in a cutting in a stiff fissured clay occurred at Bradwell in Essex in 1957. It is described in detail by Skempton and La Rochelle (1965).

The undrained strength of the clay was measured on 38 mm diameter by 76 mm triaxial (vertical axis) specimens taken from borehole samples and hand-cut block samples, with a time to failure of the order of 15 min. Although this procedure is open to criticism on the grounds that the 38 mm diameter specimens are too small to be representative of the fissured nature of the clay, it is very often followed today. No significant difference could be found between the strengths of the two types of sample, and sufficient tests were made to establish firmly the variation of undrained strength with depth. Average values of the index properties of the London Clay within the depth involved in the slides are: natural water content 33%, liquid limit 95%, plastic limit 30%, clay fraction 52%.

Five possible slip surfaces were analysed for Slide I (see Fig. 1.21) and of these the three most critical all showed the actual strength of the London Clay mobilized during the failure to be only 56% (±2%) of the average strength measured in the laboratory. For Slide II the ratio was 52%. In other words, the calculated factors of safety were about 1·8 and 1·9 respectively for the two slides. The reasons for this discrepancy are now well understood.

Thus, based on the experience at Bradwell, it would be unwise to design slopes for short term stability in brown London Clay using shear strengths

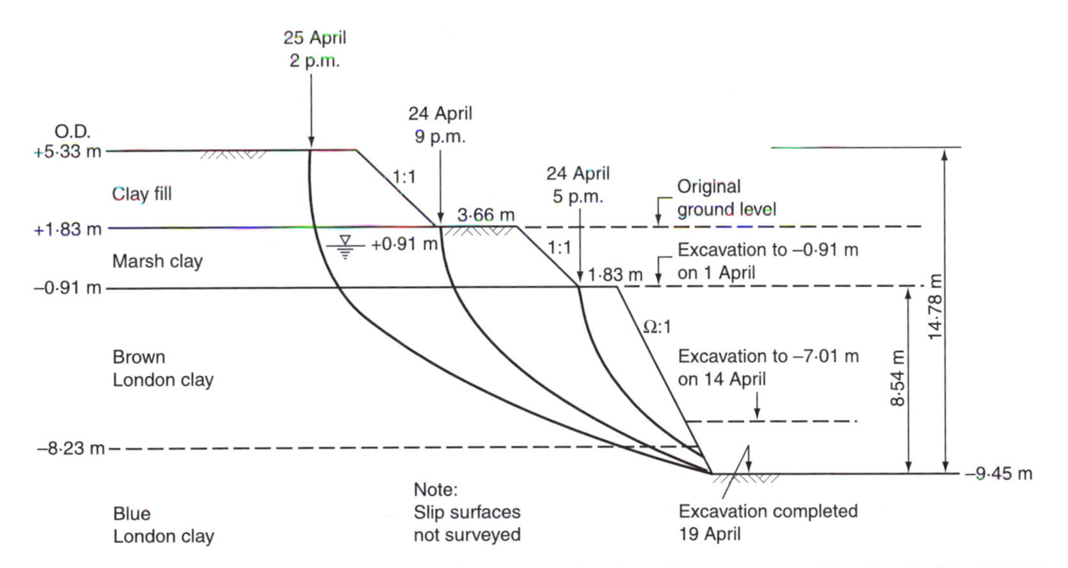

Fig. 1.21 Cross-section through Slide I, Bradwell, after Skempton and La Rochelle (1965)

exceeding 50% of the values measured in conventional undrained triaxial tests on 38 mm diameter samples.

Case study: Short term failure of a cut-off trench in London Clay, Wraysbury
This slide occurred into a cut-off trench taken down into the London Clay at Wraysbury. The slide occurred three days after the end of construction of an unsupported trench 3 m wide and 3 m deep taken down into the clay. A large proportion of the failure surface of the slip was approximately horizontal and it was therefore decided to include in the work an investigation of the variation of the undrained shear strength with orientation of the failure surface.

Undrained triaxial compression tests with a confining pressure of 207 kPa were carried out on 38 mm diameter by 76 mm high specimens with various inclinations of the axis from blocks about 250 mm cube, which in turn were hand-cut from an open excavation. The results of the tests are shown in Table 1.2 (Simons, 1967).

In an attempt to eliminate some of the scatter, the results have been corrected to a common water content of 28%, for the purpose of comparison. It can be seen that the greatest strength was obtained for samples with their axes horizontal, and least for samples with axes inclined at 45° to the vertical. Also, when the axis is inclined at 56° to the vertical (the failure surface in the test then being approximately horizontal referred to the field) the strength measured is 77% of that with the axis vertical. Thus for this particular slip where the failure surface was mainly horizontal, the normal method of measuring strength would result in appreciable error.

It is well known that undrained strength measurements made on fissured clays are influenced by the size of the specimens tested. To investigate this, triaxial specimens of sizes 305 mm diameter by 610 mm,

Table 1.2 Comparison of undrained strengths for vertical and inclined triaxial test specimens of London Clay from Wraysbury, after Simons (1962)

Axis of test	No. of tests	Water content w: %	Time to failure: min	Undrained strength s_u: kPa	Corrected undrained strength s_u: kPa (w = 28%)	Strength ratio (compared with vertical strength)
Vertical	12	28.1	7	116	117	1.00
Horizontal	12	28.1	6	124	125	1.07
45°	12	28.2	6	82	84	0.72
56° to vertical	12	27.0	4	103	90	0.77

Table 1.3 Comparison of undrained strengths for different sizes of triaxial test specimens with respect to 38 mm × 76 mm triaxial tests from U-4 sampler for London Clay from Wraysbury, after Simons (1967)

Size of test specimen: mm	No. of tests	Water content w: %	Time to failure: min	Undrained strength s_u: kPa	Corrected undrained strength s_u: kPa (w = 28%)	Strength ratio compared with 38 × 76 from U-4 sampler
305 × 610	5	28.2	63	49	51	0.62
152 × 305	9	27.1	110	51	46	0.56
102 × 203	11	27.7	175	48	46	0.57
38 × 76 (U-4)	36	26.9	8	93	82	1.00
38 × 76 (blocks)	12	28.1	7	116	117	1.43
13 × 25 (intact)	19	26.6	10	262	219	2.68

152 mm diameter by 305 mm, 102 mm diameter by 203 mm, 38 mm diameter by 76 mm, and 13 mm diameter by 25 mm were tested under undrained conditions with their axes vertical and a confining pressure of 200 kPa. The results are shown in Table 1.3. Again, to reduce scatter, the measured strengths were corrected to a moisture content of 28%. The main points to emerge are as follows.

- 305 mm diameter by 610 mm, 152 mm diameter by 305 mm and 102 mm diameter by 203 mm diameter test specimens gave approximately the same measured strength, which was about 60% of that obtained from 38 mm diameter by 76 mm specimens taken by a U-100 sampler.
- 38 mm diameter by 76 mm test specimens from hand-cut blocks indicated a strength 43% higher than 38 mm diameter by 76 mm test specimens from the U-100 sampler, showing the effect of the disturbance caused by the U-100 sampling (U-100 is shown in Tables 1.3 and 1.4 as U-4).
- The highest measured strength was obtained for the 13 mm diameter by 25 mm intact specimens.

In addition to the laboratory shear tests, undrained direct shear tests were carried out in the field, shearing a block of clay 610 mm square. Table 1.4 compares the results from an analysis of the slip, the 610 mm square in situ shear box, the 305 mm diameter by 610 mm, and the 38 mm diameter by 76 mm triaxial tests. In both sets of triaxial tests the axes of the samples were vertical.

It can be seen that measurements of undrained shear strength are related to how the measurements are carried out and great care must be exercised when choosing a particular method of test. This point is covered in more detail in Chapter 4.

Table 1.4 Comparison of undrained strengths for different types of test with respect to slip strength for London Clay from Wraysbury, after Simons (1967)

Size (in mm) and type of test specimen	Water content w: %	Time to failure T_f: min	Undrained strength s_u: kPa	Corrected undrained strength s_u: kPa (w = 28%)	Corrected undrained strength s_u: kPa (w = 28%) (T_f = 4000 min)	Strength ratio compared with slip
Slip	29.3	4000	30	35	35	1.00
610 × 610 shear box	28.1	71	48	48	41	1.17
305 × 610 triaxial vertical	28.2	63	49	51	43	1.23
38 × 76 triaxial vertical (U-4)	26.9	8	93	82	66	1.88

Long term stability

The planning of a site investigation for the long term stability of a cutting in stiff fissured clay requires a thorough understanding of the mechanics of the failure of first time slides in such cuttings. The development of railway and road systems in many countries required the construction of deep cuttings in stiff fissured clays. It quickly became clear that major slope instability problems could develop with time: steep cuttings would fail during or shortly after the end of construction; in less steep slopes, failure could occur months or years or decades after a cutting was made. It was also recognized that the failures were associated with a reduction in strength with time. Initially, attempts were made to analyse slopes using the undrained shear strength with an allowance made for a reduction in undrained strength due to 'softening'. Later on, it was felt that stability should be assessed in terms of effective stress and not total stress. It was thus necessary to:

- determine the in situ strength at failure in terms of effective stress
- find the best method of measuring or predicting this strength
- find an explanation for the long delayed failures.

The reader is referred to Professor Skempton's Special Lecture at the Ninth International Conference on Soil Mechanics and Foundation Engineering held in Tokyo (Skempton, 1977). This represented a major breakthrough in understanding the mechanics of slope instability in cuttings in stiff fissured clays.

An effective stress analysis must be used and hence it is pointless to carry out measurements of the undrained shear strength. The relevant shear strength parameters in terms of effective stress are not the peak

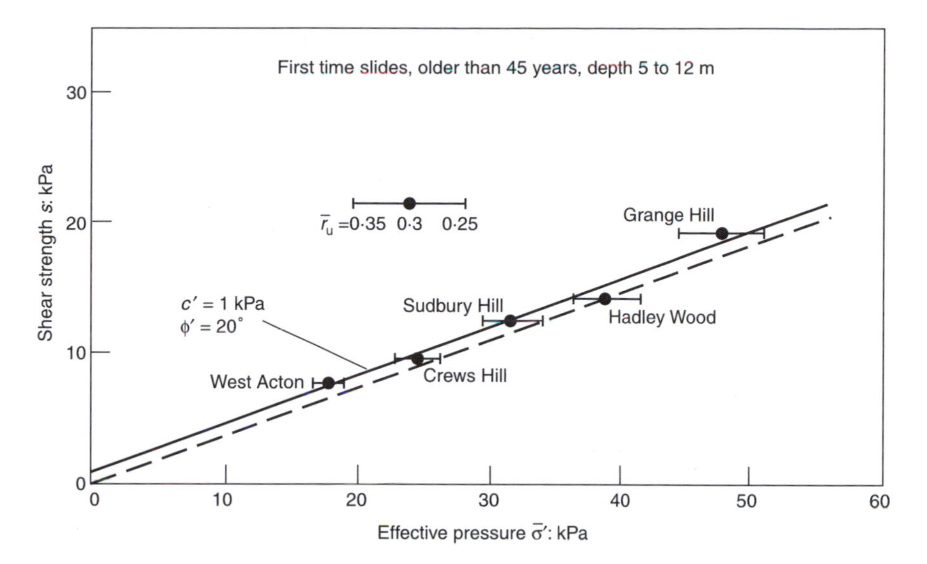

Fig. 1.22 Effective stress parameters, brown London Clay, after Skempton (1977)

parameters nor the residual parameters but approximate to the critical state values or the 'fully softened' strength. These values can be obtained from suitable effective stress triaxial or shear box tests. Effective stress parameters for brown London Clay are shown in Fig. 1.22.

It is also necessary to know the pore water pressures in the slopes at equilibration and this has to be obtained by monitoring existing long term cuts. The variation of $\bar{r}_u = \gamma_w h/\gamma z$ with time for cuttings in brown London Clay, is shown in Fig. 1.23.

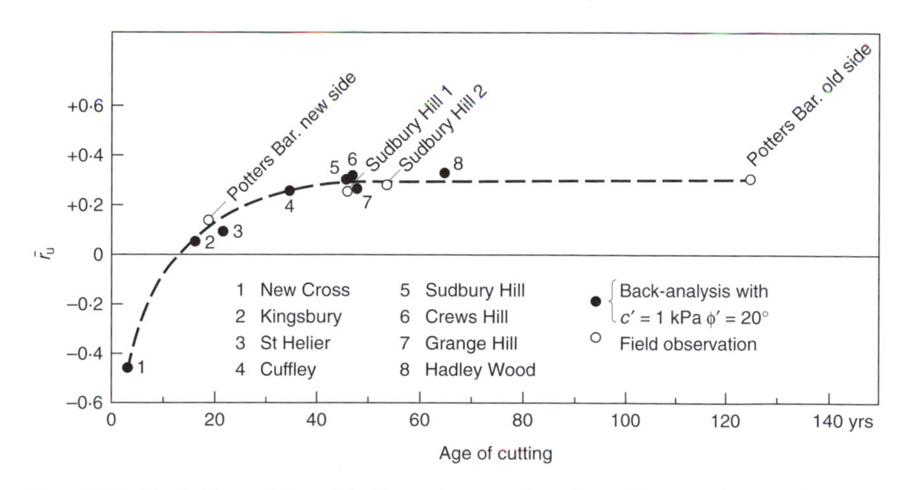

Fig. 1.23 Variation of $\bar{r}_u$ with time, brown London Clay cuttings, after Skempton (1977)

In his Special Lecture in Tokyo, Professor Skempton drew the following conclusions.

- The shear strength parameters of the brown London Clay relevant to the analyses of first time slides are $c' = 1\,\text{kPa}$ and $\phi' = 20°$.
- The peak strength even as measured on large samples is considerably higher, so some progressive failure mechanism appears to be involved.
- The in situ is given approximately by the 'fully softened' value and also by the lower limit of strength measured on structural discontinuities (joints and slickensides).
- The residual strength is much smaller than this and corresponds to the strength mobilized *after* a slip has occurred, with large displacements of the order of 1 or 2 m.
- It is a characteristic feature of first time slides in London Clay that they generally occur many years after a cutting has been excavated.
- The principal reason for this delay is the very slow rate of pore pressure equilibration; a process which in typical cuttings is not completed, for practical purposes, until 40 or 50 years after excavation.
- At equilibration, $\bar{r}_u = 0{\cdot}3$.

It should be pointed out that should a pre-existing failure surface be present in a slope, then the relevant parameters for design or analysis are the *residual* values. It is also necessary to check that when an excavation is taken down to full depth there is no danger of a bottom heave failure occurring due to water pressure in an underlying permeable layer. It is prudent to take boreholes on the line of a cutting down to a depth below the bottom of the cutting that is at least equal to the depth of the cutting and if a permeable stratum is revealed, a piezometer should be installed to record the water pressure.

Natural slopes

When investigating the stability of natural slopes, i.e. slopes that have been in existence for long periods of time (decades or more) and the pore water pressures are in equilibrium with the boundary conditions, it is vital to realize that measurements of the undrained shear strength are quite irrelevant as they do not reflect the influence of the pore water pressures acting on a potential failure surface in the field. An effective stress stability analysis is required using the strength parameters relevant to the type of clay under consideration.

These parameters can only be obtained from case records where failures have occurred and the strength mobilized in the field can be compared with that measured in the laboratory. It has been found that for *stiff fissured clays* the in situ strength falls to the residual strength as a result of progressive failure associated with down-slope creep movements over long periods of time. The residual strength can be obtained from:

- back-analyses of field failures
- ring shear tests
- reversing tests in a shear box
- triaxial tests on samples containing a failure plane at an appropriate angle, the failure plane may be formed naturally in the field following a failure, or it may be formed artificially in the laboratory.

On the other hand, back-analyses of slides in natural slopes in *intact clays*, i.e. clays without fissures, slickensides, joints or structural features of any kind, shows that the in situ strength at failure approximates to the peak strength. It is clearly of the greatest importance that clay specimens are carefully examined to determine whether or not structural features are present. It would appear that progressive failure does not take place in intact clays, but when discontinuities are found, progressive failure reduces the strength from the peak to the residual. The peak strength can be obtained from effective stress tests in the triaxial apparatus or the direct shear test (shear box).

The distribution of pore water pressure in natural slopes will be controlled by the climatic conditions prevailing at any particular site and must be obtained from piezometer observations taken over long periods of time to ensure that the must critical values are known. In the UK, for example, it is believed that for many sites the water table may rise close to the ground surface after very wet winters. For further information on this subject, reference can be made to the work of Simons *et al.* (2001).

Site investigations for slope failures

When planning an investigation of a slope that has failed, the principles outlined above should be borne in mind, but there are some important additional factors to be considered. Firstly, the geometry of an existing slope can be easily ascertained, but that of a failed slope must be reconstructed. Secondly, the pore water pressures in a stable slope can be measured but those existing in a failed slope at the time of failure have to be assessed.

Possible sources of information which may be helpful when attempting to re-establish the original slope profile of a failure include the following.

- Contoured maps, provided they are up to date and include details of recent alterations of slope geometry, e.g. excavations or undercutting at the toe of a slope or placement of fill higher up the slope.
- Making cross-sections through the stable slopes either side of the slip and extrapolating between them. An example of this is shown in

Fig. 1.24 Locations of field work, Drammen, after Kjaernsli and Simons (1962)

Fig. 1.24 for the landslide at Drammen (Kjaernsli and Simons, 1962), where two sections A and C upstream and downstream of the failure were used to re-construct a cross-section B for the slip itself. The failure having occurred, it is always possible that the cross-section for the slip was more adverse than that obtained by extrapolating from the adjacent stable profiles.

- Photographs, which may be aerial or taken from the ground. The dates on which the photographs were taken are of importance as older photographs may not reflect recent changes in slope profile.
- Witness reports from local inhabitants can provide valuable information. These should be treated with the greatest of caution as it is the experience of the writers that the recollections of different observers

may differ greatly which could lead to an incorrect and misleading assessment of the slope geometry.
- Geotechnical and engineering geology journals.
- Previous site investigation reports.
- Newspapers.

When carrying out a walk-over survey, the following features should be looked for:

- tension cracks
- toe bulges
- lateral ridges
- uneven ground
- graben features
- back-tilted blocks
- marshy ground
- streams
- springs and ponds
- cracked roads
- inclined trees
- breaks and cracks in walls
- distorted fences
- drainage paths, both natural and man-made.

Possible triggering causes of a landslide are as follows.

- Rainfall. Detailed records covering months or years preceding the landslide should be obtained and carefully considered. In some cases it is the intensity immediately prior to the slip that has a great influence and on other occasions it is the gradual increase in pore water pressure in the ground due to heavy rainfall over a long period of time that triggers a failure.
- Ground vibrations resulting from seismic activity, or underground mine working, or due to heavy trucks or trains, or from military activity.
- High wind acting on tall buildings or large trees.
- Removal of vegetation on a slope leading to an increase in pore water pressure.
- Changes in surface groundwater flow patterns. These could be due to deterioration or rupture of drains or diverting of existing flow patterns and can initiate failures in slopes some considerable distance away.
- Blockage of drains and weep holes can lead to increases in pore water pressures in the ground. It follows that drains and weep holes should be properly designed and maintained. Failure to do so could have serious consequences.

- Changes in slope geometry, e.g. undercutting near the toe of a slope either by natural causes (river erosion) or man-made excavations; placing fill near the top of a slope may initiate a landslide.
- Failure of an earth retaining structure.
- Increased pore water pressures as a consequence of pile driving.
- Loss in strength as a result of changes in soil chemistry, e.g. reduction in salt content of the pore water in marine clays by leeching by fresh water.
- Partial disturbance, for whatever reason, of a sensitive clay.
- Rapid draw-down in water level in an adjacent lake or river.

Case study: Failure of a natural slope in soft intact clay, Drammen, Norway
A comprehensive investigation into the failure of a natural slope in soft *intact* clay has been described by Kjaernsli and Simons (1962). On 6 January 1955 a rotational slide occurred in the north bank of the Drammen River at the town of Drammen in Norway, about 35 km south west of Oslo. The slide took place in soft intact marine clay of post-glacial age. The clay has occasional extremely thin seams of silt and fine sand. The investigation included five boreholes in which 54 mm diameter thin-walled stationary piston sampling tubes were used to recover samples 800 mm long at very close depth intervals. The locations of the boreholes, the eight in situ penetration vane tests, seven rotary soundings to determine depth to bedrock and two penetration piezometers are shown in Fig. 1.24 and a typical borehole log is given in Fig. 1.25. The close spacing of the samples can be seen. Liquid and plastic limit tests were carried out on most samples with many determinations of the natural moisture content. Knowledge of the plasticity index and the liquidity index can be most useful. Other laboratory tests carried out included:

- determination of salt (NaCl) content of the pore water
- determination of clay fraction
- determination of bulk unit weight
- determination of undrained shear strength using the unconfined compression test and the fall cone apparatus on both undisturbed and completely remoulded clay; the strain at failure in the unconfined compression tests is shown on the borehole log to give an indication of the degree of disturbance of a sample
- a comprehensive range of effective stress triaxial tests with isotropic and anisotropic consolidation with either drained shear or undrained shear with pore pressure measurement. The average values of the peak parameters were $c' = 2\,\text{kPa}$ and $\phi' = 32{\cdot}5^\circ$.

It should be noted that no residual shear strength determinations were made on the Drammen clay at the time the stability investigations were

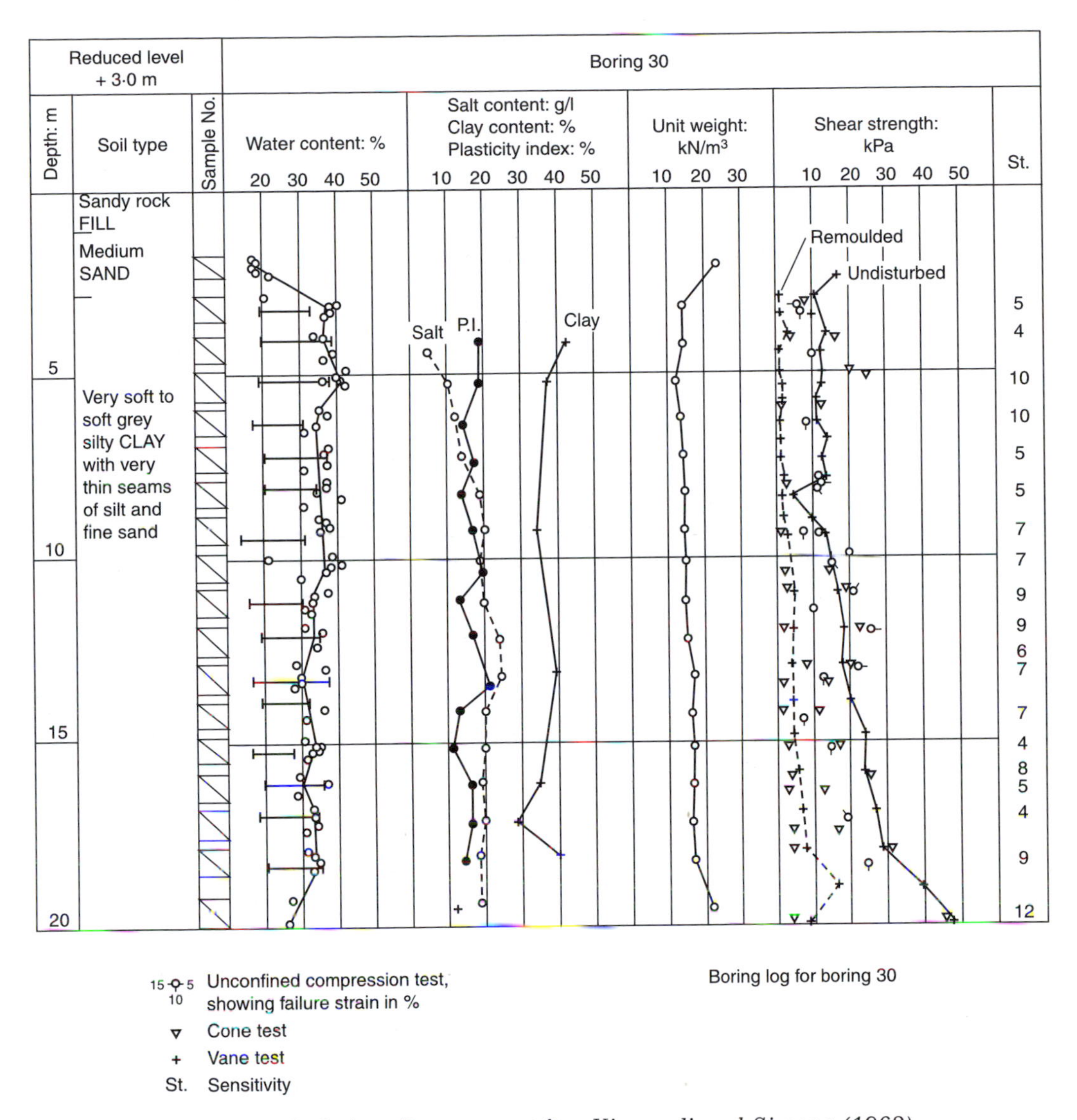

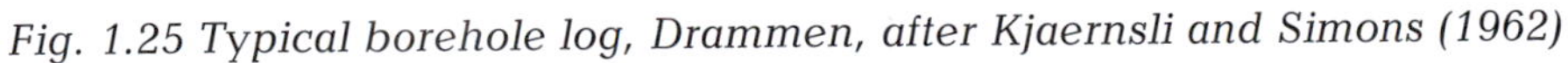
Fig. 1.25 Typical borehole log, Drammen, after Kjaernsli and Simons (1962)

carried out. We believe, however, that the peak strength is close to the residual strength and that this is why progressive failure was not a factor in the landslide.

The undrained shear strength was also measured using in situ vane tests and reasonably good agreement with the laboratory values was obtained although some sample disturbance of the deeper samples was indicated.

The results of piezometer observations at different depths and different locations are shown in Fig. 1.26 and indicate hydrostatic conditions. In

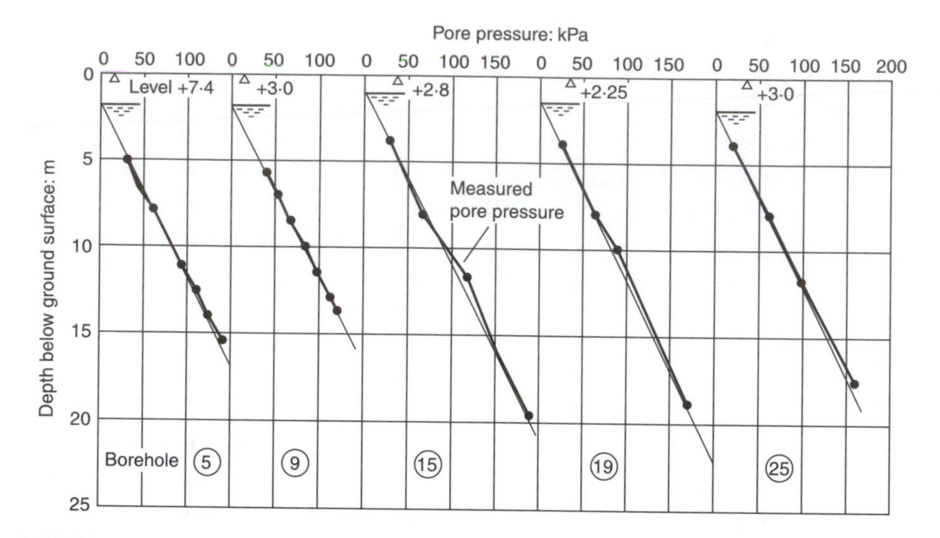

Fig. 1.26 Piezometer observations, Drammen, after Kjaernsli and Simons (1962)

the experience of the authors, this is an unusual case as very often piezometer observations at different depths in a clay do not show hydrostatic conditions. The piezometer tests were carried out by pushing the Geonor apparatus into the ground to a certain depth, waiting for equilibration to take place (which took 2 to 4 days), then taking the observation and repeating the procedure at a different depth. In this way the distribution of pore pressure with depth was quickly obtained.

The main points to note are as follows.

- The total stress stability analysis based on field or laboratory undrained shear strengths cannot be applied to determine the stability of the natural long term slopes of the Drammen River because the calculated safety factors by this method lie approximately 50% below the correct values.
- The Bishop effective stress analysis using peak effective stress shear parameters leads to satisfactory values for calculated safety factors for the *intact* Drammen clay.
- Provided that the failure criterion of maximum principal effective stress ratio is adopted for undrained tests, the values of the cohesion intercept and the angle of shearing resistance with respect to effective stress are, for all practical purposes, independent of
  - the stress path followed in the tests
  - whether the consolidation is isotropic or anisotropic
  - whether the tests are carried out drained or undrained.

Vegetation

The adverse effects of vegetation or the removal of vegetation on the behaviour of clay soils are well known. The shrinking of clay as a result of evaporation and transpiration by vegetation can cause much damage to low-rise buildings. Cutting down trees on clay soils will lead to swelling of the clay as the effective stresses in the soil decrease with time and to a heave of the ground surface and shallow foundations. The stability of slopes in clay soils will also decrease as a result of the swelling process. The growth of trees can result in relatively deep drying or desiccation of the soil which can extend to 5 m below the ground surface and horizontally up to 15 m away from the tree.

The distribution of pore water pressure with depth is illustrated in Fig. 1.27. If there is no flow of water in the soil, then a hydrostatic distribution is obtained both above and below the water table which represents the point in the soil where the water pressure is zero (or more strictly where the water pressure is equal to the atmospheric pressure). When water is lost from the soil due to evapotranspiration, large suctions can be set up in the soil profile above the water table and the soil is in a *desiccated* state. Desiccation may also be induced beneath furnaces or brick kilns. If the water loss from the soil is stopped by felling trees or covering the ground surface with a building or a road, soil will start to swell as the distribution of pore water pressure moves towards the hydrostatic line and the amount of upwards movement can be large. Cheney (1988) recorded about 160 mm of heave of a single-storey building over a period of 25 years. The amount of movement is governed by the plasticity of the

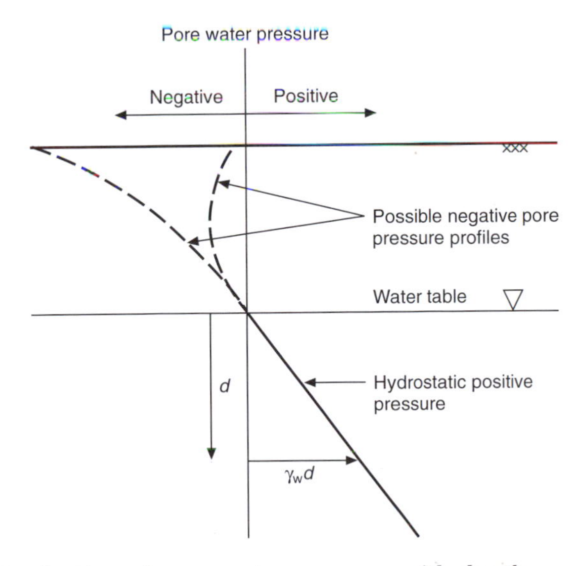

Fig. 1.27 Distribution of pore water pressure with depth

Table 1.5 Volume change potential related to plasticity index

Plasticity index: %	Potential for volume change
Over 35	Very high
22–48	High
12–32	Medium
Less than 18	Low

clay, the higher the plasticity, the greater the movement. The Building Research Establishment (1980) suggested that an indication of volume change potential can be related to the plasticity index as shown in Table 1.5. The activity chart of van der Merwe (1964) shown in Fig. 1.28 has been used successfully in South Africa.

No matter how plastic the clay, it must be in a desiccated state for heave to develop. It is of the greatest importance that a site investigation establishes whether or not a state of desiccation exists on a particular site. The Building Research Establishment (1996) has identified four main techniques for detecting desiccation:

- comparisons of soil water contents with soil index properties, e.g. the liquidity index
- comparisons of soil water content profiles
- comparisons of strength profiles
- effective stress or suction profiles.

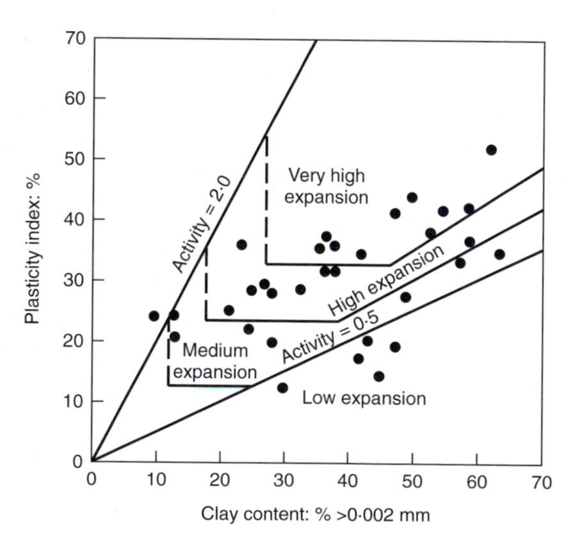

Fig. 1.28 Activity of van der Merwe (1964) for estimation of the degree of expansiveness of a clay soil. Some expansive clays from Kwa Zulu, Natal, South Africa, are shown (after Belland Maud, 1995)

The filter paper method of suction determination provides a simple, rapid and economical way of determining a suction profile (Chandler *et al.*, 1992a; Crilly and Chandler, 1993). A new instrument for measuring soil suction (Ridely and Burland, 1993) should provide more accurate measurements.

An interesting method for assessing desiccation in clay soils has been proposed by Pugh *et al.* (1995) and was originally developed by Dr A.G. Weeks. It is based on the use of the dial penetrometer which was intended to give a rough indication of the presumed bearing pressure for a foundation on a clay soil. The dial penetrometer reading gives directly the bearing pressure for a factor of safety of about three against shear failure in the ground.

The area of soil tested is small, having a diameter of 6.25 mm. This makes its use questionable in fissured clays as the results probably reflect the strength of the intact clay, while the behaviour of a large-scale foundation is influenced by the strength of the fissured clay which may be much smaller – see for example the discussion earlier in this chapter under the headings 'Cuttings in stiff fissured clays: short term stability'. Pugh *et al.* (1995) point out that the undrained strength of the intact clay between the fissures is a sensitive indicator of the mean effective stress and therefore of the suction in the clay. They present the results of many penetrometer tests which have been carried out on London Clay, and assess the depth and extent of desiccation by comparing dial penetrometer readings in a desiccated area with readings taken from a control borehole at a location where moisture conditions are at or close to equilibrium. Initially, the penetrometer tests were carried out on 38 mm diameter samples obtained from hand auger boreholes, but more recently samples taken from driven tube 'window' samplers have been used with success. Up to six 5 m boreholes with readings at 250 mm intervals can be carried out in a day.

Figure 1.29 (taken from Pugh *et al.*, 1995) shows a plot of penetrometer readings against depth adjacent to a 25 m high oak tree on London Clay (Borehole 1) and close to the former location of a 10 m high willow, but also within the zone of influence of a remaining tree. Also shown in Fig. 1.28 is the profile representing equilibrium conditions for the London Clay which has been derived from hundreds of such investigations. It can be seen that for Borehole 1 the clay is highly desiccated to a depth of up to 6 m, while for Borehole 2, although there has been some recovery, i.e. swelling at depth, the soil above 2 m remains highly desiccated.

The data presented by Pugh *et al.* relate to London Clay, but the technique should be applicable to any stiff over-consolidated clay. For example, successful applications have been recorded for Gault Clay and

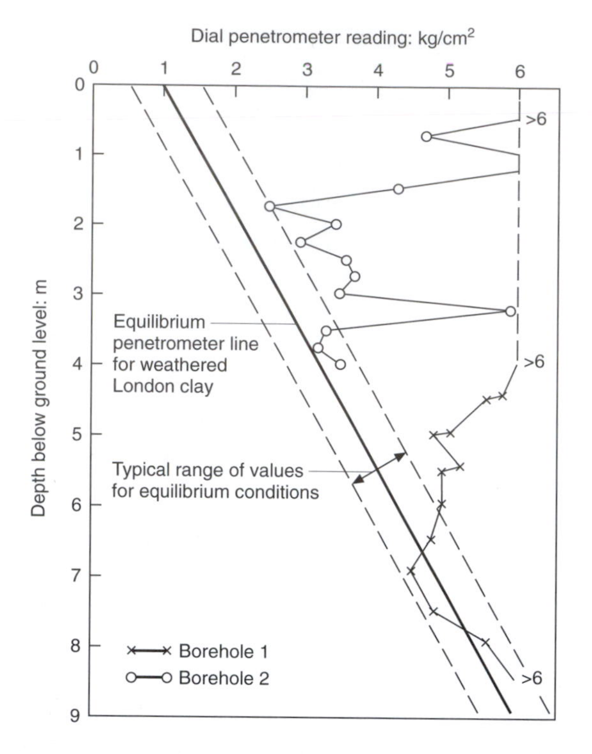

Fig. 1.29 Penetrometer readings showing desiccation, London Clay, after Pugh et al. *(1995)*

to a lesser extent for Weald and Wadhurst Clays. The technique satisfies the requirements of speed, low cost and reliability and is sufficiently sensitive to monitor moisture recovery following tree removal or pruning.

The double oedometer test described by Jennings and Knight (1957) can be used to obtain an estimate of the amount of heave which may develop. Two hopefully identical samples are tested in the oedometer apparatus. One is completely dry; the other, after application of a small seating load, is flooded with water and heave is allowed to take place. After the heave is completed, the sample is loaded up in the usual manner (Fig. 1.30). The dry test is used to give the initial voids ratio in the field at the total overburden pressure corresponding to the depth of the sample, while the flooded test is used to give the final voids ratio under an effective stress equal to the effective stress after the application of the foundation loading. The difference in voids ratios enables the heave to be calculated.

Collapsing soils

Some soils exhibit the tendency to collapse when loaded by a foundation and subsequently flooded. The collapse is usually sudden. The soils may

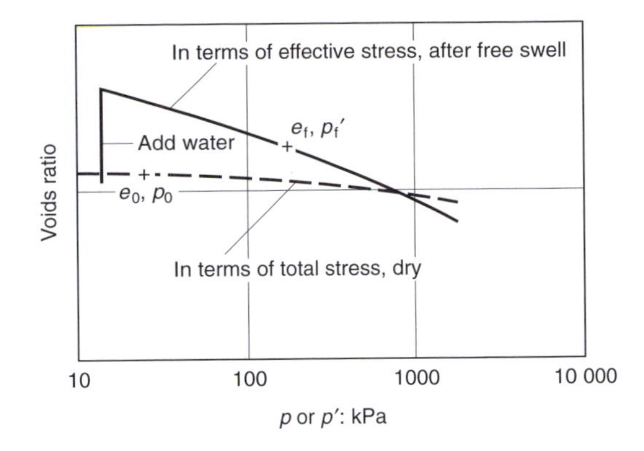

Fig. 1.30 Principle of double oedometer test. p_0 = *initial total stress;* p_f' = *final effective stress after free swell and load application*

be loess, brickearth, wind blown silts, or acid crystalline rocks such as granite or highly feldspathic sandstone which have undergone prolonged weathering to produce intensely leached residual soils. Collapsing soils show high voids ratios and relatively low densities, but in the dry state appear to be quite strong although they are susceptible to large reductions in voids ratio upon wetting following the application of foundation loading. In other words, the metastable structure collapses when the bonds between the grains break down in the presence of water.

Such soils can be recognized by placing an undisturbed sample in the oedometer apparatus and applying a loading equal to the overburden pressure plus the loading increment due to the foundation and then flooding the sample with water. A typical result of such a test is shown in Fig. 1.31.

Sands

The difficulties involved in obtaining samples of sand have been described by Clayton *et al.* (1995b) and in Chapter 4 of this book. Samples are often lost when raising a sampler to the ground surface. Very loose sands may be compacted by the sampling procedures. As a general rule it can be said that while some samplers can provide samples with the correct distribution of particle size, in situ probing may be the most reliable method for obtaining relative density. Available sampling techniques include:

- thin-wall piston samplers
- open-drive samplers
- the Bishop sand sampler
- core catchers
- impregnation
- freezing.

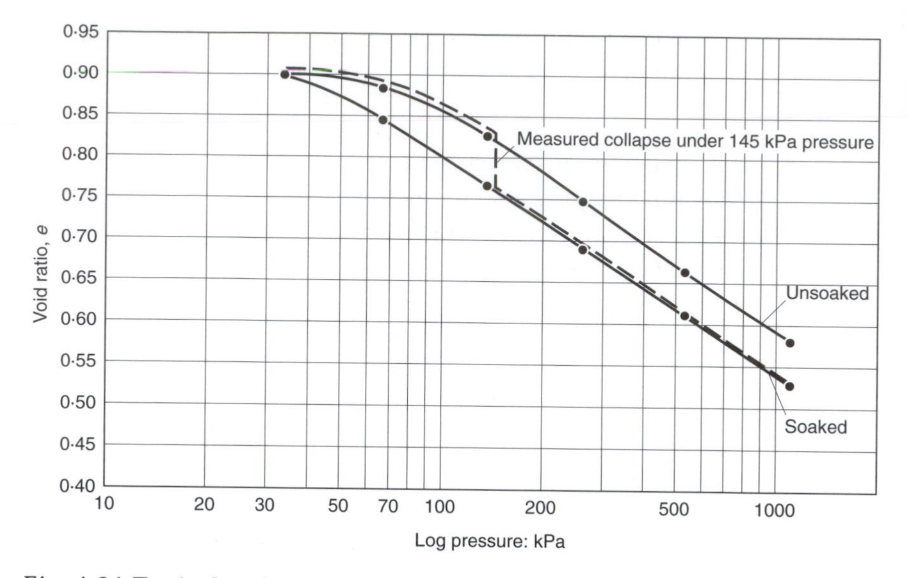

Fig. 1.31 Typical oedometer test result, collapsing soil

It is, however, vital that samples are recovered and the correct distribution of particle size is obtained. In particular, the amount of fines present must be correctly determined. Potential geotechnical problems involving sands may be:

- groundwater lowering
- groutability
- pumping of sand as fill
- response to geotechnical processes
- liquefaction.

When investigating these problems, some or all of the following soil properties are required:

- grain size distribution
- permeability
- relative density
- particle shape and angularity
- hardness of grains
- chemical composition.

The site investigation must be carefully designed so that the soil properties relevant to the particular job in hand are obtained.

Groundwater lowering

If an excavation has to be carried out below the water table some form of groundwater control will be necessary. This may be done using one of the following techniques, or a combination of two or more of these techniques:

- pumping from open sumps
- pumping from well-points
- pumping from bored wells
- pumping from horizontal wells
- electro-osmosis.

Groundwater flow may be eliminated or reduced by:

- grouting
- compressed air
- freezing
- sheet piled or diaphragm walls.

Problems of settlement of the ground surface adjacent to excavations due to piping of soil beneath sheet piles, erosion from sloping sides of excavations, infiltration of fines into unscreened pumping wells and consolidation of clay layers due to increases in effective stress as a result of groundwater lowering have to be considered. Groundwater control can have the following beneficial effects. It can:

- lower the water table and intercept inflowing seepage which would otherwise enter the excavation and interfere with the work
- improve the stability of slopes
- prevent heave of the bottom of an excavation
- reduce lateral pressures on temporary sheeting.

The choice of method used depends to a great extent on site conditions, and in particular on the particle size distribution. The range of soil types over which the various processes are applicable have been classified by Glossop and Skempton (1945) and are shown in Fig. 1.32 and Fig. 1.33. For example, the coarse gravel, soil A in Fig. 1.33, may be unsuitable for pumping because of the heavy flow through the highly permeable ground. Reference to Fig. 1.33 shows that the gravel is amenable to treatment by cement grouting to eliminate or greatly reduce the flow into the excavation. Soil B is suitable for well-pointing, but there is a risk of damage to adjacent structures due to lowering of the groundwater table. Figure 1.33 shows that chemical consolidation can be used to solidify the soil and greatly reduce or prevent inflow. The sandy silt, soil C, is too fine for well-pointing, but electro-osmosis, or freezing or compressed air could be used.

For many of the processes listed above, knowledge of the permeability of the ground is essential. It can be calculated approximately using

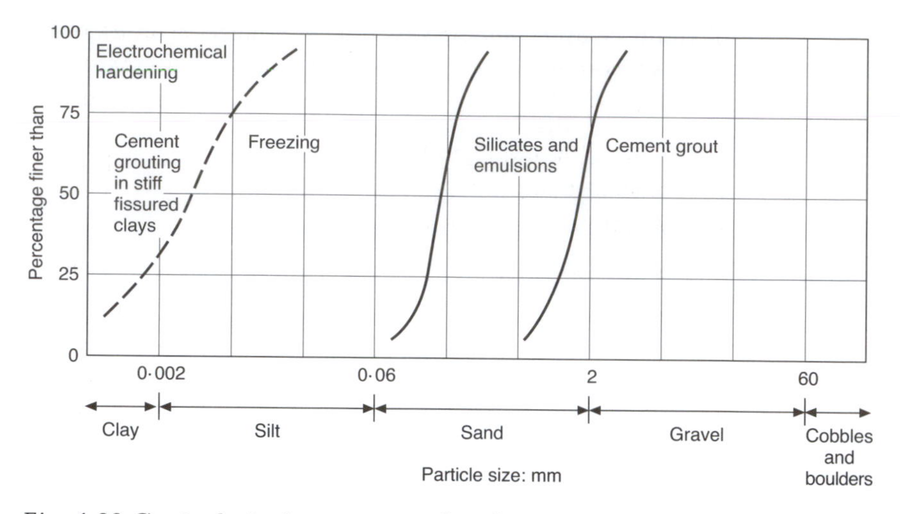

Fig. 1.32 Geotechnical processes related to soil type, after Glossop and Skempton (1945)

Hazen's formula, i.e.

$$\text{coefficient of permeability } k = C_1 D_{10}^2,$$

where k is in cm/s, C_1 is a factor varying from 100 to 150, and D_{10} is the effective grain size in cm (the effective grain size is the grain size corresponding to 10% retention on the grading curve of the soil).

In most cases, for groundwater lowering schemes it would be unwise to rely on Hazen's formula when estimating the number and diameter of pumping wells and the size of the pumps for a given size of excavation.

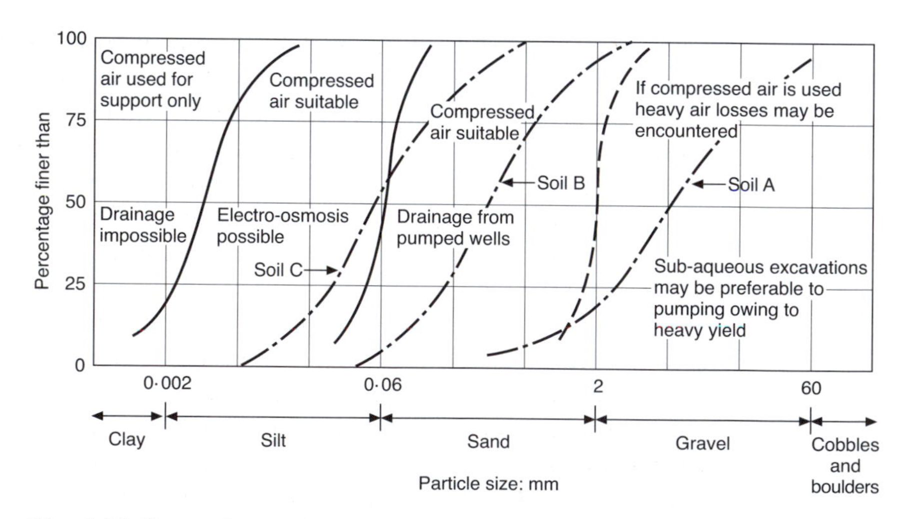

Fig. 1.33 Groundwater lowering processes related to soil type, after Glossop and Skempton (1945)

The site investigation should make provision for more reliable determinations of permeability by carrying out in situ permeability tests in open boreholes or piezometers. The three most common types of test are:

- rising or falling head tests
- constant head tests
- packer or Lugeon tests.

These tests are described in detail by Clayton *et al.* (1995b).

Pumping tests in which the groundwater is pumped from a well with observations of the draw-down in the surrounding water table using standpipes or piezometers are time consuming and costly and are usually only performed in connexion with trials for large-scale groundwater lowering schemes.

Response to geotechnical processes

If it may be necessary to improve the engineering properties of a granular soil by vibro-compaction or dynamic consolidation (heavy tamping), it is important that the grain size distribution of the deposit is determined at enough locations for the results to be representative of the whole deposit. Particular attention should be paid to the amount of fine particles (less than 0.06 mm) present which will greatly influence the efficiency of the processes.

Case study: Ground improvement by stone columns, Hodeidah, Yemen Arab Republic

An instructive case record describing the use of stone columns to improve the engineering properties of a loose silty sand is described by Hartikainen (1981). A grain silo with a capacity of 20 000 tons was to be constructed at Hodeidah in the Yemen Arab Republic. The soil conditions were assessed, prior to tendering, using five standard penetration tests (SPTs), combined with the taking of disturbed samples for both visual classification and sieve analysis. The revealed strata were mainly sands and silts to a depth of 50 m and the SPT results are shown in Fig. 1.34. The upper 10 m were very loose to loose and were described as silty sand. It was decided to compact the upper sands using stone columns, partly to reduce the settlements of the structure and partly to reduce the danger of liquefaction. The few grading curves available at tender stage indicated a silt content of less than 10% and it was planned to install the stone columns 11 m long on a 2 m by 2 m square grid. After the work started it was soon found that an acceptable level of compaction could not be achieved. Further investigations were carried out which revealed the presence of two layers of very silty sand extending from levels 0 to −1·0 m and from −4·0 m to −6·0 m with silt contents varying from 4% to

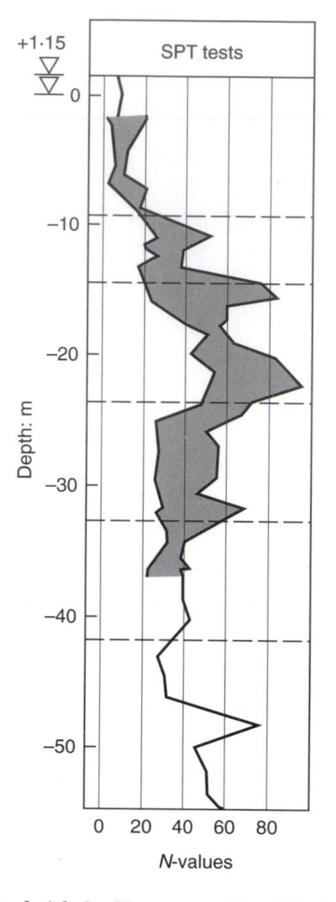

Fig. 1.34 SPT results, Hodeidah, Yemen, after Hartikainen (1981)

up to 50%, which had not been apparent from the results of the initial investigation.

It was then decided to install the stone columns in two phases. The first phase involved the original grid and column lengths and the second phase used shorter column lengths, 8·5 m, at every midpoint of the original grid, as shown in Fig. 1.35 and Fig. 1.36. The compaction was controlled using cone penetration tests (CPTs) which indicated that the specified 70% relative density was generally exceeded. The maximum observed settlement was less than 70 mm, which was considered satisfactory.

It was concluded that the low cost of the soil improvement solution in comparison with a pile solution made it possible to win the hard international competition. In spite of the difficulties encountered in the deep compaction operations and the increase of about 60% in the quantity of stone columns, the cost was still only about one quarter of the estimated cost of a piled foundation. The increase in silt content from less than 10% as initially assumed to an average as-found of 40% made the soil

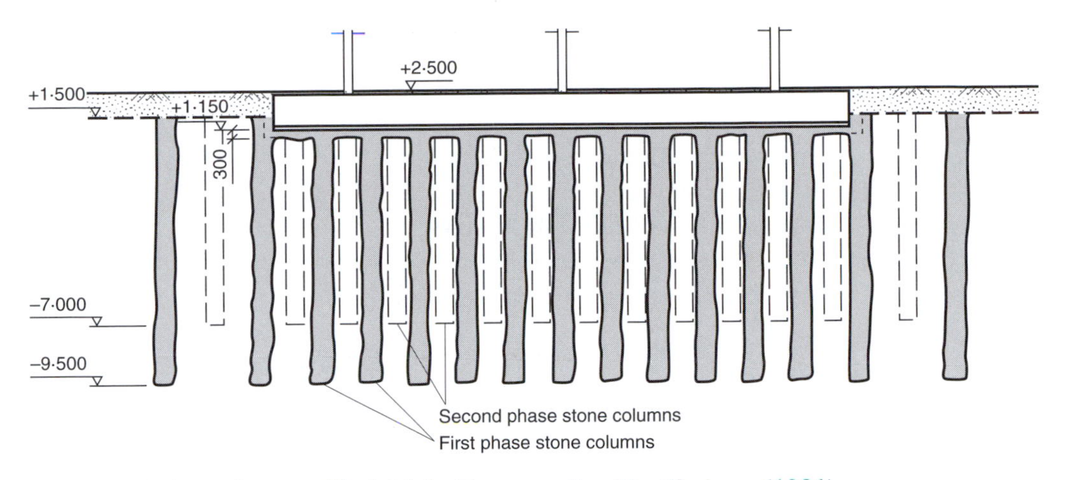

Fig. 1.35 Stone columns, Hodeidah, Yemen, after Hartikainen (1981)

considerably more difficult to compact. The inadequate initial site investigation resulted, of course, in delays to the contract.

Liquefaction

Liquefaction denotes a condition where a soil will undergo continued deformation at a constant low residual stress or with no residual resistance, due to the build-up and maintenance of high pore water pressures which reduce the effective confining pressure to a very low value; pore pressure build-up leading to true liquefaction of this type may be due to static or cyclic stress applications (Seed and Booker, 1976). Liquefaction

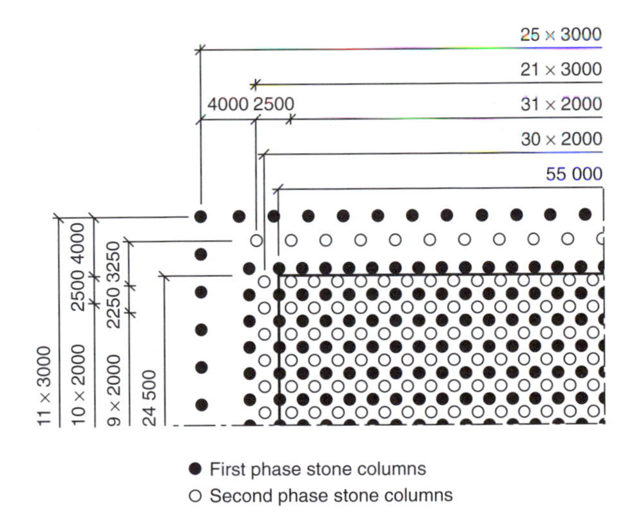

Fig. 1.36 Plan view of stone columns, Hodeidah, Yemen, after Hartikainen (1981)

generally occurs in loose, saturated fine sands and has been experienced in many countries throughout the world causing widespread devastation both on land and also under the sea. Factors affecting liquefaction include:

- grain size distribution
- relative density
- the application of static shear stress
- the application of cyclic shear stresses induced by ground motions
- the inclination of the ground surface.

A site investigation for a potential liquefaction location should provide information covering these factors. In particular the investigation should consider the following.

- A sufficient number of representative samples should be recovered to explore fully the site itself and also the environs so that a reliable picture of the grain size distribution is obtained.
- Relative density is best assessed using the in situ cone test which can also give information regarding soil type from the observations of the local friction. It has been suggested that a sandy soil is not likely to liquefy if the relative density is greater than 70%. Figure 1.37 gives a correlation between cone resistance, effective vertical stress and relative density. Figure 1.38 gives an interpretation of soil type as related to the results of cone tests, based on the work of Meigh

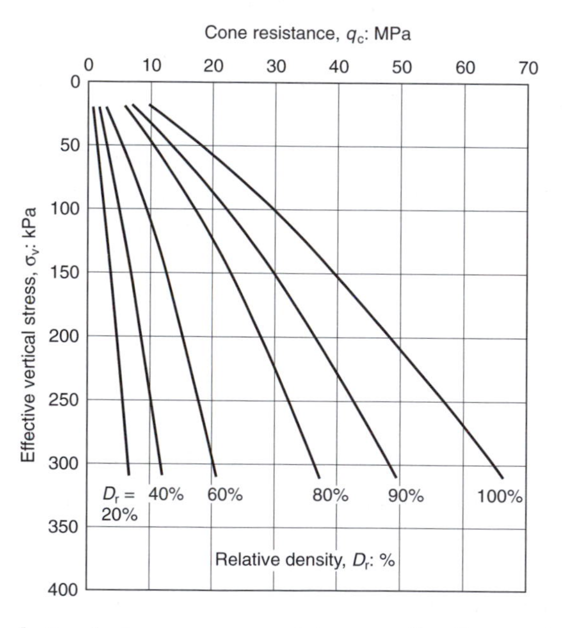

Fig. 1.37 Correlation between cone resistance, effective vertical stress and relative density

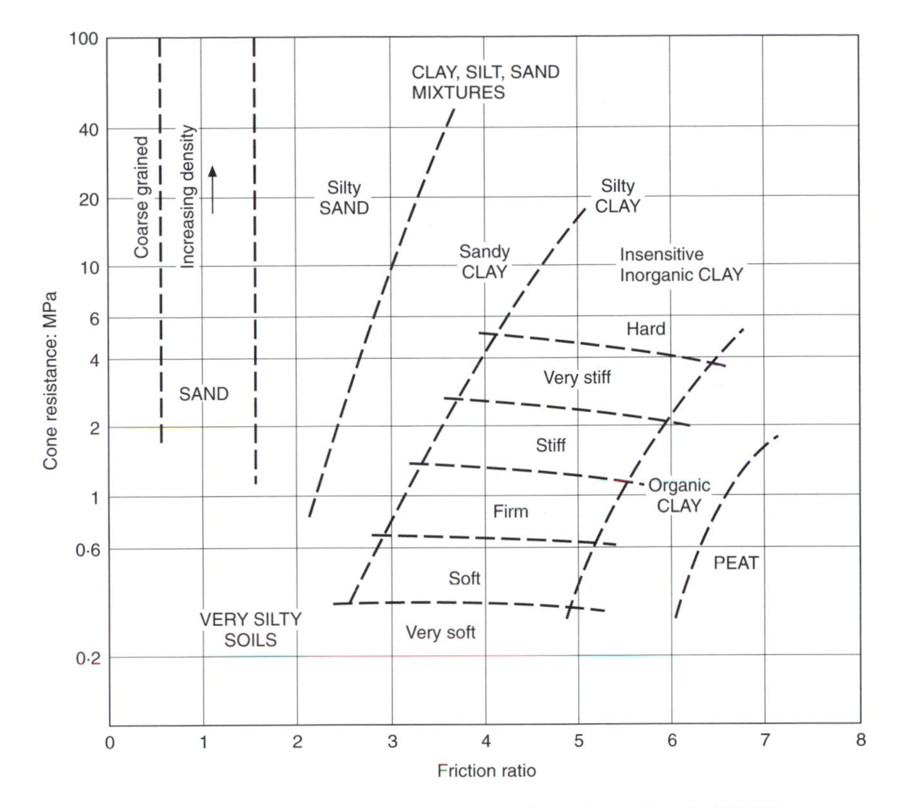

Fig. 1.38 Soil type related to cone test results after Meigh (1987)

(1987). The use of the piezocone as a probing device should be considered. Assessments of relative density can also be obtained from the SPT but the authors are of the opinion that the in situ cone test provides more reliable information.

- The history of seismic activity for the area should be obtained.
- The distribution of pore water pressure both laterally and with depth should, as always, be investigated.

Sand as hydraulic fill

If a sand deposit is being considered for fill to be placed hydraulically, then in addition to determining the distribution of grain size, the hardness and angularity of the grains should be assessed. This will then allow a dredging contractor to choose the most suitable equipment to deal with the deposit and to minimize delays to a contract due to wear and tear of the plant selected.

Foundations, footings, rafts and piles

With footings and rafts, it is generally the case that the bearing capacity from the point of view of stability is more than adequate. It can be

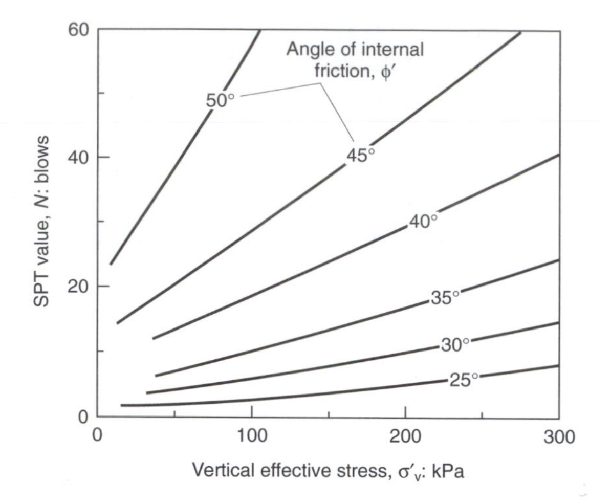

Fig. 1.39 Correlation between ϕ' and SPT N *value*

checked using bearing capacity theory. This will require an assessment of the angle of shearing resistance of the sand which may be made using the results of the standard penetration test or the dutch cone test. A correlation between ϕ' and the SPT *N* value is given in Fig. 1.39. In almost every case, however, the applied bearing pressure will be governed by settlement considerations which are usually based on the SPT or the CPT. As shown by Simons and Menzies (2000), for example, settlement predictions based on penetration tests for foundations on sands are not generally reliable since the tests on which the predictions are based do not measure the compressibility directly and do not reflect the influence of over-consolidation. For example, loose over-consolidated sand may have a low *N* value and yet have a smaller compressibility than denser normally consolidated sand with a higher *N* value.

If it is important to obtain as reliable a prediction as possible of settlement then the use of plate bearing tests should be considered preferably with plates of differing size so that the settlement of the full scale foundation can be more accurately predicted. In some cases the screw plate may be advantageous as tests can be carried out at one position at different depths both above and below the water table.

Pile bearing capacity in granular soils can be predicted on the basis of:

- bearing capacity theory
- Dutch cone penetration test (CPT)
- Standard penetration test (SPT)
- Pile driving formulae.

Correlations with special sounding tests, for example, driving rods, pushing and pulling rods, and rotating rods are available. For design considerations, reference can be made to Simons and Menzies (2000).

When planning a ground investigation for a site where one founding possibility is to use piles driven into a dense granular stratum, the soil underneath the bearing stratum should be checked to a sufficient depth to ensure that no softer layer exists that could adversely affect the performance of the piled foundation. Terzaghi and Peck (1967) report a case where a steel frame building was supported on about 10 000 timber piles 7·9 m long, driven so that their points came to bearing in the upper part of a layer of dense sand. The load per pile was 150 kN. Since the average settlement of test piles was only 6 mm for a load of 300 kN, the designers did not expect that the maximum settlement of the entire foundation would exceed this value. The observed maximum settlement was greater than 300 mm two years after the end of construction. This was due to the presence of a layer of soft clay about 50 m thick below the dense sand layer which was only 2 m thick.

Furthermore, if piles are to be driven into a bearing stratum of dense granular material, the thickness of the bearing layer should be carefully checked at a sufficient number of points to ensure that it is thick enough to allow a sufficient penetration of the piles into the bearing stratum to develop the required pile capacity, while at the same time having a suitable thickness of granular material below the ends of the piles to avoid any overstress in an underlying less competent stratum.

CHAPTER TWO

The desk study and walk-over survey

What is the desk study?

Within the construction industry desk studies are a well known but often under-used method of gathering and assessing existing information about a site that is to be used for some form of construction works. A desk study should be undertaken at the start of every site investigation. The principal objective of the desk study is to provide as much information about the following:

- probable ground conditions (topography, soil and rock types, groundwater, contaminated ground)
- previous uses of the site (made-up ground and contaminated ground)
- access to the site.

This information is used with the results of a subsequent walk-over survey to identify likely geotechnical hazards to the proposed structure. This will provide key information for the conceptual design of the structure, in particular those parts of it in contact with the ground (e.g. foundations), and provide a focus for the later more expensive and lengthy stage of physical investigation by boreholes, trial pits and in situ tests.

In addition to the features listed above the desk study should also include environmental and ecological considerations that might impose constraints on the execution of future works. For example, the site may include a wetland area containing some rare or endangered species of flora or fauna. This may impose severe restrictions on the later stages of the investigation as well as construction. The desk study may reveal that the site is of archaeological interest which may impose significant delays to the rest of the site investigation and possibly the construction work.

Why do desk studies?

eral it is cheaper to obtain and evaluate existing information than to
ate data from new investigations. This was certainly the case when
ing the route for a tunnel carrying a high-tension electricity cable
gh part of central London. The proposed depth of the tunnel

brought it within the likely depth of piles supporting the National Audit Office adjacent to Victoria railway station. Imagine the problem of finding a safe route through what can only be described as a 'forest' of piles. The critical question was: 'what is the position of the piles and how deep are they?' Initial thoughts were to look for methods such as geophysics that may be used to locate the individual piles. Any such method would have to be employed inside a working building and hence it would be difficult from the outset and the quality of the results uncertain. A comprehensive desk study was carried out in order to establish the site history and the foundations used for the National Audit Office. The records used included:

- geological maps
- geological survey memoirs
- topographic maps
- previous site investigation reports
- articles in journals including:
  - *Railway Magazine*
  - *The Architect and Building News*
  - *Civil Engineering*
  - *Proceedings of the Institution of Civil Engineers*
- Construction drawings of Victoria Station (held by Railtrack) and the National Audit Office (held by current occupants).

Sufficient information about the location and type and diameter of piles was found from the records listed above (mainly from the latter three sources listed) to recommend a safe route for the tunnel. The depths of some of the piles were estimated from knowledge of their type and diameter as well as from the general ground conditions. Although time consuming the desk study answered the key questions at a fraction of the cost of mounting a new investigation and furthermore the results were more reliable than attempting to locate the individual piles using geophysics.

In general, the better the desk study the more cost effective would be the overall site investigation. The study of geological maps, historical maps, archive aerial photographs and other documentary information can provide insights into the ground conditions, the past condition or use of a site and the hazards that such use could pose to a new development (e.g. the presence of dissolution features, ancient landslips, old coal mines, landfills or land affected by contamination from former industrial processes such as gas works).

Case study: Lewisham Extension to the Docklands Light Railway, London
The 4·6 km long Lewisham Extension to the Docklands Light Railway (DLR) runs from Mudshute, beneath the River Thames at Greenwich, to Lewisham, see Fig. 2.1. The collation and evaluation of existing information

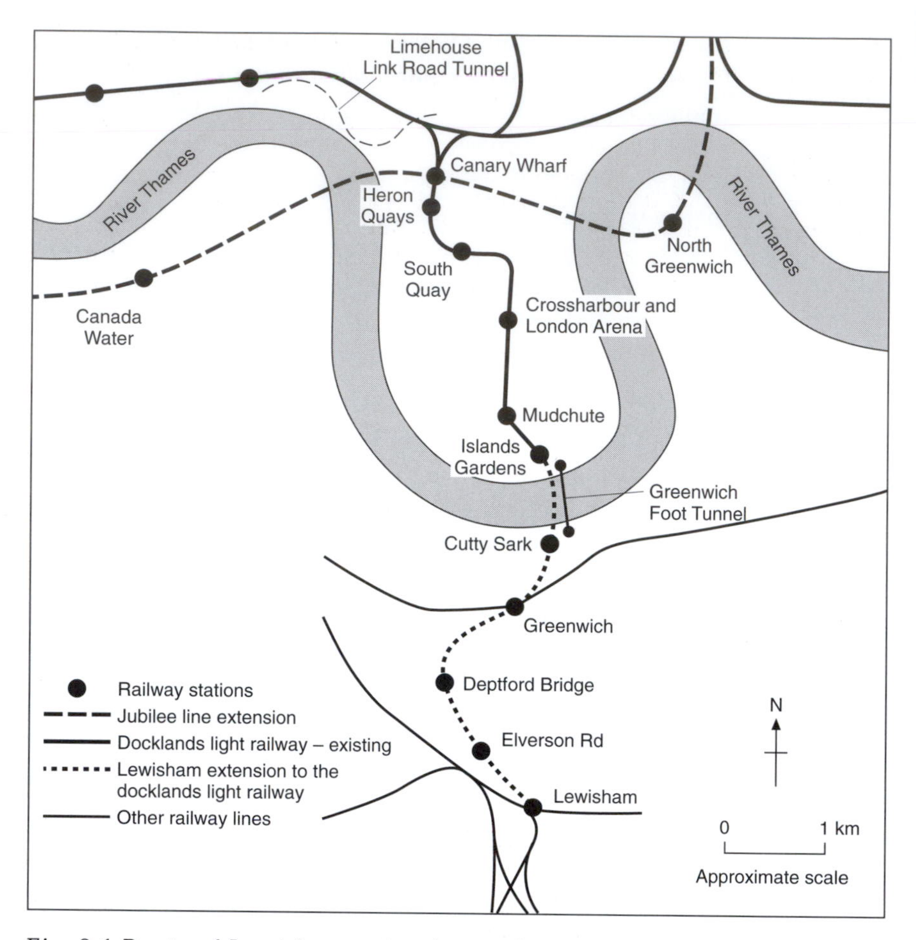

Fig. 2.1 Route of Lewisham extension to the Docklands Light Railway, London (after Shilston et al., *1998)*

by means of a desk study was, therefore, central to the design of the additional investigations and to the execution of the required geotechnical tasks (Shilston *et al.*, 1998).

The desk study comprised a review of existing geological and hydrogeological data, site history and construction precedent. The principal sources of information examined during the desk study and their main benefits to the project are listed below and summarized in Table 2.1.

- *Geology.* The geology within the area of interest comprises made ground and alluvium, Terrace Gravel, overlying sediments of the Lambeth Group (formerly known as the Woolwich and Reading Beds), Thanet Formation and Upper Chalk.

  Of particular benefit to understanding the geology was the work by Howland (1991), who presented a review of the geology of the

Table 2.1 Summary of the principal sources of information examined during the desk study and their main benefits to the project (after Shilston et al., 1998)

Classes of information	Description of information	Examples	Benefits
Geology and hydrogeology	1 : 10 560 scale geological map shows Greenwich fault and associated folding and disturbances	British Geological Survey (BGS) maps and memoirs	Better understanding of local geological structures including faults, folds and unconformities and of potential hazards such as swallow holes
	BGS has records of numerous boreholes in the vicinity of the site	Borehole records from previous site investigations in the area	Enhanced appreciation of ground conditions, thus limiting the required amount of additional geotechnical investigation
	Howland (1991): *The engineering geology of the London Docklands*, which is based on 4500 boreholes, provides a useful overview	Published geological papers and literature	As above
	CIRIA (1989) indicates the site is in an area which may be critically affected by rising groundwater table in the deep aquifer	Published hydrogeological papers and literature	Better understanding of groundwater regime including influence of rising groundwater table
Site condition, land use and history	Historical maps from 1703 onwards and Ordnance Survey maps from 1869 to the present day	Historical maps	Identification of potential hazards associated with previous land use including contaminated land, old foundations (piles, footings, etc.) and previous construction (old river walls, basements, etc.)
	Department of the Environment (1991): *Review of mining instability in Great Britain* records mining activities in the Greenwich area	Mining records	Identified potential hazard of encountering shafts/deneholes

Table 2.1 – Continued

Precedent	Limehouse Link project experienced problems with de-watering part of the Lambeth Group which contained perched water and had high permeabilities. Records of the construction of the Greenwich foot tunnel and shafts (Copperthwaite, 1901)	Published case histories of construction projects in the area, particularly those involving deep excavations and tunnels	Obtain details of previous construction experience, particularly identification of potential construction problems and appropriate solutions – in other words 'Don't reinvent the wheel!'

Docklands area based on a study of 4500 boreholes collated by the London Docklands Development Corporation. This comprehensive study provided a more detailed interpretation of the stratigraphy and structural geology of the site than could be obtained from geological maps alone and enabled the subsequent subsurface investigation to be more focussed and allowed savings to be made on the number of boreholes employed.

- *Hydrogeology.* Hydrogeological conditions in the Docklands area of London are complex and are in a state of transition as the groundwater level in the Chalk aquifer beneath London is rising (CIRIA, 1989). The results of numerical modelling of the rising groundwater in the London Basin giving predictions of groundwater levels in the Chalk up to 2040 have been published by Lucas and Robinson (1995). They concluded that the piezometric groundwater level in the Chalk under parts of London is rising at rates in excess of 2 m per year. This information and other data evaluated during the desk study provided a sound basis for determining the groundwater conditions to be assumed in the geotechnical design of the project.
- *Site history.* The Docklands and Greenwich areas of London have undergone numerous phases of development and re-development dating back to medieval times. A review of historical maps of the area allowed assessments to be made of the possible influences of previous construction activities (such as old foundations and excavations) on the proposed new works. Where possible, construction records for prestigious or particularly sensitive structures that might be affected by the works were obtained as part of the desk study. The existing Greenwich mainline station was built in the 19th century and is a Grade E listed historic building. Drawings of both the superstructure and foundation were found which proved invaluable in assessing the impact on the station of the proposed tunnelling and surface works. The historical study also identified a number of other potential hazards including:

  - unexploded bombs: the Docklands area of London was heavily bombed during the Second World War
  - mining instability: chalk and sand were mined in the 19th century in the Greenwich area
  - contaminated land: ground and groundwater contamination are present from previous industrial activity
  - building foundations: possible piles from existing and demolished buildings in the vicinity of the surface sites and bored tunnels.
- *Construction precedent.* Information on construction precedent in the area was important in assessing the geotechnical hazards faced by the project and in the determination of engineering properties for the design of the works. The close proximity of the Lewisham Extension to several major engineering projects with well documented case histories, such as the Limehouse Link road tunnel, Jubilee Line Extension and Canary Wharf, was extremely valuable (see Fig. 2.1). The information obtained on these projects by the desk study highlighted construction difficulties and particular geotechnical hazards. These included, for example, difficulties encountered with de-watering the excavations for the Limehouse Link project (Cruickshank, 1993).

  Also of benefit were the records of much older structures such as the Greenwich Foot Tunnel, which was opened in 1902. This tunnel runs beneath the Thames and connects Island Gardens to Greenwich close to the proposed route for the DLR bored tunnels (see Fig. 2.1). There is good documentation for this tunnel, including a contemporary paper published in the *Proceedings of the Institution of Civil Engineers* which describes the method of construction and the ground conditions encountered (Copperthwaite, 1901).

Overview of how a desk study is done

As mentioned above a desk study involves collecting and evaluating existing information about the site and the surrounding area. The Engineer acts like a detective collecting and sifting through clues to find the person or persons who committed the crime. In the case of the desk study the outcomes are a geological model of the site, the past condition and use of the site and the identification of potential geotechnical hazards. The analysis and evaluation of evidence must be carried out with the same care and attention to detail as used by a detective. For example, a minor road marked on a map may be called Chalk Pit Lane and yet there is no chalk pit marked on the current revision of the topographic maps of the area. The road name implies that it once passed a chalk pit. This evidence should lead to further investigation of historical maps and aerial photographs to find the chalk pit and its location in relation to the site under investigation. The

Engineer must treat all the evidence with caution until it can be supported with corroborating evidence from historical maps, reports and other sources.

In the UK and USA it is generally possible to find a great deal of information about a particular site. In other parts of the world the only data that may be available at the desk study stage are aerial photographs or possibly satellite images. Shilston *et al.* (1998) provide a useful generic checklist for sources of information which can be augmented for individual countries or regions (Table 2.2). For the UK more detailed information on many of the sources mentioned in Table 2.2 is given by Guy (1992), Clayton *et al.* (1995b), Reeves (1996) and Perry and West (1996).

Table 2.2 Aide memoire to classes of information for geotechnical desk studies (after Shilston et al.*, 1998)*

Classes of information	Examples
Topography	• Maps • Aerial photographs
Geology and hydrogeology	• Maps, memoirs and reports • Aerial photographs • Published papers and records • Thematic databases (e.g. in UK for landslides, natural cavities, etc.) • Previous ground investigations
Environment and Planning	• Planning maps • Archaeological site and historic building records • Contaminated land records • Landfill and wastes disposal records • Environmental statements • Meteorological records • River and coastal information
Site condition, land use and history	• Historical maps • Historical documents • Aerial photographs • Land use and planning maps
Local knowledge and experience	• Local history societies • Neighbours • Previous site users • Construction records • Building control office
Precedent	• Case histories • Construction records
Codes, standards and guidance	• Professional bodies and institutes • Government departments • Research organisations and universities

Table 2.2 shows that topographic maps, geological maps and aerial photography provide key data to any desk study and hence should be regarded as the minimum data requirement.

Table 2.3 gives a general checklist outlining the key information that the desk study should provide together with the sources of data. Again map information (topographical, historical and geological) and aerial photography are highlighted as key sources of desk study data.

What to look for in a desk study

Geology

The steps that should be taken in assessing the geology of a site at the desk study stage of a site investigation are the following.

- Identify the geological units outcropping under the site together with any superficial deposits.
- Determine the nature and thickness of the rocks and soils present beneath the site.
- Determine the regional geological structure (regional dip, folding and faulting patterns).
- Determine the local geological structure, e.g. local dip, faulting, cambering and valley bulging.
- Identify topographic features and their association with the geology, e.g. escarpments, landslip features.
- Determine whether there are any mine workings below the site or nearby.
- Determine the depth, extent and type of weathering.

As the geological information about the site is collected it is important that a geological model of the site in three dimensions is constructed. At the very least this should involve drawing a number of sections through the site. At this stage these may only be sketch sections but they will help in understanding the subsurface structure of the site and as more data are collected during the desk study, walk-over survey and the later more detailed investigations the model can be refined, as suggested by Fookes (1997).

The principal sources of information on the geology of the site are geological maps and associated memoirs or reports. Geological literature, such as learned journals and books, can often provide useful information but of course coverage is likely to be limited – particularly if the site is not in an area of special geological interest. The information given on geological maps, although extremely valuable, is often difficult to interpret to those not trained in the science of geology. The main problem is with understanding the names given to the strata shown on the map. Geologists do

Table 2.3 Desk study checklist

Feature	Source of data								
	Maps				Photography	Written records			Internet
	Topo-graphical	Historical	Geo-logical	Specialist	Aerial photographs	Published memoirs and reports	Specialist reports	Magazine/ newspaper/ journal articles	
Geology									
What geological strata lie below the site? Define in terms of principal constituents and whether they may be considered as rock, weak rock or soil			***		*	***	*(SI)	**(J)	*?
What is the geological structure? Define in terms of regional and local dip magnitude and direction, folding, faulting and igneous rocks (intrusive and extrusive). A geological section must be prepared			***		**	***	*(SI)	**(J)	*?
Are there any coal seams under the site?			***			***	**(M)	**(J)	*?
Is the site covered by superficial deposits such as alluvium, boulder clay or loess? If so attempt to determine likely thickness, principal constituents, structure and extent			***		**?	**	**(SI)	**(J)	
What is the depth of weathering? Can a typical weathering profile be established?						**	**(SI)	**(J)	
Is the site covered by residual soil deposits? If so determine type, extent and thickness			**?		*	**	**(SI)	**(J)	

Are there any special conditions associated with the geology, for example, mining, dissolution, landslipping or swelling ground?			**	***	**	**(SI)	**(J)	
Topography and land cover								
Does the site lie on sloping ground? If so what is minimum, maximum and average slope?	*** (DEM) [1 : 25 000]			**	*	**(SI)		
Are there any significant changes in slope which may indicate a change in rock type or landslipping?	*** [1 : 25 000]			***	*	**(SI)		
Is there evidence of changes in ground level, for example by placement of fill, subsidence or the demolition of old structures?	**	**		***	**	**(SI)	**	
Are, or were there, trees or hedges growing in the area of the site?	*	*		***				
Surface water, groundwater and drainage								
Are, or were there, any springs, seepages, ponds, lakes or water courses on or near the site?	***	***		***	***	**(SI)	*	
What are the water bearing strata? Are there any major aquifers under the site?			***		***	**(Hy)	**(J)	*?
What is the depth to the water table?	**	**	**	**	**	***(W)	*(J)	*
Are there any perched water tables within the area of the site?			**		**	**	**	*?
Is there any likelihood of artesian conditions within the area of the site?			***			***(Hy)		

Table 2.3 – Continued

Feature	Source of data								
	Maps				Photography	Written records			Internet
	Topo-graphical	Historical	Geo-logical	Specialist	Aerial photographs	Published memoirs and reports	Specialist reports	Magazine/ newspaper/ journal articles	
Surface water, groundwater and drainage – continued									
What is the likelihood of flooding?	***	***	***		***	**	***	***	***(UK)
Will changes in groundwater level have any adverse effects such as subsidence?			**			**	**(SI)	**	
What is the likelihood of the proposed construction resulting in pollution of groundwater or surface water			**			***	**(Hy)	**	
Site history									
What were the previous uses of the site?	***	***			***	**	**	***(N)	***(UK)
Are there any disused quarries or pits within the area of the site?	***	***			***	**	**	**	**(UK)
Is there any landfill within the area of the site? If so what type of waste was deposited in it?	*	*		*	**?	**	**	**	**(UK)
Are there any landfill sites nearby? If so establish any geological connection with site	*	*	***	*	**?	**	**	**	**(UK)
Is there, or has there ever been, mining or quarrying activity in this area?	***	***	***	*	***	**	**(M)	**	**(UK)

Is there any evidence of made ground under the site? If so determine extent, type and thickness	*	*			***	**	**(SI)	**	
Are there any services running under the site, e.g. electricity cables, gas pipes, water pipes, etc?	*	*		***(S)	**		***(S)		
Are there any archaeological features within the area of the site?	*?	*?		**(A)	***?	**	**(A)	***(M/J)	**?
Are there any archaeological features nearby? This may give an indication of the likelihood of similar features within the area of the site	*?	*?		**(A)	***?	**	**(A)	***(M/J)	**?
Ground conditions									
Is there available information on the strength, compressibility and permeability of the ground?						*?	***(SI)	**?(J)	*?
Is the subsoil a shrinkable clay?						*?	***(SI)	**?(J)	*?
Does experience suggest that groundwater in these soil conditions may attack concrete?						**	**	**	*?
Is there any evidence of landslipping either on or adjacent to the site or on similar ground nearby?			**		***	**	**	**	*

*** Best source of data
** Intermediate source
* Worst source
SI Site Investigation reports
Hy Hydrogeological reports
M Mining records
A Archaeological records

S Records/maps held by service providers, e.g. electricity, gas etc.
W Well records (held by BGS in the UK)
J Journal article
M Magazine article
N Newspaper article
DEM Digital Elevation Model

not name strata according to material type as may be considered sensible from an engineering standpoint. They will use age, type area (the town or region where the material was first described) or some feature of the rock. It is often impossible to determine whether the material shown on the map may be considered as a granular or cohesive soil or a weak or strong rock from the stratigraphic name alone. Some examples of such names from UK stratigraphy are:

- Gault (mainly clay)
- Reading Beds (mainly clay)
- Cornbrash (limestone)
- Hythe Beds (sandstone)
- Bracklesham Beds (interbedded sands and clays).

For example, the 'Gault' may vary in composition from one part of the country to another. When the Gault was being deposited millions of years ago the landmass from which the material was derived lay to the west of the present day outcrop and hence in eastern areas such as Dover the Gault has a high clay content. As one moves westward the dominant grain size increases such that in Dorset the Gault is sandy. The Reading Beds also vary in composition from one location to another. The Reading Beds are generally thought to consist entirely of highly plastic, highly shrinkable clays, but in some parts of the country they are predominantly sandy. Even in cases where the stratigraphic name indicates a material type such as 'London Clay' it does not necessarily mean that it is entirely composed of clay. In the case of the London Clay the lower parts of the formation contain layers of weakly cemented sand. The only way of discovering the material types and variations is by consulting written geological records for the local area.

Another problem faced when attempting to interpret geological maps is the manner in which superficial deposits are shown. Superficial or drift deposits are the youngest materials shown on the map and generally form relatively thin (in geological terms) but often patchy layers covering the much older rocks which are referred to as 'solid' on UK geological maps. They are often of critical importance to engineering works since they occupy the top 10 m of ground. For example, some of the most common superficial deposits shown on geological maps are the following.

- *Alluvium* – material deposited by rivers and streams, often a mixture of clay, silt and sand; found in valleys and may well be associated with the groundwater close to the surface.
- *Head (or Coombe Rock)* – material produced as a result of solifluction, often rock fragments in a finer-grained groundmass; found on gently sloping ground and in valleys.

- *Terrace deposits* – materials associated with a previous route of a river or stream. Renewed erosion has resulted in the river cutting through its original flood plain leaving the deposits at a higher level and sometimes in a different location than the present day flood plain. These deposits are often granular and may be associated with perched water tables.
- *Boulder clay* – this is a generic term used to cover a wide range of glacially derived deposits. The principal characteristic of these deposits is that they can be expected to be highly variable vertically and laterally and may contain large boulder-size fragments of rock. The fact that clay is used in the material name does not necessarily mean that the material contains clay.

These deposits share one thing in common – they tend to vary considerably in thickness and composition over short distances. This is true of many other superficial deposits. It is for this reason that their thickness or range of thicknesses is not shown on the key to geological maps. Unfortunately, not all geological maps, particularly those at 1 : 63 360 or 1 : 50 000 scales, give information about the presence or extent of all superficial deposits. In the UK geological maps are divided into the following types.

- *Solid* – shows only the solid geology in detail; superficial deposits are not shown in colour and are depicted by symbols only.
- *Solid and drift* – both solid and drift deposits are shown in colour.
- *Drift* – the superficial or drift deposits are shown in colour.

In most cases a solid and drift or a drift map will be most useful. In some areas where drift deposits are extensive separate solid and drift maps may be required.

In order to determine the nature and thickness of local superficial deposits it is necessary to consult geological records such as regional guides[1] or the memoirs[2] associated with the geological map covering the site or other records. It is unlikely that these records will provide detailed information on the thickness of superficial deposits. This type of detailed information can only be gained from borehole or well records or existing site investigation reports for the site or nearby sites.

In general the most important information that can be found from geological records is:

- the type of material forming the ground below the site (e.g. sand, clay, weak rock or strong rock)

[1] *British Regional Geology* guides are published by the British Geological Survey. They cover a wide area (e.g. 'Central England'), so descriptions of geological formations and superficial deposits will be rather general.

[2] *Sheet Memoirs* are published by the British Geological Survey (formerly the Institute of Geological Sciences). They give more specific descriptions of the materials found in the area shown on a given 1 : 63 360 or 1 : 50 000 scale geological map.

- the thickness of the different soil or rock types below the site
- difficult ground conditions associated either with particular soil or rock types (e.g. subsurface cavities may be present in limestone or chalk or in coal-bearing rocks as a result of mining), or a combination of the soil type and the form of the site (e.g. clay soil plus trees may result in swelling ground if trees are removed; clay plus sloping ground may mean unstable slopes)
- depth of weathering
- presence and type of residual soils in tropical regions.

In order to deduce the regional geological structure it is necessary to use relatively small-scale geological maps. In the UK the 1 : 63 360 and the more recent 1 : 50 000 scale geological maps are ideal for this purpose. The key features to look for are the regional dip, unconformities and the general trend and type of folds and faults. This will aid in the interpretation of larger scale geological maps of the site. In some countries it may not be possible to zoom in any further than the regional level because larger scale maps may not be available. For extended sites such as road alignments these small-scale maps are very useful.

In regions where intense folding and faulting are not present the dip direction and dip of sedimentary strata can be determined on a regional level by looking at the outcrop pattern and relative ages of strata and the width of each outcrop in relation to the thickness of that stratum. This is illustrated in Fig. 2.2. Figure 2.2(a) is a sketch map of the geology of the Guildford area. It can be seen that the outcrop of the London Clay and the Reading Beds form nearly parallel-sided bands across the map. They 'V' slightly to the North as the outcrops cross the valley of the River Wey, indicating a northerly dip. In fact these beds dip at about 9° towards the North. The outcrop of the Chalk displays a different geometry. The outcrop is very wide East of Guildford. To the West of Guildford the width of the outcrop reduces significantly. This is due to the fact that the chalk is dipping very steeply here (70°–80°) as shown in the geological section (Fig. 2.2(b)).

In order to obtain a detailed picture of the geology of a relatively small site a larger-scale map is required. In the UK the base scale for most geological mapping used to be 1 : 10 560 (6 inches to 1 mile); more recently it has become 1 : 10 000. Hand coloured copies of original maps can be viewed at the offices of the British Geological Survey or black and white photocopies may be purchased from them. Since most of the geology in the UK was mapped in the late 19th and early 20th centuries the topographic base maps used form a valuable historical record that can be used in assessing the history of the site which is discussed later. These maps also contain useful notes written by the field geologists relating to the rocks or features such as landslips.

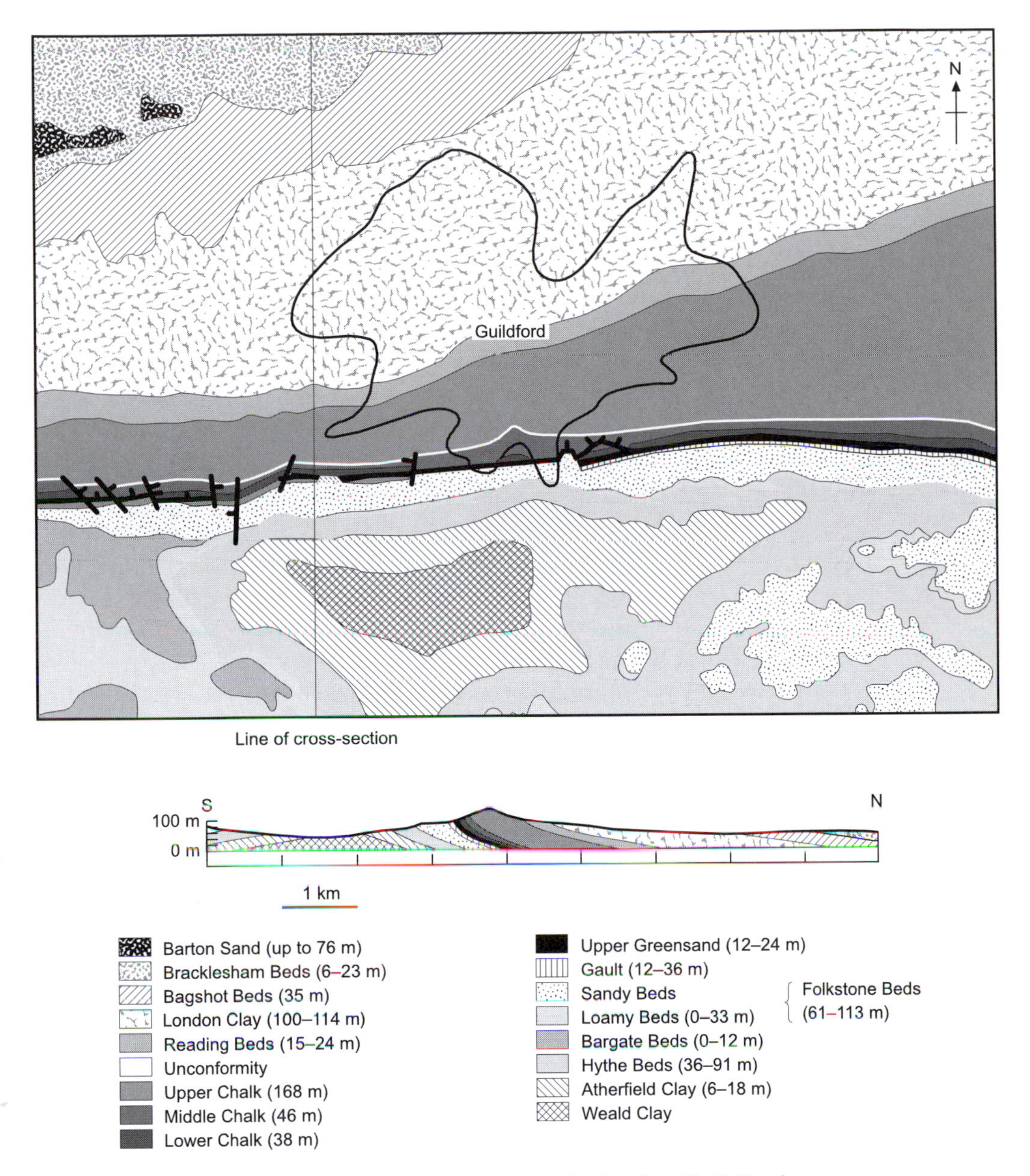

Fig. 2.2 Geological sketch map (a) and section (b) for the Guildford area

On any site, present or past mining activity could present a hazard to construction works. Mining is not confined to the traditional coal fields in the UK but is widely scattered in places across much of the rest of the country. In particular, in the South East mining problems are possibly not considered because traditional coal fields are not associated with this area.

Mining instability is associated with five mineral types:

- metalliferous mining
- rock mining (that includes limestone, sandstone, etc.)
- coal and associated minerals (which obviously relates to the traditional coal field areas)
- evaporates
- ironstone mining.

The West Midlands is well known for coal mining activities. There are however scattered metalliferous mining activities in Shropshire, Staffordshire, Herefordshire and Worcestershire. There is even a small gold mine in Worcestershire where placer deposits in the Old Red Sandstone have been worked for gold.

The first step in any mining desk study is to build the geological model using the published geological maps, geological records and journal articles where relevant. The geology in a mining desk study is the key to understanding the mining activities and aids formulating a strategy for searching for relevant information. The old 1 : 10 560 scale county series geological maps are the primary source of information particularly as there is sometimes mining information and notes on the old maps which have not been transferred to the more modern maps. In addition to the published maps another useful but less accessible source of information is the original field notes (known as field slips) that the geologists used when they were mapping the area. These are available for viewing at the British Geological Survey's library. The field slips contain observations made by the geologists during the late 19th century and early 20th century indicating the location of old dumps, old workings or in some cases unrecorded shallow coal mining, dating from Elizabethan times. The sheet memoirs (for the relevant 1 : 63 360 or 1 : 50 000 scale maps) will often provide a good insight into the extent of the mining problem but the amount of detail varies considerably from memoir to memoir both in terms of area covered and when the memoir was written.

The next source of information for mining desk studies is old or historical maps. For mining areas these maps can indicate the extent and intensity of mining. They indicate shaft positions, areas of spoil from mining operations, etc. There are occasional errors in them as to the mineral that was being worked. Other types of old maps which are often useful are deposited railway plans. During the railway boom years of the 19th century surveyors were surveying plots of land for proposed railway developments. Those surveys predate the first edition 1 : 2500 scale Ordnance Survey maps by up to 40 years and they can be very detailed. They are also accompanied by a book of reference which gives some information about the land use and the ownership.

One of the principal sources of information when carrying out a mining desk study in the UK is the Coal Authority Mine Records Office. Located at Bretby near Burton-on-Trent, the office holds a centralized archive of coal mine plans which also includes associated minerals such as iron stone, fireclay and ganister.

Aerial photography is extremely useful for identifying ancient, unrecorded shallow outcrop mine workings and unrecorded mine shafts. In some cases aerial photographs will be the only source of confirmation that an unrecorded shaft exists. For example, you might see a circle on a late 19th century mine plan, without any annotation. An examination of the aerial photographs reveals a characteristic ring of debris, and this indicates the location of an unrecorded mine shaft.

The British Geological Survey (BGS) has statutory rights to copies of records of boreholes deeper than 30 m. As a result the BGS holds over 600 000 borehole records. The BGS internet site (www.bgs.ac.uk) allows searches to be made for borehole records for any site in the UK and copies of the records can be purchased. The BGS also holds geophysical logs, site investigation reports, road reports, mine and quarry plans and sections, field notebooks and unpublished survey reports.

Use of geological maps

Case study: Redevelopment of a site, Milton Keynes, UK

Figure 2.3 shows a section of the 1 : 25 000 geological map (solid and drift) for Milton Keynes (Sheet SP83 and Parts of SP73, 74, 84, 93 and 94). The area shown is dominated by the valley of the River Ouzel or Lovat that runs approximately North to South on the East side of the map. The terrace deposits can be seen above the flood plain. Although the nature of these cannot be determined from the map, an examination of a more recent topographic map (see Fig. 2.4) shows a lake ('Willen Lake') occupying the outcrop of the terrace deposits near Willen village. This suggests that these are granular deposits which have been exploited for aggregates. Boulder Clay occupies the high ground above the Ouzel valley. The lower slopes of the Ouzel valley are covered by Head deposits. Head is also found in small valleys between the interfluves cut into the western slopes of the main Ouzel valley. The nature and thickness of the Head cannot be determined from the map, except that it is a drift deposit and may, therefore, be loose or soft. The 'solid' deposits comprise:

- *Oxford Clay* – bluish grey mudstones about 67 m thick
- *Kellaways Beds* – fine grey sands overlying bluish grey clay 4·9–5·2 m thick

LIVERPOOL JOHN MOORES UNIVERSITY
LEARNING SERVICES

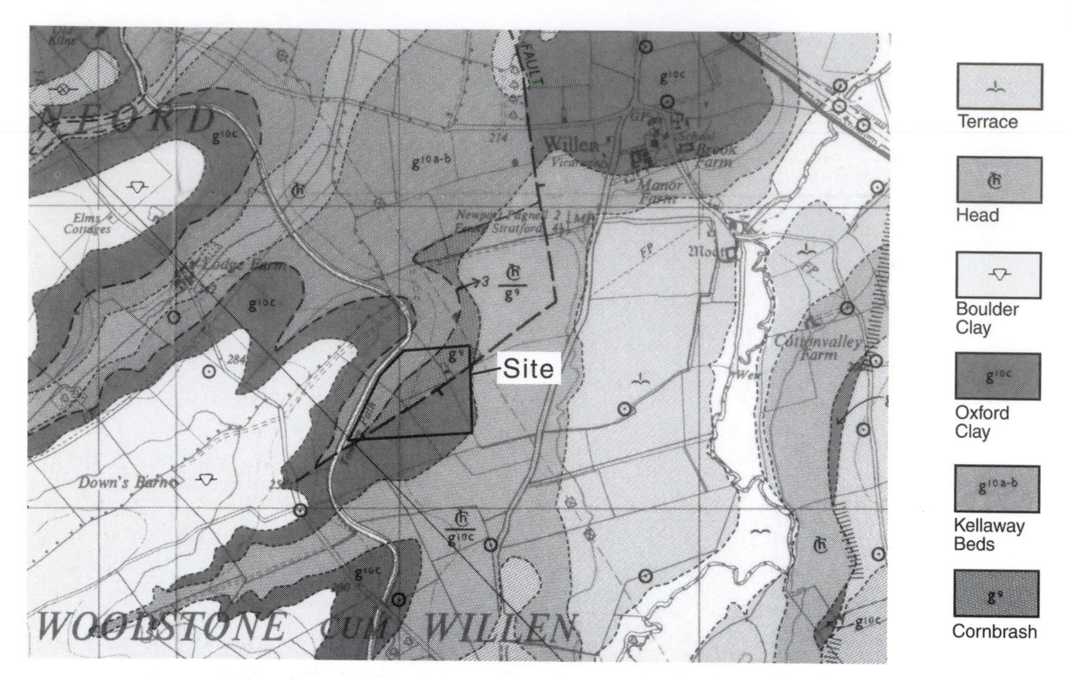

Fig. 2.3 Part of the 1 : 25 000 geological map of Milton Keynes (Sheet SP83 and parts of sheets SP73, 74, 84, 93 and 94) (Reproduced with permission of BGS)

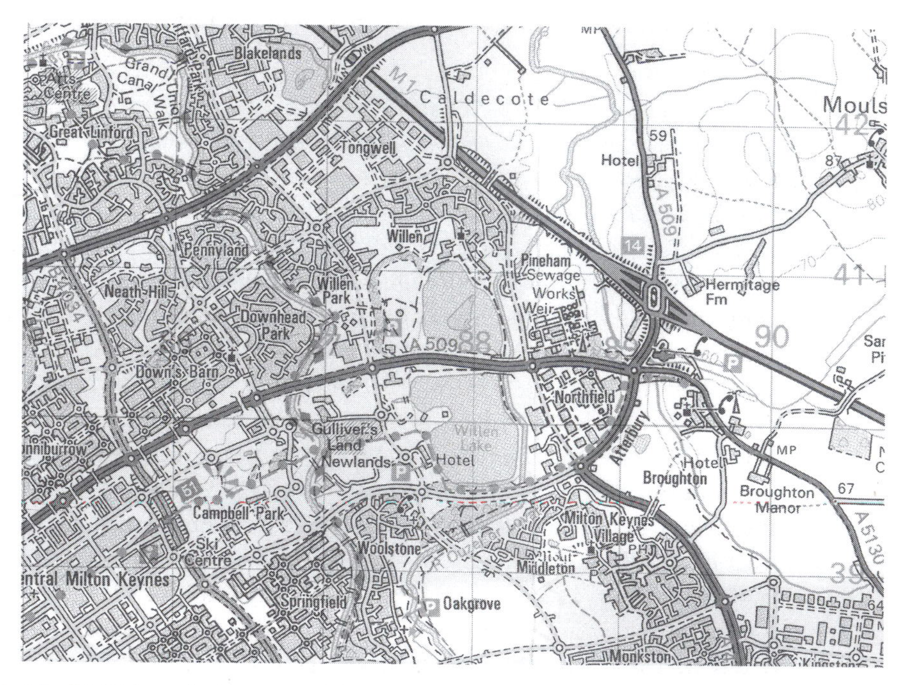

Fig. 2.4 1 : 50 000 scale topographic map of the Willen Park area of Milton Keynes (grid lines shown at 1 km intervals) (Reproduced with permission of Ordnance Survey)

- *Cornbrash* – limestone rock 1·0–2·1 m thick.

The dip is clearly shallow as the outcrops are almost parallel to the contours. This is confirmed by the dip arrow shown on the outcrop of Cornbrash which indicates a dip of 3° towards the East.

A fault ('Willen Fault') is shown as a dotted line on Fig. 2.3 trending NS from the top of the map and changes direction to NE–SW at the centre of the map. The small tick marks perpendicular to the dotted line indicate the side of the fault which has moved down relative to the other side. This is known as the 'downthrow'.

The area within the box was the site proposed for a low-rise school complex. The geology of the site comprises Kellaways Beds and Cornbrash on the North side of the fault. The throw of the fault has brought the Oxford Clay down to the same level as the Cornbrash on the South side of the fault. The eastern edge of the site is on the edge of the outcrop of Head. The extent of superficial deposits is difficult to map and hence it would be reasonable to expect to find Head covering at least the eastern part of the site.

Extensive trial pitting and borehole excavation carried out on this site showed that the Head extended further West than shown on the geological map. The trial pits confirmed that, under the proposed building area, the Cornbrash was at a suitably shallow and uniform depth to provide a foundation for spread footings, with the exception of a small area in the south-east of the site. In this area the Cornbrash was found to be 10 m below ground level as a result of the fault.

Case study: Redevelopment of a site, Bristol, UK

Figure 2.5 shows the published 1 : 10 000 geological map of an area of south-east Bristol. The area marked was the site of a jam factory. A proposal was put forward to redevelop this site as a superstore. The geology shown on the map is relatively complex with Mercia Mudstone (formally Keuper Marl) shown as MMG on the map laying unconformably on coal measures strata which dip towards the South (note how the outcrops 'V' towards the South in the valley which runs across the map from the NE corner towards the SW corner). Numerous coal seams are shown outcropping on or near the site. The 'Pott Seam' outcrops at the northern end of the site but the southerly dip means that it is likely to be at a shallow depth below the proposed structure. In many parts of the UK thin coal seams have been mined. A special mining desk study using the information sources discussed earlier was carried out and revealed that the seam had been mined and was at a maximum depth of 10 m below the ground surface within the area of the site. The mined areas of the seam were subsequently filled with grout.

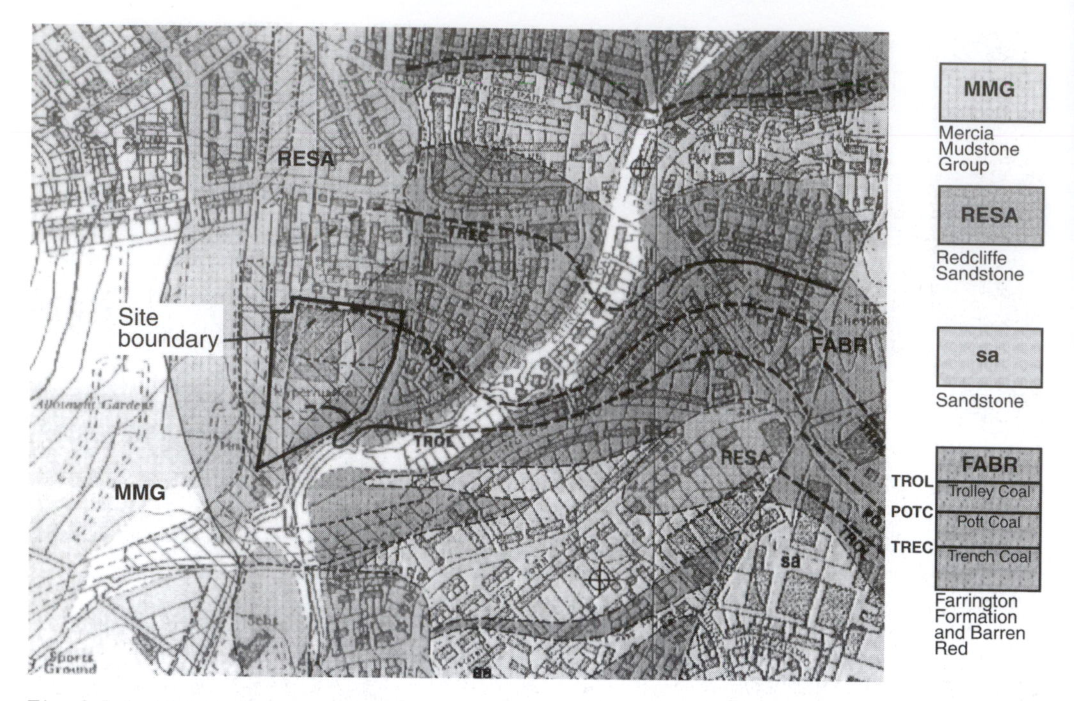

Fig. 2.5 1 : 10 000 scale geological map – part of Sheet ST67 SW (Drift) of south-east Bristol (Reproduced with permission of BGS)

Topography and land cover

The form of the land, particularly when it is combined with the geology, can provide valuable information about potential geotechnical hazards. For example, existing slopes of more than about 4° on clay subsoil could contain pre-existing slip surfaces which could lead to instability if the slope were to be loaded or steepened during construction. Uneven, hummocky and poorly drained ground on hill slopes is normally a sign of instability. The land cover is equally important. Again where the subsoil is clay the removal of vegetation to make way for construction may in time result in damage to structures on shallow foundations as a result of the clay undergoing expansion as its moisture content increases. The principal sources of information about topography and land cover are topographic maps and aerial photography.

Topographic maps are available in many scales and formats. The most common format is a thematic map depicting roads, railways, rivers, buildings, contours, etc. printed on paper (see Fig. 2.4). In some countries (such as Italy for example) topographic information (e.g. contours) is superimposed directly on aerial photography (Fig. 2.6). Such maps are more objective than conventional thematic maps since no information has been omitted. Such maps, however, require more experience to interpret

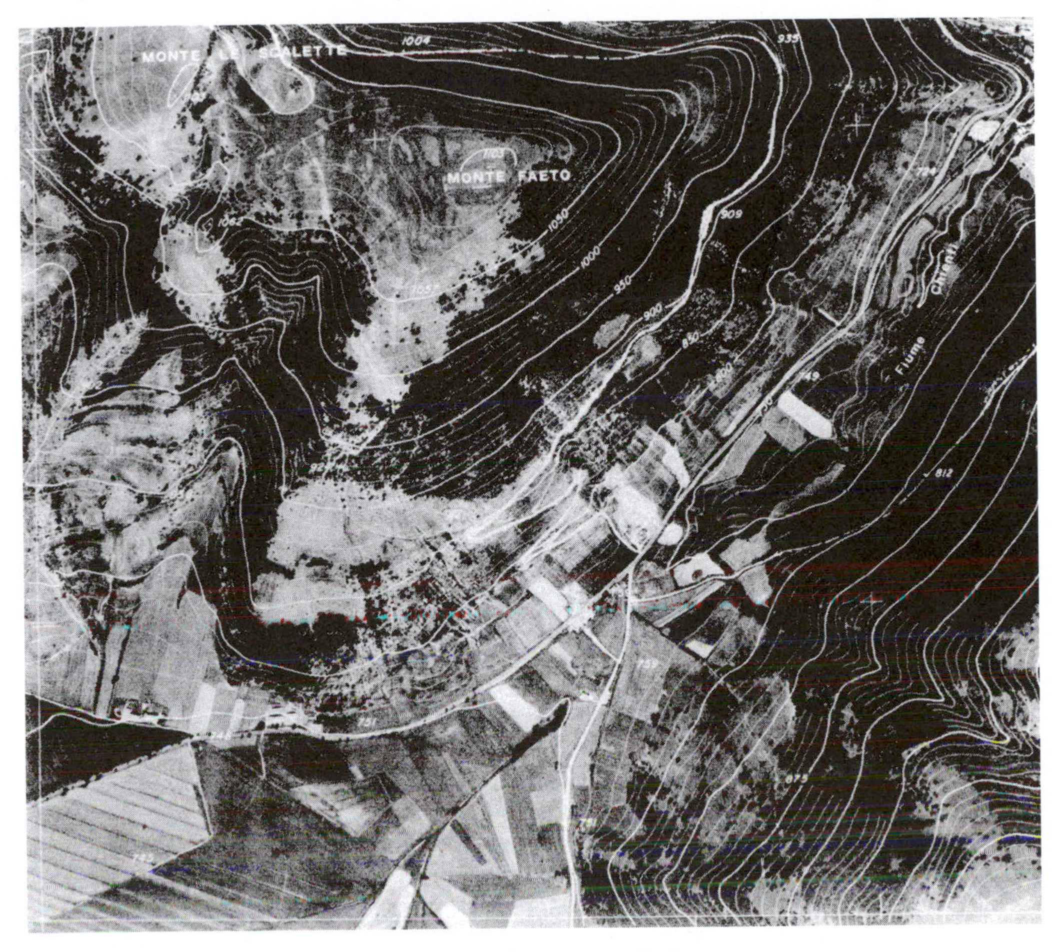

Fig. 2.6 Topographic map of the Monte Faeto area, Italy

as a result. The availability of digital map products is growing rapidly. In the UK digital map products are available at most scales, particularly at large scales. The advantage of digital maps is that they can be incorporated into a Geographic Information System (GIS). The most common scale of map available in the UK is 1 : 50 000 (2 cm to 1 km Landranger Series, e.g. see Fig. 2.4). These maps cover an area of 40 km × 40 km and generally contain too little detail for most site investigations where the site is of limited extent. They are more useful for extended sites such as highways, where existing road and footpath access may be complex.

The 1 : 25 000 map (4 cm to 1 km) is rapidly replacing the 1 : 50 000 map in the UK as the general purpose map product (Fig. 2.7). These maps show field boundaries. This allows ground features such as dolines (dissolution features) or disused pits or quarries to be correlated more easily with other data sources such as historical maps and aerial photography. The maps

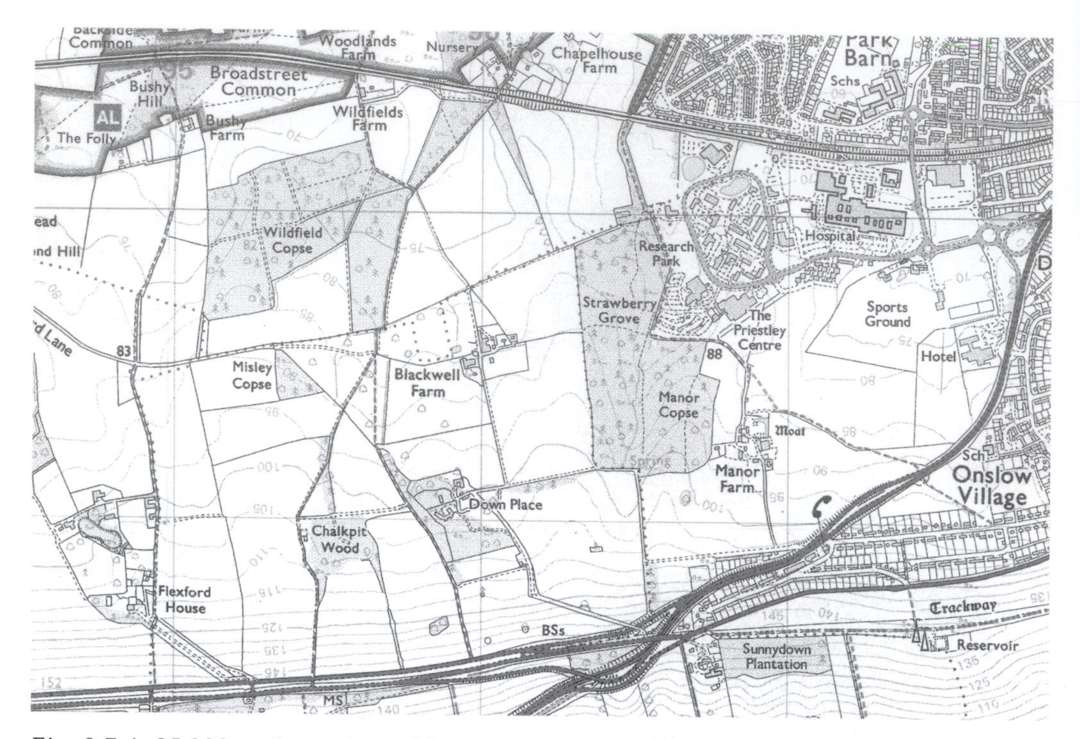

Fig. 2.7 1 : 25 000 scale topographic map of Manor Farm, near Guildford, Surrey (gridlines at 1 km intervals) (Reproduced with permission of Ordnance Survey)

show contours at a 5 m interval which permits relatively minor topographic features to be identified. Steep slopes which may suffer from, or be liable to, instability may be identified and targeted for further investigation. In addition to surface water features such as rivers, ponds and lakes, springs seepages (marshy ground) and wells are marked on these maps. The locations of springs and seepages provide advanced warning of a high groundwater table.

Large-scale maps of the UK are available at scales of 1 : 10 000 (5 km by 5 km), 1 : 5000 (3 km by 3 km), 1 : 2500, 1 : 1250 and 1 : 500. For site investigation purposes, the 1 : 1250 scale map is useful for noting the positions of any features seen during the walk-over survey. All the large-scale maps have the positions and levels of benchmarks shown and hence will be useful during the later subsurface exploration stage of site investigation when the ground level at boreholes is required.

Topographic data for the UK can now be obtained from the Ordnance Survey in a digital format either as contours or as a Digital Terrain Model[3] (DTM) at a scale of 1 : 50 000. These data can be incorporated

[3] A DTM is a digital representation of a topographic surface interpolated from surveyed contours. It will not include surface features such as buildings and trees.

into a GIS and used for flood potential studies as well as the identification of landforms that may be subject to instability. For more detailed topographic information high-resolution height data captured using airborne Light Detection And Ranging (LIDAR) systems is now available. The LIDAR systems work by sending a laser pulse from an aircraft to the ground and measuring the speed and intensity of the return signal. A Digital Elevation Model[4] (DEM) is created by interpolating the distance travelled by each laser pulse and the exact position of the aircraft when the pulse was transmitted. The resolution of LIDAR systems is such that subtle topographic features that indicate slope instability for example may be identified.

Surface water, groundwater and drainage

Water can present significant problems during construction if its presence is not predicted at an early stage. Level ground in valleys (usually flood plains) is likely to have a high groundwater table and may be subject to periodic flooding from surface water, groundwater or both. Springs, seepages, ponds or rivers are also a sign of a high groundwater table. The movement of plant over such a site may prove difficult and problems will be encountered with excavations.

The principal sources of information relating to surface water, groundwater and drainage are topographic maps, geological maps (the extent of flood plains is clearly depicted on geological maps from the outcrop of alluvium) and aerial photographs. In addition to these sources there are specialist records and maps such as well records held by the British Geological Survey and flood potential maps available on the internet from the Environment Agency. Well catalogues held by the British Geological Survey can provide useful information about groundwater levels as well as the depths of different soil types.

Site history

A knowledge of the previous uses of a site can provide valuable information relating to potential geotechnical hazards such as the presence of made up ground, landfill or contaminated ground. The desk study must include a search of all available large-scale topographical maps (i.e. 1 : 10 560 or larger) particularly those that are now out of print (referred to as historical maps in Table 2.3). In the UK the most useful maps are the 'County Series', produced from the mid 19th century up until the

[4] A DEM is a generic term describing a digital representation of a topographic surface. This may be represented by contours, spot heights on a regular grid or triangulated irregular network (TIN). A DEM may include surface features such as buildings, trees, etc.

Second World War, and the Ordnance Survey National Grid mapping which replaced the County Series. In addition to these plans, county libraries, muniment rooms, and archives will contain valuable historical information, for example:

- old mining records
- trade directories
- the Victoria County History
- specialist archival materials on local industries
- old newspaper files (also available from newspaper publishers)
- Public Health Act plans (c. 1850 prepared for the installation of early Victorian sewer systems) and
- early maps, for example, related to property ownership.

Another important source of information about site history is aerial photography. Since the Second World War there has been very good coverage of most of the UK with small- and large-scale aerial photography (mainly panchromatic, i.e. black and white). Local authorities generally commission aerial photography of their areas on a regular (usually every 10 years) basis. It is generally not difficult to build up a good pictorial record of a site covering the past 50 years. The advantage of aerial photography in the UK is that the frequency of photography is greater than the revision of topographic maps. This can be extremely useful in tracing the extent of landfill since the final size of quarries before they fell into disuse can be readily established using aerial photography.

Sites that were once used for industrial purposes may be contaminated. Table 2.4 gives the common contaminants associated with various industrial uses. Simply describing historical maps and aerial photographs (or reproducing extracts from them) in a report does not exploit the full range of information they are likely to contain. Knowledge of the historical features one is trying to investigate is essential for a reliable interpretation of historical information (for example, old waterfront structures, military installations, mines and quarries and the industrial activities in dockyards). A good knowledge of geology, geomorphology and landscape history is also essential for the interpretation of historical information.

Use of historical records

Case study: Biddulph Moor, UK

This case study illustrates the use of historical maps to detect made ground. Figure 2.8 shows extracts from 1:2500 scale topographic maps of the Biddulph Moor area of Shropshire. The maps shown are the revisions of 1876, 1899, 1925, 1960 and 1968. The 1899 and 1925 maps show a broken line marking the route of a small stream North of the road that runs WSW to ENE across the centre of the area shown in Fig. 2.8. The

Table 2.4 Contaminants associated with various industrial sites (from DD175: 1988)

Industry	Examples of sites	Likely contaminants
Chemical	• Acid/alkali works • Dyeworks • Fertilizers and pesticides • Pharmaceuticals • Paint works • Wood treatment plants	Acids, alkalis, metals, solvents (e.g. toluene, benzene), phenols, specialized organic compounds
Petrochemical	• Oil refineries • Tank farms • Fuel storage depots • Tar distilleries	Hydrocarbons, phenols, acids, alkalis and asbestos
Metal	• Iron and steel works • Foundries, smelters • Electroplating, anodizing and galvanizing works • Engineering works • Shipbuilding/shipbreaking • Scrap reduction plants	Metals (especially Fe, Cu, Ni, Cr, Zn, Cd and Pb), asbestos
Energy	• Gasworks • Power stations	Combustable substances (e.g. coal and coke dust), phenols, cyanides, sulphur compounds, asbestos
Transport	• Garages, vehicle builders and maintenance workshops • Railway depots	Combustable substances, hydrocarbons, asbestos
Mineral extraction Land restoration (including waste disposal sites)	• Mines and spoil heaps • Pits and quarries • Filled sites	Metals (e.g. Cu, Zn, Pb), gases (e.g. methane), leachates
Water supply and sewage treatment	• Waterworks • Sewage treatment plants	Metals (in sludges), micro-organisms
Miscellaneous	• Docks, wharfs and quays • Tanneries • Rubber works • Military land	Metals, organic compounds, methane, toxic or flammable or explosive substances, micro-organisms

stream is shown running approximately NS under the road and emerges on the South side from a culvert where it continues its course southwards. The part of the stream North of the road is not shown on the 1960 and 1968 editions of the map. The stream has been infilled and levelled to make way for the housing develepment. The stream, however, is still shown flowing on the South side of the road. Considerable structural damage

LIVERPOOL JOHN MOORES UNIVERSITY
Aldham Roberts L.R.C.
TEL 0151 231 3701/3634

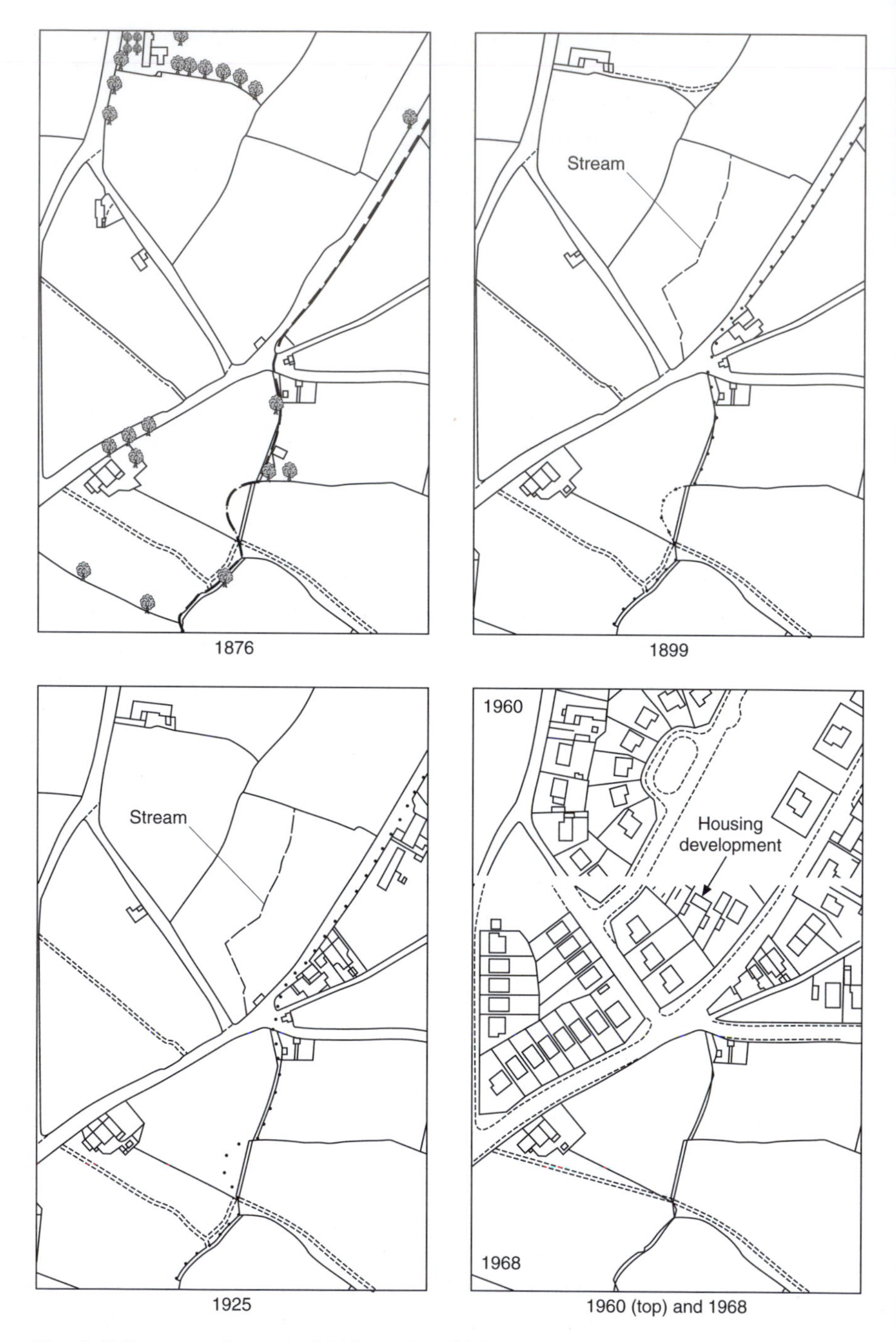

Fig. 2.8 Extracts from 1 : 2500 scale of Biddulph Moor, Staffordshire (after Clayton et al., *1995) (maps are all to same scale)*

occurred to the houses in the area of the infilled stream. Subsequent investigations revealed that the houses were built on made ground. The maps should have given warning of this.

The 1876 map does not show the stream. This indicates the importance of collecting all available archive material, particularly different editions of maps and air photographs.

Case study: Basildon, UK

This case study illustrates the use of aerial photography to investigate houses damaged by heave, following the removal of trees (Clayton *et al.*, 1995). Figure 2.9 traces the history of a site in Basildon, Essex using aerial photography. Figure 2.9(a) shows the site before construction. The site comprises areas that are densely wooded and open areas covered with grass. The geological maps of this area reveal that this site is underlain by London Clay which is known to be shrinkable clay. Figure 2.9(b) shows the construction phase of a housing development and Fig. 2.9(c) shows the completed development. The trees shown in Fig. 2.9(a) have been removed to make way for the housing development. Following construction, some of the houses in the terraces began to crack and investigations revealed that parts of these structures were undergoing heave. Large differential movements were found to occur where these long structures crossed from the once wooded areas into the once open grassed areas. The movements resulted from the changes in moisture content of the clay due to the removal of the trees. This is a common geotechnical hazard that can be readily predicted from knowledge of the geology and the position of the trees, both of which can often be found from a desk study and walk-over survey. Aerial photography provides the key information since it provides a permanent record of the size and position of trees and other vegetation which may be removed some years before such developments are envisaged. In addition, aerial photography also provides a valuable permanent record of construction.

Case study: Paddington Station, London

In the desk study carried out for the redevelopment of Paddington Station, London, industrial archaeology and the use of specialist archive material played a key role (Shilston *et al.*, 1998). The proposed redevelopment work would involve disturbance to the built fabric both above and below ground. This highlighted the need to bring together existing geotechnical and historical information for the station. Historical information was obtained about the original construction of the various parts of the station (including, for example, details of foundations, cellars or basements and underground railways) and their subsequent piecemeal modification. As anticipated, this information was found to be particularly

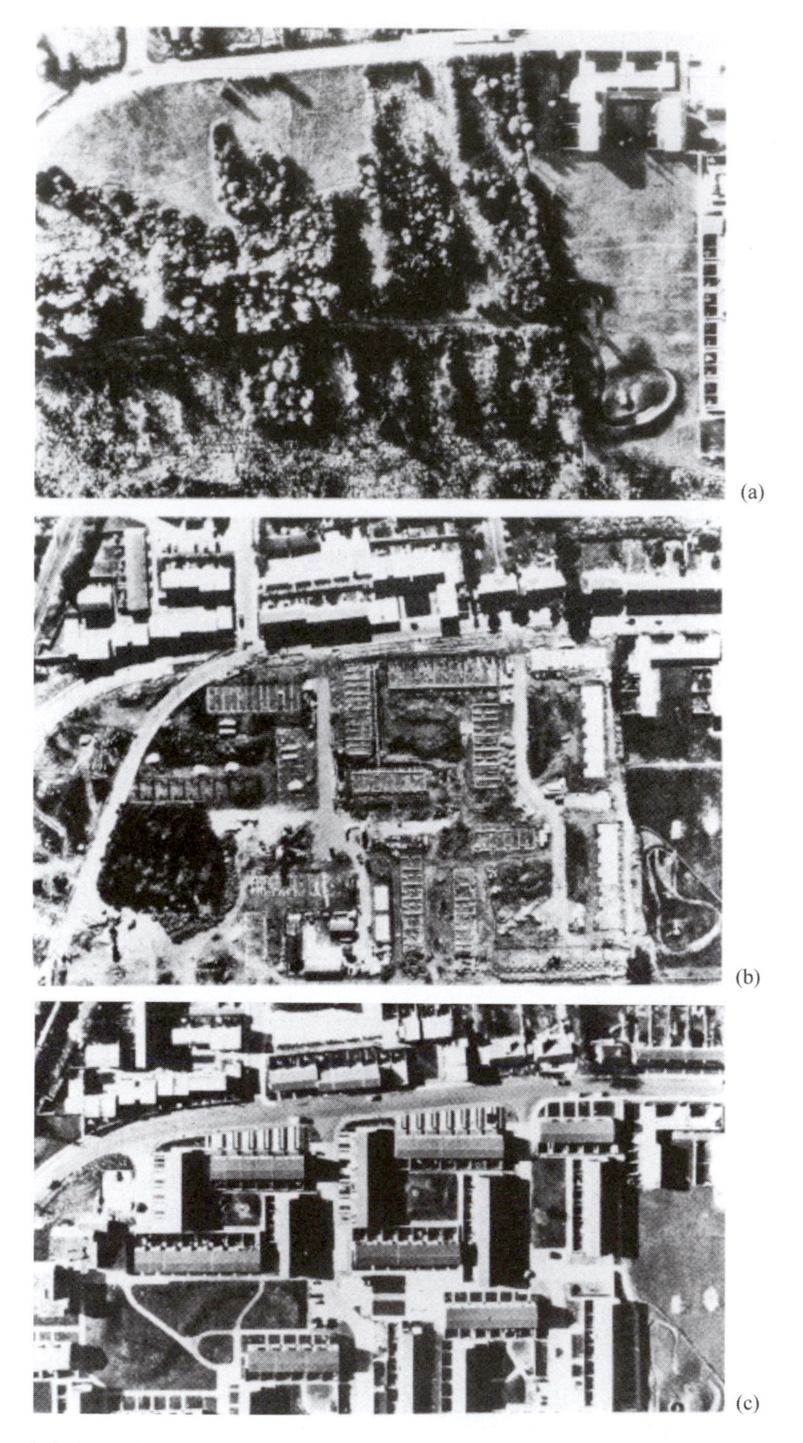

Fig. 2.9 Aerial photographs showing the development of a housing area in Basildon

valuable for the interpretation of the building's current and former states and for interpreting the results of borehole and other subsurface investigations.

Paddington Station was designed and built by Isambard Kingdom Brunel between 1850 and1854 for the Great Western Railway. It replaced a temporary station also built by Brunel which had been constructed in 1838. The Great Western Railway Royal Hotel was designed by P.C. Hardwick and also opened in 1854. This structure bordered the station on its East side, with a waiting area in between called The Lawn. Between 1854 and 1916 Paddington Station underwent a series of piecemeal expansions and alterations to accommodate the increasing number of passenger and goods traffic. In 1916 an additional span was added to the train sheds and in 1924 all the original columns were replaced by steel columns. During the 1930s two modern high-rise office blocks were built to the North and South of the station (Tournament House and Macmillan House) and by 1984 more alterations had been carried out to improve passenger access and the condition of the station structures. Two separate underground railway systems have lines beneath and adjacent to Paddington Station.

The principal sources of historical information examined during the desk study and their main benefits to the project are summarized in Table 2.5. One particular point of interest was the use of historical photographs taken during the replacement (in 1924) of the original Brunel columns. The photographs clearly showed the foundations of the columns and the method of their replacement. If the detailed historical investigation had not been undertaken, then details such as these could have been overlooked, potentially creating increased uncertainty and additional work for the engineering and architectural design teams.

Ground conditions

Ground conditions relate to knowledge of the physical and mechanical properties of the geomaterials present under the site (e.g. strength, compressibility, permeability and, in the case of clays, swelling and shrinkage characteristics), and the existence of features such as chemically aggressive groundwater, slope instability and dissolution features. Information on the physical and mechanical properties of the soils and rocks and chemically aggressive groundwater can only really be gained from reports from previous site investigations carried out on or near the site. Information on groundwater quality may be obtained from the local water supply company particularly if they are abstracting significant quantities of water for public supply.

Evidence of slope instability may be gained from geological maps and aerial photography, previous site investigation reports and articles in

Table 2.5 Summary of principal sources of historical information examined during the desk study for the redevelopment of Paddington Station (after Shilston et al.*, 1998)*

Classes of information	Description of information	Examples	Benefits
Topography	Ordnance Survey First Edition (19th century); 18th and 19th century estate maps	Historical maps	OS maps held by public libraries and therefore inexpensive to access; topographic information quite accurate, even on earliest editions. Estate maps sometimes show features not mapped later because destroyed
	Post-Second World War aerial surveys held by the National Library of Air Photographs, Swindon	Aerial photographs	Show features not recorded by mapping, particularly where buried remains are suspected
Environment and planning	Sites and monuments records held by local authority. Scheduled ancient monuments records held by English Heritage. Listed buildings information held by local authorities	Records held and updated by the statutory national and local planning and heritage organisations	Identification of legally protected buildings and remains required to avoid breaking planning law. Vital information for preliminary identification of heritage potential. Opportunity for opening dialogue with statutory consultees which indicated no archaeological site within station. Also established which parts of the station were listed
Site condition, land use and history	Map sequences from earliest reliable sources allowed development of station footprint to be traced	Historical maps	Map 'regression' exercise allowed identification of major changes in layout indicating areas of potential fill and ground disturbance
	Greater London Record Office/Metropolitan Archives hold books on the station's history; Royal Commission on the Historical Monuments (England) have a substantial railway archive section. Other materials such as railway and engineering magazines, newspapers, local history society pamphlets and specialist publications also available	General archive sources	Detailed understanding of development of station construction (expansion/contraction) allowed areas of potential disturbance or those which required further detailed ground investigation to be identified. Specific design details of station construction also found which were later used in design of the new foundations. Details of bombing during the Second World War

	I.K. Brunel archive held by Bristol University. Railway Heritage Museum and Great Western Railway Museum in Swindon hold material on Brunel's Paddington Station, such as back issues of the *Great Western Railway Magazine* dating from the late 19th century	Specialist museum archives.	As above.
Engineering and architectural records	These archives hold contemporary accounts of station construction and details of later expansion. They also hold information on the 1930s office buildings, which had not been investigated since their construction, and on the underground railways and stations	Institution of Civil Engineers, London Underground and Post Office Railways	Engineering information obtained allowed more cost effective design of foundations; also assisted with the investigation of the construction techniques and materials of the 1930s office buildings
	Architectural and historical survey carried out by English Heritage in 1993 assisted with finer points of design details, such as spandrel design by M. Digby-Wyatt in 1850s	English Heritage	As above. The information was also useful to architects designing the new additions to the station – sympathetic design in historic contexts is currently a very important issue for designers
	As above	Specialist and museum archives	As above

learned journals. It is important when checking the site for slope instability to check for evidence of instability in the area around the site where the geology, slope and slope aspect (dip direction of slope) are similar to that of the site. Large landslips are often marked on geological maps. The key source of information, however, is aerial photography. The important features to look out for are hummocky ground, ponding, deranged drainage, rear scarp features and lobate features.

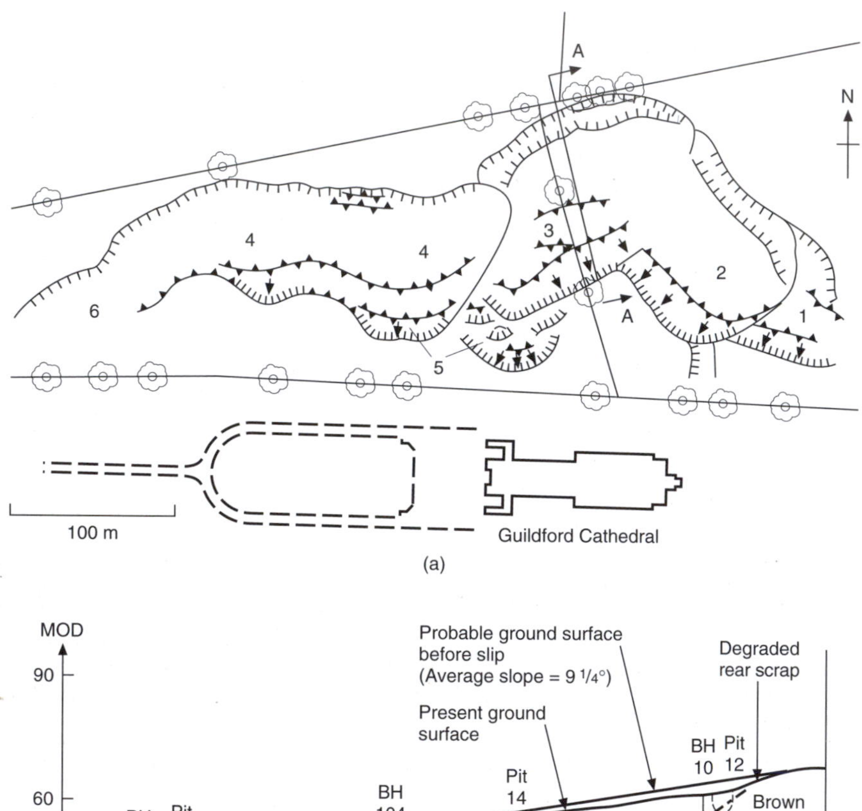

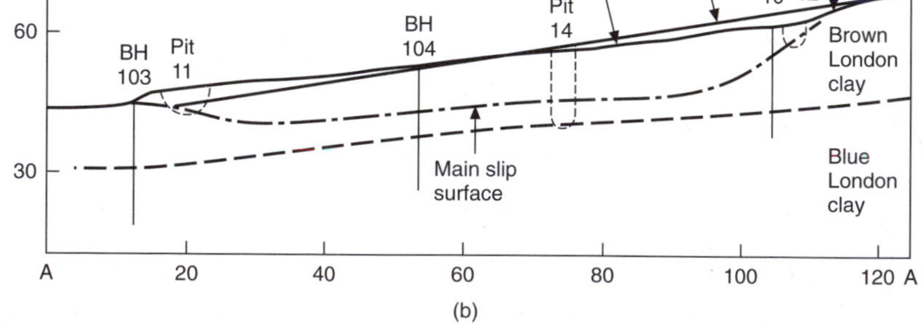

Fig. 2.10 Landslip at Stag Hill, Guildford. (a) Extent and sequence of landslides on the North slope of Stag Hill prepared from oblique air photographs. (b) Section A–A through landslip at Stag Hill (after Skempton and Petley, 1967)

Case study: Stag Hill, Guildford, UK

This case study illustrates the identification of landslip features using vertical and oblique air photography. The North face of Stag Hill near Guildford, Surrey, was the site chosen for the University of Surrey. The reason why a site so close to the town centre of Guildford had not been previously developed was because of a large landslip occupying the North face of Stag Hill. This landslip occurred on a 9° slope in brown London Clay (Fig. 2.10) and its extent is clearly visible on vertical aerial photographs taken before construction of the university began in 1967. An example of such a photograph (taken during 1961) is shown in Fig. 2.11.

The landslip is identified on Fig. 2.11 mainly by shadow, relief and vegetation. The rear scarp and toe of the landslip form a small step in the slope. These topographic features are enhanced by shadow allowing the limits of the slip to be estimated readily from a single photograph (Fig. 2.12). It can be seen from Fig. 2.11 that a line of trees and bushes marks the position of part of the rear scarp and toe. The topographic expression of these features appears to be more pronounced here than elsewhere, making cultivation across them very difficult. Thus natural vegetation has been become established on these parts of the rear scarp and toe.

Several coalescing landslips of different ages give rise to the overall feature seen in Fig. 2.11. Many subtle landslip features are, however,

Fig. 2.11 The Stag Hill landslip, Guildford

Fig. 2.12 Oblique air photo of Stag Hill landslip, Guildford

difficult to identify even though the contact scale of the photographs is relatively large (1 : 4000). The minimal shadow normally required for vertical air photography can severely limit the detection of such minor features. Oblique photographs (when available) can often overcome this problem. The oblique air photograph shown in Fig. 2.12 was taken with a low sun angle. Minor topographic features associated with the landslip are clearly visible because of the long shadows.

Case study: Sevenoaks by-pass, UK

This case study illustrates the use of aerial photography to identify solifluction lobes near Sevenoaks, Kent (Clayton *et al.*, 1995). The construction of the Sevenoaks by-pass faced problems with earthwork failures. Subsequent investigation (Weeks, 1970) revealed that the route crossed several solifluction lobes which are in a state of limiting equilibrium. These solifluction lobes are underlain by an extensive solifluction sheet which contains several principal slip surfaces located at the base of the

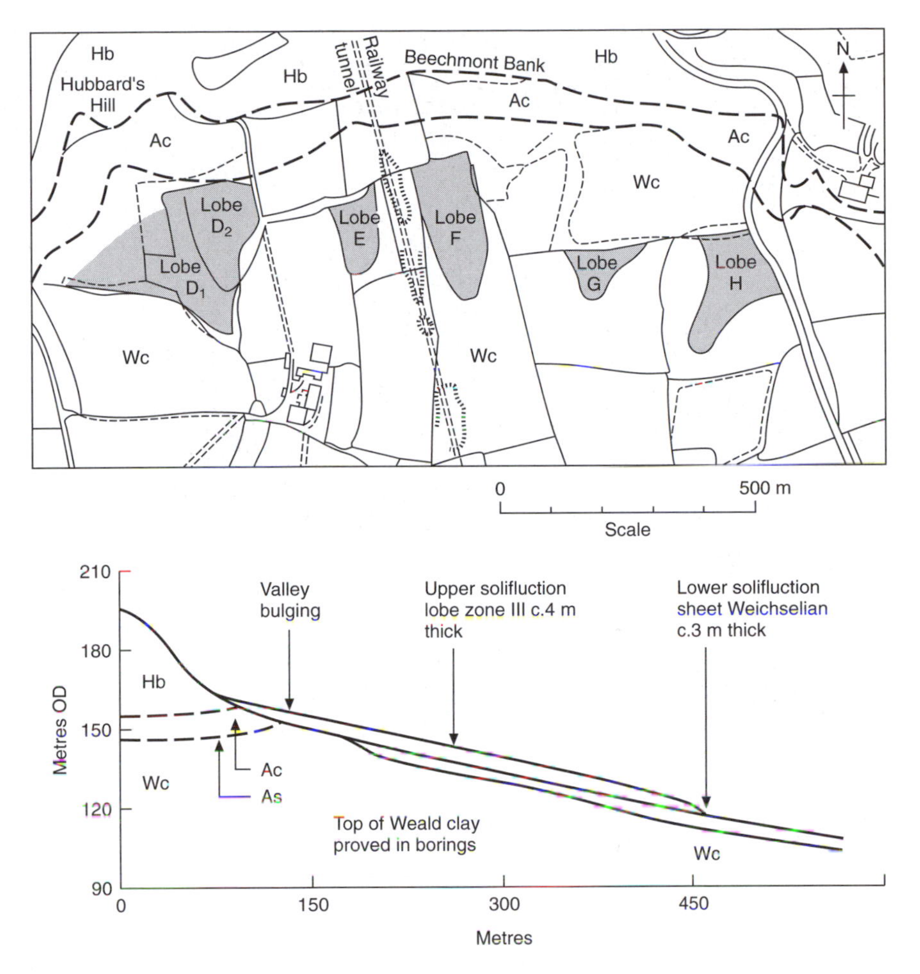

Fig. 2.13 Geological map and section of solifluction lobes near Sevenoaks, UK (section from Weeks, 1970). Ac, Atherfield Clay; As, Atherfield silt; Hb, Hythe beds; Wc, Weald Clay

Fig. 2.14 Mosaic of vertical air photography, showing the landslip complex of the Sevenoaks by-pass site (after Clayton et al., *1995)*

sheet (Weeks, 1969; 1970). The lobes appear to be more unstable than the underlying sheet.

The geology of the area was determined from 1 : 50 000 and 1 : 10 560 scale geological maps and transferred to a base map shown in Fig. 2.13. The Hythe Beds which comprise sandy limestone, limestone and sandy silty clay layers form a steep escarpment. These beds are underlain by the Atherfield Clay and the Weald Clay, both of which outcrop at the base of the escarpment as indicated in the geological cross-section shown in Fig. 2.13. The ground at the base of the escarpment slopes towards the South at angles between 3° and 10°.

Aerial photography of the area in front of the escarpment revealed a series of lobe-shaped features which are characterized by hummocky ground (see Fig. 2.14). The hummocky ground gives rise to a distinctive photographic texture which is referred to as 'turbulent' by Clayton *et al.* (1995). The position and extent of the lobes identified from Fig. 2.14 is shown on Fig. 2.13.

Lobe H is most noticeable. The field in which it is situated is covered by a series of light-toned lineations which are disrupted by the lobe and help enhance its characteristic 'turbulent' texture. Lobe D is not easily identified because the field boundaries follow the edges of the lobe to a large extent. The upper part of the eastern edge of the lobe occurs downslope of the field boundary and is easily identified by a change in slope and characteristic texture.

From ground level the lobes appear as areas of hummocky ground (Fig. 2.15). The change in slope which marks the edge of the lobes

Fig. 2.15 Ground photographs of lobe E, Sevenoaks, UK

(particularly the 'nose' of each lobe) is a most noticeable feature at ground level. It would be impossible, however, to appreciate fully the extent and shape of these features simply from a walk-over survey. Aerial photography has the advantage of providing the necessary overview which allows a rapid assessment of ground features such as these.

Aerial photography

Aerial photography represents one of the most important desk study data sources for the reasons detailed below.

- *Improved vantage point.* It is often difficult to piece together observations made in the field in order to fully appreciate their pattern and extent. This is because we cannot usually see the whole picture at once. The 'bird's eye view' provided by aerial photography allows the patterns made by, say, vegetation or landforms to be recognized more easily and their engineering significance to be deduced.
- *Permanent record of existing conditions.* Unlike maps, which are subjective, aerial photographs represent raw data. The detail presented on a map represents only what the cartographer chose to include. No detail is excluded from an aerial photograph. Detail, however, may be obscured by vegetation or it may be too small to be recognized due to the scale of the photography.
- *Permits detailed study of site history.* The frequency with which an area is photographed from the air is generally greater than the frequency with which maps are revised. Therefore aerial photography can provide a more detailed record of the history of a site (e.g. the extent of quarrying at different times, the position of mine shafts, and the extent of landfill). Dumbleton (1983) gives a detailed account (with case histories) of how aerial photography can be used in the study of site history. There was little air photo coverage of the UK before the Second World War. The best coverage is found after about 1950. County Councils generally commission aerial photography on a county-wide basis every five or ten years. These photos, together with the Ordnance Survey and air photo libraries (e.g. Royal Commission on the Historical Monuments of England and Aerofilms Ltd), provide a comprehensive source of historical data.
- *Broadened spectral sensitivity.* Most aerial (and ordinary) photographs contain information within the visible part of the electromagnetic spectrum and a small part of the near-infrared portion of this spectrum. This means that there is generally more information in the photograph than you would be able to see by looking directly at the ground. Patterns relating to different ground conditions (e.g. disturbed ground) may show up more clearly on aerial photography. It is also

possible, by using special film and filters, to enhance the view of the near-infrared part of the spectrum. This is useful in looking at subtle changes in vegetation characteristics which may be associated with geotechnical hazards such as landfill.

- *Availability*. We are fortunate in the UK to have excellent air photo coverage. Only a very few out-of-the-way places have no coverage at all. Most of the available photography is panchromatic, however, but colour is being increasingly used. Between 1998 and 2001 every part of the UK was photographed in colour from the air using conventional 9 inch film from an average height of 1524 ft as part of the Millennium Map Project. These photographs have been scanned and transformed into digital images. They are available for viewing (at low resolution) and purchase (at full resolution) on the internet (www.getmapping.com and www.multimap.com). The Millennium Map Company has published two aerial photographic atlases covering London and England. These books provide a valuable source of data for desk studies. They are not a substitute for purchasing the photography since in many cases the scale at which the photographs are shown in these books is too small for detailed interpretation. It is the intention of the Millennium Map Company to update the aerial photography of the whole of the UK on a regular basis (e.g. every five years). In the future this will provide an extremely valuable source of site history information.

  In general the availability of aerial photography is usually good in developed countries. In developing countries, however, aerial photography may have to be specially commissioned.
- *Cost*. Aerial photography represents an extremely cheap source of data in terms of the amount of information it contains. The unit cost in some cases may be relatively high. You should perhaps compare this with the cost of a digital map of the same area of ground and then consider the fact that the aerial photograph contains raw data whereas the map is accurate but subjective. The cost benefit of the photography is greater provided you have the ability to interpret the data adequately.

Photographic parameters affecting interpretation

Regardless of the skill of the interpreter or the methods of viewing the photographs, the amount of information that can be obtained will depend largely on the photographic image. The parameters which affect the image include:

- geometry of image (orientation of camera axis)
- emulsion and filter combinations
- scale of image

- image medium
- time of day of photography
- season of photography.

Geometry of image

On the basis of the orientation of the camera axis, aerial photographs may be classified as vertical (camera looking vertically downwards, Fig. 2.11) or oblique (camera axis at some angle with the vertical, Fig. 2.12).

Vertical photographs are used more extensively because they can be used for mapping purposes. The scale changes that occur across the field of view in an oblique photograph make it very difficult or even impossible to make accurate measurements. Oblique aerial photographs, however, are more easily interpreted, particularly by the inexperienced interpreter, since they give a more familiar view of objects than that provided by a view looking vertically downwards. The value of oblique photography is often underestimated. Obliques are particularly useful when viewed in conjunction with vertical photographs. The great advantage of oblique photography is that it can be obtained rapidly and cheaply by mounting a 35 mm camera in a radio-controlled model fixed-wing aircraft or helicopter or below a kite or tethered balloon.

Emulsion and filter combinations

The main types of film used in aerial photography are black and white, true colour and false colour infrared. Of these, true colour is the only one that gives a true picture of the ground surface. Objects seen in true colour photographs are made readily identifiable since the eye is capable of separating at least one hundred times more colour combinations than grey levels. Colour films, however, have less definition than the corresponding black and white films. Furthermore, it can often be more difficult for the interpreter to assimilate all the information contained within a colour photograph. The result is that subtle textural variations and geometric patterns may be easily missed, whereas they are usually more distinguishable on black and white photographs.

Infrared films record reflected radiation in the visible part of the spectrum but are also sensitive to reflected infrared radiation (up to wavelengths of about 0·9 μm), which is invisible to the naked eye. The photographs produced using colour infrared film are referred to as false colour photographs. This is because the colours are shifted to enable the infrared information to be displayed on the image. It is important when interpreting false colour infrared photographs to understand how the colours are shifted (Table 2.6).

Vegetation reflects a significant amount of radiation in the near-infrared part of the electromagnetic spectrum and hence appears as different

Table 2.6 Colour shift for false colour infrared photographs

Reflected electromagnetic energy	Colours produced on the image
Blue	Not represented (filtered out)
Green	Blue
Red	Green
Infrared	Red

shades of red on the false colour image. False colour infrared photography is useful for detecting disturbed ground, springs, seepages and contaminated ground (through changes in type of vegetation). The disadvantage of false colour infrared photography is that it requires specialist film and processing and hence is expensive. As a result existing coverage of false colour infrared aerial photography is very limited.

Scale of image

The amount of detail that can be seen on an aerial photograph will depend to a large extent on the scale of the image. Unlike a topographic map the scale of a vertical aerial photograph varies in relation to the terrain elevation. The contact scale, S, given for an aerial photograph is related to the average terrain elevation h_{av}, the altitude of the aircraft H and the focal length of the camera lens f by the expression:

$$S = f/(H - h_{av})$$

The scale of aerial photographs varies from about 1 : 1 000 000 to about 1 : 1000. The typical scale for vertical aerial photographs in the UK is about 1 : 10 000. Although this scale is suitable for topographic mapping, some small-scale features of geotechnical interest (such as tension cracks, seepages, sinkholes and small landslips) may not be seen easily. A more suitable scale is between 1 : 5000 and 1 : 8000. However, since the area of ground seen on an aerial photograph is inversely proportional to the scale, by increasing the scale it becomes difficult to place large features such as major landforms or drainage patterns into an environmental setting. Ultimately the scale of image required will depend on the type of project. For route location studies small-scale photography is likely to be more appropriate. For the investigation of a site of limited extent larger scale photography may be used.

Image medium

The most commonly used photographic product is the familiar positive paper print. This may be a contact print or an enlargement. The standard format of aerial survey prints is 230 mm × 230 mm (9 in × 9 in). It is not normal practice to enlarge standard aerial survey photographs since not only are they more expensive but secondary processing tends to reduce the resolution of the original.

The photographic image often appears sharper when printed with a glossy finish. Often aerial photographs are printed with a non-reflective matt silk finish which tends to reduce image sharpness and tonal contrast. Special photographic paper may be employed to increase the tonal contrast of black and white photographs. During a walk-over survey it is more than likely that photographs will be taken into the field. The prints should therefore be printed on thick paper to increase durability.

An alternative to the opaque paper print is the positive transparency. This may be the initial film product, as is the case with colour reversal film or infrared Ektachrome (false colour infrared) or it can be produced directly from the negative (diapositives) as is often the case with black and white photographs used for photogrammetric purposes. Film transparencies are more difficult to handle and work with than paper but in some instances colours can be more true and the image sharper.

It is possible to enhance the quality of the photographic image in terms of contrast and brightness by scanning it and using image processing software. Such raster-based digital images can be georeferenced and annotated (after the walk-over survey) so that they can be incorporated into a Geographic Information System (GIS).

Time of day of photography

The time of day at which aerial photographs are taken has a great influence on the appearance of ground features. The controlling factor is the sun's elevation. At low elevations (i.e. during early morning or late afternoon–early evening) long shadows are cast by objects, which in some cases can aid the recognition of certain surface features (e.g. see Fig. 2.12).

While shadow can aid interpretation, it can also be a hindrance particularly in areas of high relief, since important detail may become hidden. It is for this reason that most vertical aerial photography is taken during late morning or early afternoon when the sun's elevation is at or near maximum, and hence the amount of shadow is minimal. Aerial photographs with low sun elevation are not readily available for most of the UK. The most common use of such photography is for archaeological surveys. The photographs used in this work are generally oblique and can be very useful in geotechnical desk studies.

Season of photography

The time of year in which aerial photographs are taken can be a critical factor in detecting features of geotechnical interest such as springs, seepages, instability and made-up ground. Springs and seepages may not be visible on photographs taken when the water table is low during the summer months. Moisture differences show up best in spring. Variation in soil types or moisture content may be seen directly as soil marks or indirectly as crop marks. These variations may be natural or due to disturbances such as trenching for buried services or the presence of land drains, buried foundations or archaeological remains. Soil patterns representing such features are best seen during spring and autumn, particularly when the ground is bare.

Crop marks are used extensively by archaeologists to detect disturbed ground associated with ancient settlements and field systems. They occur where there are differences in the water retention properties of the soil which, in the right conditions, affect the growth rate of certain crops. Crop marks may also occur in association with contaminated land. In limestone areas the moisture-retaining nature of the soil infill within some sinkholes causes the vegetation to be more lush over the feature, thus aiding detection (e.g. Edwards, 1969). Crop marks are, however, transitory features, being visible only for a few days, during the summer months when the soil becomes increasingly dry. The best conditions are during periods of drought. Aerial photographs taken during the UK drought of 1976 provide some very good examples of crop marks. Crop marks are best seen in barley, sugar beet and peas between June and August. They may also be seen in open grassland, but only during the driest weather. Moorland vegetation shows the best effects in late autumn when it is drying back and in spring when differences in reflectivity from dead vegetation are greatest (Dumbleton, 1983).

Aerial photographs taken after a period of snowfall can be very useful in locating features of geotechnical interest. Snow cover has the effect of removing distracting detail and leaving only information about the general topography (beware of snow drifts!). Differential snow-melt can also be helpful in locating buried services which are relatively close to the ground surface.

In general, aerial photographs taken during spring and autumn will normally provide sufficient data for identification of most geotechnical features. Of course the choice of season is very much limited by the availability of existing photography. In the UK periods of good weather which are favourable for aerial photography (i.e. minimum cloud cover and good visibility) amount to less than 500 hours per year. The summer offers the most settled conditions but haze may restrict visibility to some extent. Spring and autumn, however, offer the highest proportion of days with good visibility and hence photographs taken during these seasons should be readily available.

Interpretation of aerial photography

The examination of any type of photograph from the family snapshot to a false colour infrared aerial photograph involves interpretation. You may be unaware that you are performing interpretation when viewing the family photo album but in fact features such as height, build, facial expression, etc., help you distinguish one person from another. This information is fed into your internal database which finds the names to put to the faces. Interpretation therefore involves two stages: recognition and deduction.

The basic knowledge or reference level required to interpret a family snapshot is of course different from that required to perform interpretation of aerial photography for geotechnical purposes. For desk studies, the basic reference level required by the interpreter is knowledge of physical and cultural features, their relationship with geology, geomorphology and land use, their engineering significance and their common form of appearance on aerial photographs (both vertical and oblique). The development of such knowledge requires substantial experience. The starting point is a basic knowledge of engineering geology.

During each desk study your basic reference level will be extended partly through the study of the other sources of data such as maps and geological memoirs and from the overall experience gained from that particular project. In general, interpretation demands keen powers of observation, imagination and above all patience.

Principles of image interpretation

Every image must be treated on its merits and hence there are no rules to define how an image should be interpreted. The people in the family snapshot, for example, are recognized in photographic terms largely on the basis of shape, tone, colour and size. These recognition elements are used subconsciously in this case because the reference level of the interpreter is high. In the case of an image of the ground surface the situation is different in terms of the features to be identified and of course, the reference level of the interpreter. The interpreter, particularly an inexperienced one, will make a more conscious and systematic use of recognition elements. These relate mainly to the characteristics of the image alone and include the following.

- *Tone.* Tone refers to the colour or reflective brightness of features. Tone is the most important recognition element since most features will be distinguished on the basis of tone. In general, on black and white photographs dark tones are indicative of dark-coloured materials or wet conditions (e.g. poorly drained soil or seepages) and light tones indicate light-coloured materials or dry conditions. The fossil river channels shown as soil marks in Fig. 2.16 were recognized

Fig. 2.16 Air photograph showing infilled channels, Fenlands, UK

on the basis that they displayed a lighter tone than did the ground either side. These are composed of fine sand and silt whilst the ground on either side is composed of soft clay and peat. The dark tones characterising these materials are marked 'A' in Fig. 2.16.

- *Shadow.* As a recognition element, shadow depends primarily on the use of shape and tone. The shape of objects and ground relief are enhanced by shadow. With vertical aerial photographs shadow permits an effective profile view of certain features. Mounds may be distinguished from depressions, and embankments from cuttings using shadow. One of the most important uses of shadow is in aiding the recognition of subdued topography. For example, such subdued topographic features are often associated with old landslips and other forms of disturbed ground. These features are best seen in photographs taken with low sun elevations (i.e. long shadows). A good example of the use of shadow in recognizing landslip features is shown in the oblique photograph of the North side of Stag Hill (the site of the University of Surrey prior to construction). The long shadows picked out the rear scarp and toe features of the landslip very clearly (see Fig. 2.12).

- *Texture.* Texture may be defined as the frequency of tone change within the image. It is produced by an aggregate of unit features too small to be clearly discerned individually, and is a composite of several image characteristics such as tone, shadow, size, shape and pattern. For example, leaves are too small to be recognized individually on most aerial photographs, but the size and shape of many leaves, together with the shadows they produce, combine to give a distinctive texture which allows one type of tree to be distinguished from another, particularly coniferous and deciduous trees. Texture is also useful in the identification of slope instability. The hummocky ground (often enhanced by shadow), and the impedance of drainage give rise to variations in tone which produce characteristic textures. Clayton *et al.* (1982) describe a 'turbulent texture' which is commonly associated with old mudflows and solifluction lobes. An example of this texture is shown in Fig. 2.14. This texture can also be seen in Fig. 2.6.
- *Pattern.* Pattern refers to the orderly spatial arrangement of geologic, topographic, vegetation or man-made features. The most noticeable patterns seen on aerial photographs of the UK are either a 'patchwork' of fields in rural areas, or a network of streets in urban areas. Land use patterns are essentially a ground characteristic and are described later. Variation in near-surface soil types may give rise to soil or crop patterns. The manifestations of such patterns are dependent upon the season of photography discussed earlier. Patterns of natural drainage are generally related to the geological structure. Man-made drainage measures produce distinctive regular patterns on photographs (e.g. see the linear features in the lower left corner of Fig. 2.6). These features are indicative of poorly drained ground or slope instability.
- *Shape.* Shape refers to the form of features seen from the air. Man-made features are normally characterized by straight lines or regular

curves, and hence are often recognized by shape alone. Many natural features have distinctive shapes. For example mudflows are generally lobate and sinkholes are commonly circular or elliptical. Natural features, however, may not always be recognized by shape alone.

- The *size* of objects can be used to aid recognition. Size, however, must be considered in relation to the scale of the photograph, or in relation to objects within the photograph of known size. Vertical exaggeration associated with viewing aerial photographs stereoscopically will make objects appear much taller than they are in reality.

These recognition elements allow the interpreter to classify the image on the basis of ground characteristics. This is normally done using overlays. Commonly used ground characteristics include the following.

- *Landforms* such as hills, scarps, valleys, flood plains, etc. are usually easily identified from the air. Photographs can be divided up into a number of major and minor landforms depending on the contact scale. Identifying the landform commonly identifies the natural process that formed it, and often gives an indication of the types of material that may be expected by association. Knowledge of geomorphology is required to gain the most from studying landforms.
- *Land cover* refers to the materials covering the ground surface, the most common of which are different forms of vegetation. Land cover is often confused with land use. Grass is an example of land cover. An area covered with grass may be a golf course or a field, both of which are examples of land use. Vegetation in cultivated areas is controlled by various environmental factors, the most important of which are soil type, slope and the availability of water. Local differences in soil characteristics, depth and moisture condition can be detected from the resulting differences in growth of crops or natural vegetation and ripening of cereals (crop marks have already been discussed). These data may be used in conjunction with other ground characteristics such as landforms and drainage to interpret soil and rock boundaries together with other features of geotechnical interest. Crop marks and different types of vegetation show up more clearly on false colour infrared photographs than on conventional black and white panchromatic photographs.
- *Land use* may be divided into five broad categories:
    - agriculture
    - urban development
    - forest
    - uncultivated (e.g. grassland, moorland and rock slopes)
    - materials extraction (e.g. quarrying and mining).

In general the most useful categories for interpreting ground conditions are agriculture, forest and uncultivated land.

- *Drainage patterns* are usually identified from aerial photographs. The type of drainage pattern and the density of the drainage network are often indicative of geological structure as well as rock/soil type.
- *Lineations* – many features, both natural and man-made, appear on aerial photographs as lineations. Such features are commonly the linear expression of characteristics such as tone, texture, landforms, drainage and vegetation. In some cases, these characteristics may be disrupted in a linear fashion, thus giving rise to a different linear feature. Natural lineations can be used to interpret soil and rock boundaries, together with structural features such as faults and bedrock jointing.

The recognition elements and ground characteristics are combined to deduce the significance of features seen in the photographs in terms of:

- geology
- geomorphology
- instability
- made ground
- contaminated land
- site access
- site drainage
- site history
- archaeology
- flood potential.

The above list is by no means complete and only represents the more common applications of aerial photography.

Interpretation procedure

When interpreting aerial photography it is necessary to have a systematic approach. The following stages are recommended.

- Preliminary examination:
  - broad overview
  - recognition of large-scale features
  - identification of areas which require special attention.
- Detailed examination:
  - classification of features and production of overlays
  - recognition of small-scale features.
- Deduction:
  - analysis of the geotechnical significance of features identified in the above steps.

- Compilation:
  - transferring information onto maps
  - identification of features which require inspection during walk-over survey.

It should be remembered that every time you look at a set of aerial photographs you see something new.

Sources of aerial photography

The largest collection of aerial photography is held by the Royal Commission on the Historical Monuments of England (RCHME), National Monuments Record, Air Photographs. The RCHME holds over three million photographs in its library; these include:

- oblique air photographs illustrating architectural, archaeological and landscape subjects
- vertical air photographs
  - RAF 1940–1965, scale 1 : 2500–1 : 60 000 (including National Survey 1946–1947 at 1 : 10 000
  - Ordnance Survey 1952–1970, scale 1 : 5000–1 : 23 000
  - Meridian Airmaps collection 1952–1984, scale 1 : 3000–1 : 30 000
- the RCHME has a GIS which enables rapid searching for aerial photographs based on six-figure OS grid references for the area of interest. The RCHME will send you photocopies (for a fee) of results of a search so that you can decide what to order.

Other sources of existing aerial photographs include:

Ordnance Survey
Air Photo Sales
Romsey Road, Maybush
Southampton SO9 4DH

Simmons Aerofilms Ltd
Gate Studios
Station Road
Borehamwood WE6 1EJ

Small air photo companies often have useful libraries of photographs. It is worth looking in the *Yellow Pages* for the area relating to your site to find the local companies. Unitary authorities (old County Councils in particular) fly their areas on a regular basis (five or ten year intervals). It is worth contacting the authority for the area associated with your site to find out what photography is available. The purchase of such photographs is often done through the air photo company which was commissioned to take the photographs since it normally holds the negatives.

Overview of the walk-over survey

The walk-over survey is an important and necessary supplement to the desk study. It should include:

- an inspection of the site and surrounding area on foot
- the examination of local records concerning the site not included in the desk study
- the questioning of local inhabitants about the site.

The primary objective of a walk-over survey is to check and make additions to the information already collected during the desk study and hence it is essential that all information concerning the site is studied thoroughly before any visits are made. This will allow a greater understanding of the significance of features seen on or around the site and enable more effective research of local records.

The site inspection involves walking across the whole site and surrounding area making full use of all the information obtained during the desk study. It is useful to go prepared with maps (topographic and geological) and aerial photographs (suitably protected from the elements) and to prepare a special map at a suitably large-scale (not smaller than 1 : 10 000, preferably 1 : 2500) to record features seen during the inspection. On this map should be marked by hand the geology and other features of special interest (e.g. small topographic features, isolated clumps of trees in fields, springs, seepages, etc.) that have been noted during the desk study. During the site inspection the position of any further features (e.g. hedges, pits, exploratory holes, shallow depressions, mine waste, damaged structures, etc.) should be marked on the map and the existence of previously identified features confirmed.

Equipment needed for a walk-over survey

A number of simple tools are required for the site inspection. These include:

- 20 m or 30 m long tape to measure the position of features of interest
- compass to orientate map
- Global Positioning System (GPS) to locate features of interest on map
- pocket penetrometer or hand vane to measure the strength of soil
- Abney level or Rabone Chesterman 'angle finder' to measure ground slope angles
- posthole auger and spade to probe and expose soil for examination and sampling
- polythene bags, labels and waterproof marker pen for taking samples
- camera for visual records
- waterproof note pad and pencil for making notes and sketches

- hand lens for examining soils and rocks
- binoculars for viewing nearby features
- geological hammer for breaking open rock specimens
- penknife for scraping and cutting soil and weak rock for examination
- *Identification and description of soils* and *Identification and description of rocks* – laminated card supplied with this book.

Feature identification

Many features may be observed during the site inspection. Only with experience can the significance of these features, particularly with respect to potential geotechnical hazards, be interpreted. The following types of information should be recorded in a notebook and/or on the site plan.

Slope angles

These can be measured more accurately and with better resolution during the site inspection than from topographic maps during the desk study. Slope angles and changes in slope angles should be recorded. The slope angles will give an indication of the underlying material types and changes in slope changes in material types. For example, rocks stand steeper than clays. Very flat slopes associated with streams or rivers probably mark the extent of the flood plain and hence areas which are likely to be underlain by alluvium and may be prone to periodic flooding.

Exposures

Exposures of rock or soil may be found in cliffs, stream and river beds, quarries, pits and road and railway cuttings. The material seen in such exposures should be described in detail (BS 5930, 1999). Notes should be made of discontinuities in rock and soil exposures. These should include the types of discontinuity (e.g. bedding), orientation (dip and dip direction), average spacing, persistence and aperture of each main set of discontinuities. Samples of soil and rock from such exposures should be taken for index tests and petrological examination.

Vegetation

If the site is on clay, make a record of the position of all trees and shrubs on the site, together with their approximate sizes, heights and girths and where possible their species. If in doubt about the species take leaf samples and photographs. The position and size of hedges should be recorded. Compare the current pattern of trees, shrubs, hedges, etc. with the most recent aerial photographs (e.g. Fig. 2.9) or other records and note where vegetation has been added or removed. If the site is underlain by soluble rocks (e.g. chalk, limestone, rock salt) either at some depth or at the surface note the location of any isolated clumps of trees and shrubs and whether these are associated depressions in the ground surface. These may be associated with dissolution features.

Instability

The presence of hummocky broken or terraced ground on hill slopes should be noted since these features are normally associated with land-slipping. Other features that are indicative of instability are trees with bent trunks (except in very windy areas, trees normally grow vertically – ground movements will cause them to change direction), kinks in hedgelines (e.g. Fig. 2.15) and kerblines, tension cracks, ponding (e.g. Fig. 3.3(b)) and deranged drainage. These features should be noted on the site plan. The alignment of hedges and fences should be compared with recent maps and aerial photographs. The extent and type of landslip should be noted on the site plan. Aerial photographs will aid the identification, classification and mapping of landslip areas. The relative age of the landslip should be noted where possible. In the case of rotational slips the amount of degradation of the rear scarp may give an indication of relative age. Where there is more than one landslip the relative age of each slip should be determined by examination of how the more recent landslip features cross-cut and obscure older features. Structures situated on or adjacent to a landslip should be inspected for structural damage. Evidence of soil creep is particularly noticeable where a hedge or other barrier traps the creep material on the uphill side.

Structures

Examine structures in the area of the site and record signs of damage. Methods of assessing damage to low-rise buildings are described in BRE Digest 251. The pattern and extent of cracks in damaged structures should be recorded using detailed sketches and notes. Measurements of cracks may be made and the damage classified using a system as described in BRE Digest 251 (Building Research Establishment, 1981). Other signs of distortion, such as non-verticallity of walls should be noted. Where possible, information should be obtained on the types of foundation commonly used in the area and the history of damage should be studied. The common causes of structural damage to buildings include clay heave or shrinkage, excessive differential consolidation settlement, settlement due to made ground, slope instability, groundwater lowering, soil erosion, structural failure of foundations, subsidence due to mining or dissolution features, vibration and chemical attack.

Made ground

Note the position of any infilling being carried out at the time of the site inspection. Look for areas that may have been previously infilled, by comparing the available topographic maps and aerial photographs with what can be seen on the site.

Mining and quarrying

Look for signs of mineral extraction in the area. These may include old mine buildings, derelict or hummocky land, surface depressions, evidence of infilling or spoil heaps.

Groundwater

Note the positions of springs, ponds and other water. Water-loving vegetation such as reeds, rushes and willow trees may indicate seepages and wet ground conditions. The absence of features noted above which are shown on topographic maps and aerial photographs may indicate that fill has been placed on the site. If hand auger holes are made the presence or absence of groundwater should be noted. Shallow wells on or near the site may give a reasonable indication of groundwater levels.

Surface water and erosion

Note should be made of ponds and streams which are not shown on the maps of the site. The likelihood of flooding should be assessed for the site and surrounding area. Any history of flooding in the area may be obtained through local enquiries. Any evidence of active soil erosion by surface water, such as gullies, should be noted.

Dissolution features

Land underlain by soluble rocks (e.g. chalk, limestone, rock salt) may contain naturally-occurring cavities which may be filled with soft soils. These can collapse or settle beneath a structure. This type of ground is associated with dry valleys and surface depressions, and with areas where streams disappear into the ground.

Access

Check the ease of access for drilling rigs or excavators which may be required for detailed ground investigation work. Record and take photographs of the condition of gates and tracks that plant might use, so that any damage caused can be properly quantified.

Local enquiries

Local enquiries involve talking to local people and visiting sources of reference material. These include the following.

- *Local builders and civil engineering contractors.* They may provide information on ground problems and types of foundation used in the area.

- *Local authority engineers and surveyors.* They may have extensive experience of building in their area and will be able to comment on general ground conditions, the possibility of flooding, any occurrences of structural damage associated with ground movements and previous site use.
- *Public utilities.* Utilities such as gas, electricity, telecommunications and water will give information on the position of their services in and around the site. Clearly these must be avoided during the ground investigation and may require re-routing during the development of the site. The engineers attached to the public utilities may provide information on any ground related problems encountered in laying the services.
- *Local archives.* Libraries, muniment rooms and county archives will hold old maps which may provide information on areas of fill or previous works on the site. Records of flooding, landslipping and mining activity may be found in these archives.
- *Local inhabitants.* Local inhabitants who have lived in the area for some time are often a useful source of information concerning previous uses of the site, structural damage to buildings on or near the site, mining subsidence, the location of old mine shafts, flooding and landslipping. A certain amount of caution should be employed when assessing any information given by local inhabitants as it is sometimes exaggerated, vague or ambiguous.
- *Local clubs and societies.* Local clubs and societies can often provide valuable information concerning the site and surrounding area. Such clubs and societies include archaeology societies, industrial archaeology societies, local history societies, natural history societies, caving clubs and geological societies.
- *Schools, colleges and universities.* Local educational establishments, particularly colleges and universities, are often a valuable source of local information. Many colleges and universities have departments of geology, geography and/or civil engineering. It is likely that these departments have carried out detailed studies of various local areas. These studies may provide valuable information concerning the site and surrounding area.

CHAPTER THREE

Geotechnical hazards and risk management

Overview

Engineering works will typically involve loading or unloading the ground or a combination of both. The response of the ground to these changes needs to be predicted with a reasonable degree of confidence and accuracy. This requirement presents a dilemma to the engineer. Because we are dealing with natural materials (most of the time) it is not possible to make such predictions without removing all the ground within the zone of influence of the proposed structure. Clearly this is impractical as well as being prohibitively expensive. It is necessary to compromise on both confidence and accuracy in providing information about the ground. In attempting to optimize these parameters the geotechnical engineer endeavours to identify potential hazards associated with a site at an early stage of the investigation. These can be targeted for more detailed investigation at a later stage. In general the hazards the ground poses can be divided into five distinct types.

- The ground may be very much weaker than the materials (such as concrete and steel) that we typically use for construction.
- Its properties can change with time (for example due to weathering) so that ground conditions become worse during the life of a structure or during a short term event such as an earthquake or rain storm.
- The ground is often so variable, both vertically and horizontally, that we can never know the complete variation of properties under the site.
- It can contain substances hazardous to human health, or which can attack construction materials.
- It contains water, which can both weaken the ground and apply large loads to structures.

The data collected during the desk study and walk-over survey provide key information for the preliminary identification of potential hazards. It is often the combination of factors that indicates the existence of a hazard. A number of examples are given below.

- The combination of clay soil (from geological maps/reports, etc.) plus trees (from topographic maps, aerial photography and walk-over survey) can present the hazard of swelling if the trees are removed. This is of particular importance in the construction of low-rise buildings.

- The combination of clay soil (from geological maps/reports) plus sloping ground (from topographic maps, typically slopes $>5°$) can mean unstable ground. The hazard becomes more acute if there is evidence of instability on similar slopes in the same geology in the area or if features such as ponding, deranged drainage and 'turbulent texture' are seen on aerial photographs of the site.
- The combination of chalk (from geological maps/reports) and a cover of sand or clay (from geological maps/reports) indicates the possible existence of dissolution features (sometimes referred to as 'sinkholes'). The hazard becomes particularly acute if bowl-shaped depressions are seen on aerial photographs of the site and/or its surroundings.
- A quarry that is shown on historical maps but cannot be found on more recent maps and aerial photographs or from the walk-over survey is indicative of the existence of made ground. It also raises the question of contaminated ground.
- The combination of loose granular soils (from geological maps/reports) and a seismically active area (from seismic vulnerability maps/reports, etc.) indicates a hazard from liquefaction during earthquake events.

In all the above examples the combinations do not necessarily provide conclusive evidence that a particular hazard exists. In most cases further investigation is required to prove the existence of the hazard. Combinations like these are important, however, in raising the engineer's awareness of the likelihood of a particular hazard being present.

The use of combinations like these is one way of identifying potential hazards. Another way is to consider, systematically, all the types of geotechnical problems that have ever been known to occur, and to ask whether they might happen on one's own site. To do this you need a catalogue of previous experience. This now follows.

Hazards from natural and man-made materials

Naturally occurring and man-made materials can, in themselves, represent a hazard to building and construction, for two primary reasons:

- they may react with construction materials, weakening them to such an extent that collapse occurs
- they may present a health and safety risk to human beings, either during construction, or once completed.

Reactions with construction materials

The most common hazards of this type occur because of acidic ground and groundwater, sulphate attack on concrete foundations or reactive aggregates. Each of these is now discussed in turn.

Acidic ground and groundwater – the problem

Ground and groundwater exist naturally at different levels of acidity. For example, peaty and sandy ground tends to be acidic, whilst limestone tends to be slightly alkaline. Thus careful consideration of the geology of the site and the surrounding area should give a good indication of the likelihood of this hazard.

In almost all cases the threat to construction comes from acidic groundwater, because this attacks the concrete used in foundations. Cement is a carbonate material, but in addition it is common to use limestone aggregates, which are also dissolved by acid water, in concrete. The risk from acid ground conditions should be assessed not only from pH values, but also based on the depth of foundation, depth to water table, likelihood of water movement, thickness of concrete, and pressure conditions. Examples of low and high risk are:

- *low risk* – pH of groundwater 5·5–7·0, stiff un-fissured clay soil with groundwater table below foundation level
- *high risk* – pH of groundwater <3·5, permeable soil with water table above foundation level and risk of groundwater movement.

Sulphate attack on concrete – the problem

In the UK and many other parts of the world, sulphates are widespread in near-surface geological deposits. Aqueous solutions of sulphates will attack the hardened cement in concrete, leading to chemical changes and the creation of new compounds associated with a large volume increase. This increase of volume causes cracking and spalling. If fresh sulphates can readily move to the concrete, and penetrate the cracks, the speed at which deterioration takes place will be accelerated. If fresh sulphates cannot travel to the concrete then the reaction will gradually cease, and for this reason it is soluble rather than insoluble sulphates that are a problem.

The rate at which sulphate attack can occur is therefore a function of:

- type of sulphate
- concentration of sulphates
- amount of groundwater movement around the concrete
- type of cement
- the shape of the structure.

Reactive aggregates in concrete – the problem

Aggregates, in the form of sand, gravel or crushed rock, are used as an essential part of concrete, and in road construction. As the best sources of aggregates are used up less desirable aggregates may then be used, but these may contain materials which react undesirably with Portland cement.

Typically, such reactions cause expansion to occur, thus cracking the concrete, weakening it, and leaving it (and any reinforcing bar it contains) open to further damage by water ingress. Poor aggregate may also iron-stain the concrete, making it unsightly, or it may weather badly.

Undesirable reactive aggregates may include:

- clay minerals
- micas
- feldspars
- opal
- chert
- pyrite and marcasite.

Reactions vary from poor weathering of the concrete as a result of the inclusion of clays or weak stone, to alkali–aggregate reaction causing dramatic and extensive surface cracking. Inclusion of small quantities of pyrite and marcasite will lead to ugly iron-staining on concrete finishes.

Hazards presented to human beings

A number of significant hazards to health and safety are associated with both naturally occurring and man-made materials. Any construction or new development may bring carcinogens (cancer-inducing substances) into contact with human beings. Toxic and explosive gases may move from the ground into buildings, tunnels, sewers, trial excavations, and other confined spaces, causing risks of explosion, an increased possibility of cancer, or immediate death as a result of asphyxiation.

Common causes are:

- asbestos particles
- radon gas
- methane, carbon monoxide and carbon dioxide gas (natural or man-made)
- hydrogen sulphide gas

Asbestos

'Asbestos' is widely used as a term to describe fibrous minerals, generally obtained from metamorphic rocks, and used in heat resistant applications. Asbestos has a reputation as a cancer-inducing substance, but in fact only two types of mineral (out of a much wider range that has been used) are now thought to cause cancer:

- sheet minerals (chrysotile), which constitute 95% of the asbestos used
- amphiboles [amosite (brown asbestos), crocidolite (blue asbestos), anthophyllite, tremolite, actinolite, ferroactinolite], which together make up only about 5% of commercial asbestos.

Exposure to specific types of asbestos has been found to cause fatal lung diseases. Any fibres that are inhaled and retained in the lung do not dissolve or disperse, and because of their shape cause a constant irritation. The body's reaction over tens of years is to produce cancerous cells at the site of the irritation. Once these cells overcome the immune system, death follows rapidly. Materials that are currently in use pose minimal health risk, except under conditions of very heavy industrial exposure.

Radon

Radon is unique amongst naturally occurring gases because it is radioactive. Radon has not been recognized as a hazardous gas until relatively recently. Radon gas (radon-222) is a radioactive isotope produced as uranium-238 decays, eventually becoming stable lead-206. Mylonites, which are metamorphic rocks produced typically in fault zones, where rocks are sheared and ground together, are now known to be very important sources of radon, but uranium is also enriched in granites, gneisses, schists, slates, some sandstones, and some glacial deposits, as well as being highly concentrated in black shales and phosphate deposits.

Radon risk is usually associated with the level of content of uranium in the underlying bedrock. However, the residual soils left by the dissolution of limestones and dolomites can also have a high uranium concentration, even though the underlying bedrock does not present a hazard.

Methane, carbon monoxide, carbon dioxide and hydrogen sulphide gas

Toxic gases can occur naturally, or can be produced by human activity (for example the landfilling of municipal solid waste). Methane, carbon monoxide and carbon dioxide are all produced as part of the decay of organic matter. Many geological materials have been formed in this way (coal, lignite, peat) so that there is always the possibility that underground excavations (of whatever size) may encounter these materials, and that high concentrations may build up in the absence of adequate ventilation.

These three gases are traditionally termed 'fire-damp' by coal miners, who fear them because they cause many deaths by fire, explosion or suffocation. They are colourless, and methane is combustible. Excessive concentrations of carbon dioxide and carbon monoxide lead to suffocation. Carbon monoxide is particularly insidious, with no taste, no smell, and no sense of loss of air during death.

Hydrogen sulphide occurs in conjunction with some oil bearing deposits, and is more lethal than cyanide. It has a 'rotten egg' smell that makes it easy to sense, but even so whole oil-drilling crews have been killed instantly by this gas suddenly coming out of oil wells.

Geotechnical hazards

Geotechnical hazards are many and diverse. The most common hazards are due to:

- groundwater
- slope instability
- subsidence
- compressible soils
- swelling soils
- weak soils
- weathering of soil and rock during the life of the construction

Groundwater

Nearly all geotechnical hazards are associated in some way with water. Water in soils and rocks can, through reducing effective stresses, bring about a reduction in strength which may lead to slopes becoming unstable. Changes in groundwater conditions can cause the collapse of subsurface voids that can result in subsidence. Changes in moisture content in certain clay-rich soils can result in swelling and shrinkage leading to structural damage of buildings founded in them. The movement of water through certain rocks and soils can result in weathering at a rate that may result in changes in the mechanical properties during the life of a structure placed on or in them. The problem of acidic groundwater has already been discussed above.

Case study: Subsidence from sinkholes, Fontwell, West Sussex, UK
This case study illustrates collapse subsidence due to groundwater. The area around Littleheath Road, near Fontwell, West Sussex was subject to severe collapse subsidence in November 1985 after a water pipe burst, and again in January 1994 following a period of exceptional rainfall (Rigby-Jones *et al.*, 1993; McDowell and Poulsom, 1996). The mechanism of collapse is thought to be associated with granular soils subsiding gradually into dissolution pipes in the underlying chalk. Arching allows voids to form in the granular soils at the interface with the chalk. In time these migrate upwards, eventually reaching the ground surface. The collapse features observed at the ground surface were typically cylindrical holes, 1 to 4 m in diameter (see Fig. 3.1) and 0·5 to 3 m deep. They are typical of the 'plug collapse' described by McDowell and Poulsom (1996) and are referred to as sinkholes. These features occurred suddenly with little or no warning. Prior to the collapse there were no ground surface features which might indicate the presence of a subsurface void. As part of a study described by Matthews *et al.* (2000) a trench was dug across part of the site shown in Fig. 3.1 to reveal voids that had not collapsed. Figure 3.2 shows

Fig. 3.1 Air photo showing extent of collapse features at Littleheath Road, West Sussex, UK (courtesy of Sealand Aerial Photography)

one of the voids encountered by the trench. The voids were hemispherical in shape and approximately 1 m high and 2·5 m in diameter. The floors of the voids were characterized by loose gravel which had fallen from the roof.

The geology of this area comprises Upper Chalk with a highly irregular surface overlain by 2·5 to 5·8 m of fine yellow marine sand belonging to the Slindon Sand formation. This forms part of a raised beach cut into the Upper Chalk (Mottershead, 1976; McDowell and Poulsom, 1996). The sand is overlain by between 4 and 5 m of dense to medium dense flint gravel in a silty clay matrix, known locally as Head Gravel. Brickearths comprising silty clay are also found in this area. The Head Gravels and Brickearth have been transported into the area by solifluction, wind or flood waters in a periglacial environment (Mottershead, 1976; McDowell, 1989). The water table in the Chalk is approximately 30 m below ground level in this area, but perched water can be held within the Slindon Sand (McDowell and Poulsom, 1996).

Fig. 3.2 Section through a void in granular soil exposed in the side of a trial trench at Littleheath Road, West Sussex, UK, after Matthews et al. *(2000b)*

Slope instability

Slope instability is a generic term that refers to the movement of soils and rocks under the influence of gravity. Terms such as landslides, slips, slumps, mudflows and rockfalls are commonly used to describe different types of slope instability. In general slope instability can be divided into the following three major classes based on the way in which the movement takes place.

- *Slides* – the moving material remains largely in contact with the parent or underlying rock or soil during the movement which takes place along a discrete shear surface.
- *Flows* – the moving material becomes disaggregated and can move without the concentration of displacement along a discrete shear surface.
- *Falls* – involve the immediate separation of the moving material from the parent material with only infrequent or intermittent contact thereafter, until the debris comes finally to rest.

Slides

Slides are generally translational in nature. In rock, sliding generally involves the translational movement of joint bounded blocks. The orientation of the discontinuities in the rock mass relative to the orientation of the rock face (natural or excavated) will determine whether blocks will slide out of the face or not. The build-up of water pressure on joint bounded blocks will aid sliding and result in movement of blocks on relatively low-angle surfaces. In highly fractured rock and relatively homogeneous soils shearing will take place through the rock and soil and will tend to follow a curved shear surface resulting in rotational movement. The soil or rock tends to move as a block with most disruption occurring at the bottom or toe of the landslip. In the presence of different lithologies within a stratified deposit slips often follow the bedding and hence have a relatively flat basal shear surface for some distance. The principal mode of movement will be translational. The curved rear part of such slips results in some rotation and causes great internal distortion of the sliding mass, which may be reflected by breaks in the ground surface.

An example of such a landslip is the site of the University of Surrey (described in Chapter 2). The University of Surrey site slopes gently northwards, at about 9° to the horizontal. Beneath it lies a major landslide, about 700 m long, 100 m from rear scarp to toe, and with a slip surface up to 10 m deep (Fig. 2.10). This huge mass of soil was probably de-stabilized during the last ice age, when permafrost conditions would have occurred in the south-east of England. The base of the landslip is approximately parallel to the bedding in the London Clay which dips northward. Clearly when investigating sites on northward facing slopes in London Clay in this area such landslip hazards should be considered. Using air photographs like those in Fig. 2.11, we can build up a map of the unstable ground. Such information is extremely useful in planning future developments on a site. These slopes typically have only a very small margin of safety, so that any development on the site requires careful design. Construction can easily de-stabilize the slope, either by excavation at the toe, or by loading the upper part of the slope.

In terms of morphology, movements such as mudslides, flow slides and debris flows (Hutchinson, 1988) may be classified as flows. In terms of mechanism, mudslides predominantly slide rather than flow, whereas flow slides and debris flows probably exhibit varying degrees of sliding and flowing. Mudslides are especially well developed on slopes containing stiff fissured clays. Disturbed soils are very susceptible to the infiltration of rainwater. The backtilted elements or grabens in the slides described above often cause ponds to form by intercepting surface water flows. The discharge of these frequently results in the formation of mudslides. Mudslides can also be formed by the careless discharge of water on to a slope.

Flow slides occur in loose cohesionless materials, very lightly cemented, high-porosity silts and high-porosity weak rocks. In the latter case a high degree of disturbance to the rock fabric is required to induce metastability. In all the cases of flow slides in weak rock, the necessary disturbance has been provided by the collapse of a steep cliff. Debris flows are potentially highly destructive and involve rapid flow of wet debris. They are generally associated with mountainous areas, where sudden access of water, usually from heavy rain or melting snow, can mobilize debris mantling slopes and incorporate it into a debris flow.

Solifluction is defined as '*the gradual movement of wet soil down a slope*' (Oxford English Dictionary). During the last ice age large areas of land existed under periglacial (i.e. adjacent to the ice sheet) conditions, which meant that the ground was permanently frozen to some depth. During the summers the upper layers of soil would thaw, and because water could not drain through the permanently frozen soil below, a wet soil layer would gradually creep downslope, only to be frozen during the next winter. This continual movement led to the formation of polished 'residual' shear surfaces at modest depths (typically 3–5 m) below the ground surface. Construction on these soliflucted materials is almost certain to suffer from major instability problems. Many cases of solifluction have been recognized on the clays of southern England, and there have been a considerable number of construction problems, principally associated with highways crossing sidelong ground on scarp slopes.

Two classic cases involved the re-alignment of the M4 motorway near Swindon, and the Sevenoaks by-pass, to the south-east of London. Figure 2.14 shows part of the instability on the Sevenoaks by-pass. A number of solifluction lobes of this type exist, but were not detected before construction. The extent of the instability can be clearly seen on air photographs, but these were presumably not studied during initial design work.

Flows

A flow is a form of mass movement that involves much greater internal deformation than a slide. Typically the water content in a flow is so high that it behaves as a fluid. This is the case in clay soils when the moisture content is greater than the liquid limit. A classic case is the Norwegian quick clays. These relatively young clays have a metastable structure resulting from deposition in a marine environment and subsequent leaching by fresh water. Disturbance through for example excavation results in collapse of the structure at a moisture content greater than the liquid limit. As a result the material flows as a liquid. Extensive destruction of property has resulted from the initiation of flows in this material.

Rock falls

Rock falls involve the rapid movement of boulder-sized material down steep slopes. They generally occur only in mountainous terrain, and may be initiated by freeze–thaw action, earthquakes, or perhaps by construction. Rockfalls may threaten highway traffic, or new housing developments.

Landslip identification

The desk study and walk-over survey can provide a great deal of valuable information about unstable slopes. The value of aerial photographs cannot be overemphasized. Air photos taken at different times of the year and at different times of the day can be most revealing and can indicate down-slope movements that may not be obvious to the naked eye at ground level (e.g. see Figs 2.11 and 2.12).

The following features should be looked for when carrying out the walk-over survey:

- tension cracks
- toe bulges
- lateral ridges
- uneven ground
- graben features
- back-tilted blocks
- marshy ground
- streams
- springs and ponds
- cracked roads
- inclined trees
- breaks and cracks in walls
- distorted fences
- drainage paths, both natural and man-made.

Some of these features are indicated in Fig. 3.3(a). Tension cracks and ponding in the scarp of a slip is shown in Fig. 3.3(b).

Walk-over surveys, carried out during and immediately after heavy rainfall, can be most revealing and can show adverse drainage conditions that may not be apparent under dry weather conditions. Weep-holes in walls should be carefully studied to see whether they are functional or blocked up, as is often the case.

Just before a failure like this occurs, tension cracks and bulging at the toe may be noticed. After the failure the rear scarp will degrade, and mudflows may help transport clay down the sides of the slip towards the toe. The ground will become very wet and 'boggy'.

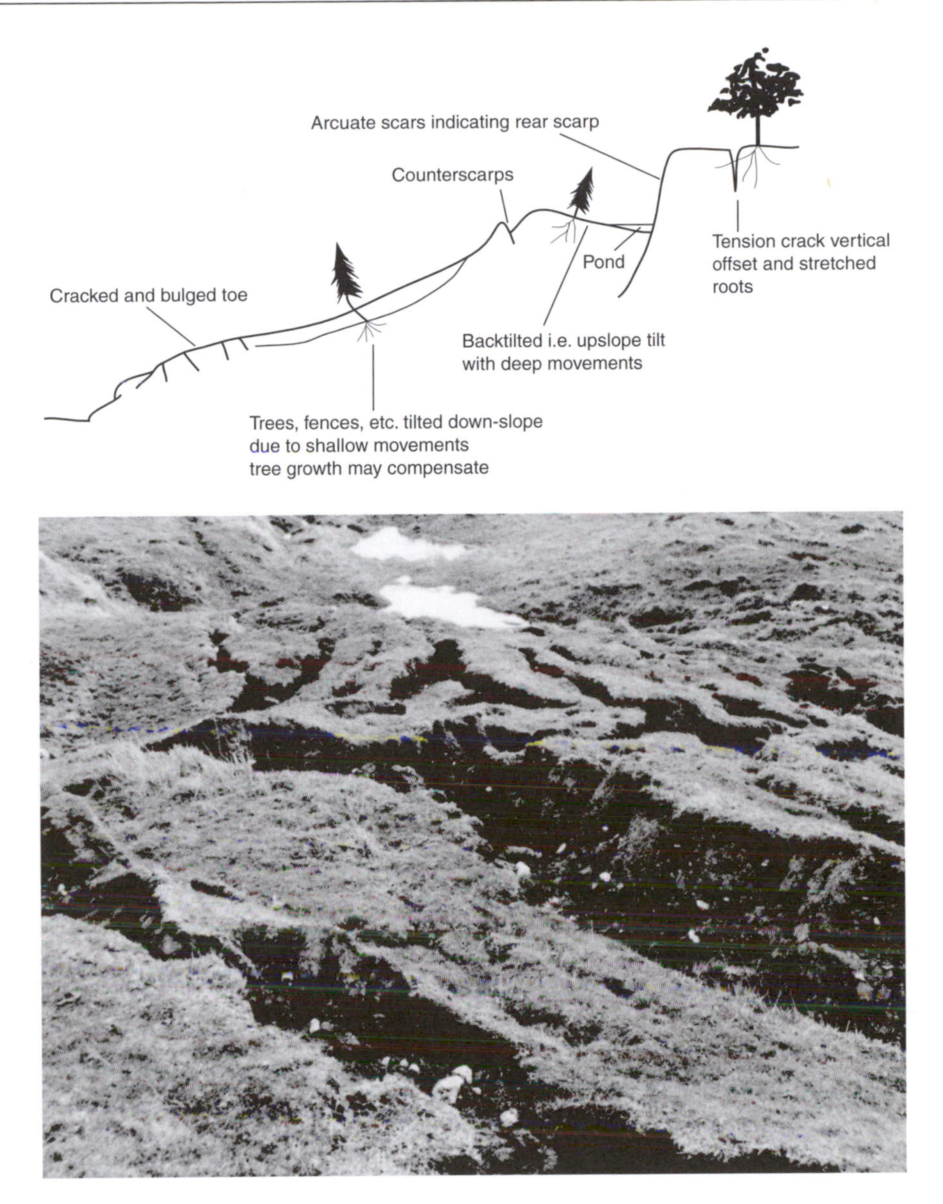

Fig. 3.3 (a) Surface indications of slope instability, after Bromhead (1979). (b) Ponding in the scarp of a slip

Slope failures may be triggered by:

- changes in slope angle (re-grading during construction)
- excavation or natural erosion at the toe
- loading the upper part of the slope (e.g. with new buildings)

- raising the groundwater table (heavy rain, or discharge from roofs, roads, drains, etc.
- vibrations from earthquakes.

Ground movements

Buildings, bridges and other civil engineering structures can be damaged by a number of ground-related mechanisms as follows.

- Shear forces transmitted by earthquakes may cause damage to columns, or settlement of foundations.
- Ground movements due to other causes (e.g. slope instability) may take foundations with them, cracking the buildings above.
- Structural loading may cause sufficient settlement of foundations to damage the superstructure above.
- Structural loading may lead to failure of the soil. The very large movements that occur during foundation failure are certain to damage the superstructure.

Broadly, the movements which occur below foundations can be classified as:

- *settlement* – downward movement of a foundation caused by structural loading
- *heave* – upward foundation movement caused by unloading (e.g. excavation) or swelling soils
- *subsidence* – downward foundation movement caused by external factors, such as mining, dissolution, desiccation of clay soils (e.g. by tree roots) and de-watering.

Uniform settlement would not pose a great problem for a structure, although services (pipes for gas, water, sewerage, cable television, etc.) might well be adversely affected. On the other hand, relatively small amounts of differential foundation movement can be very damaging to certain types of structure. Unfortunately it is very unusual for structures to settle uniformly, even when both the loading and the ground conditions appear uniform. Typically, differential settlements will exceed 50% of the maximum total settlement.

Heave and settlement, respectively, may occur as a result of the precipitation or removal of soil/rock components by groundwater. Heave may occur if the addition of water during the hydration of certain minerals (e.g. anhydrite, $CaSO_4$) causes expansion. Conversely, reduction in volume resulting in settlement may occur during dehydration although this appears to be less significant in practice. Evaporite minerals (e.g. rock salt, gypsum) are the main cause of ground movements associated with hydration and dehydration. The active deposition of such minerals typically occurs in arid environments such as the Middle East.

Subsidence

'Subsidence' (from the Latin words 'sub' and 'sidere' meaning to settle down) is the sinking or caving in of the ground or of buildings on top of it. In geotechnical engineering it is considered distinct from 'settlement', which occurs when loads are placed on the ground (for example by foundations).

Subsidence may be the result either of natural or man-made processes. Although subsidence does occur in remote areas, it is most significant when there are buildings or roads over the area that collapses. The typical sequence of subsidence, as it affects construction, is as follows:

- ground conditions are modified to make the soil or rock 'meta-stable', i.e. prone to collapse
- new construction takes place
- collapse is triggered, usually by water ingress, but occasionally by earthquake vibrations.

The most common natural process to cause metastable conditions is the dissolution of soluble rocks. Dissolution is a common weathering mechanism in rocks such as rock salt, limestone and chalk. Typical features associated with dissolution in chalk include an irregular interface between the chalk and the cover material, downward tapering pipes infilled with compressible material, and voids. If undetected, dissolution features pose a risk to construction through differential settlement or collapse. The risks associated with foundations which arise from dissolution features include:

- differential settlement due to variations in chalk level
- large settlements due to loose and metastable materials overlying dissolution pipes or collapses or loose and metastable chalk caused by collapse (e.g. cave collapse)
- collapse settlement associated with metastable materials, caves, cavities overlying dissolution pipes or man-made openings within the chalk
- reduction in load-carrying capacity associated with dissolution widened discontinuities.

Weathering and dissolution of carbonate rocks, such as limestone, gypsum, dolomite or chalk, are caused by acid rainwater. As rain falls through the atmosphere it becomes weakly acidic, because of its reaction with CO_2. As the groundwater percolates the ground surface it dissolves the rock, and leaves it in a fragile state. In chalk and limestone, preferential solution takes place, and caves and sinkholes are produced along with the characteristic rough karst terrain. In some areas such as Florida in the USA and the Transvaal in South Africa the risk of sudden collapse resulting from dissolution is well known.

Dissolution features similar to those in limestone areas have been reported in areas underlain by gypsum and anhydrite. The existence of subsurface dissolution channels in gypsum was found in New Mexico and sinkholes associated with gypsum opened up in Rapid City, South Dakota, USA. Dissolution features (surface depressions) associated with rock salt are common in Cheshire in the UK.

Mining activities can produce meta-stable conditions similar to those associated with dissolution. Old mine workings are typically left open, and allowed to collapse down in their own time. Additional loads, and especially the disposal of water into meta-stable ground, can lead to sudden collapse. Mining in the UK has extended from Neolithic times (about 4000 years ago), through the origins of the Industrial Revolution in the late 18th century, to the present, so there is much to be wary of!

Compressible soils

If foundations are built on the more compressible clays, peats, or uncompacted made ground then total settlements will be very large, and as a result differential settlements will cause damage to the structure. It is normal to avoid this problem by piling through poor ground.

Swelling soils

Swelling soils are also known as 'expansive soils'. When mobilized they expand under buildings, causing the buildings to heave differentially and suffer structural damage. Four types of ground commonly cause this type of problem:

- high-plasticity (potentially expansive/shrinkable) clays
- calcium sulphates (anhydrite and gypsum)
- black shales (which typically contain iron sulphides) or other shales
- fill composed of industrial materials (certain types of slag).

An expansive clay is an unsaturated clay with a high liquid limit. When fed with water the clay will swell. Clays which have previously been desiccated (for example by a dry climate or by vegetation, e.g. tree root action), can exert pressures of up to 1000 kPa on structures above them, lifting foundations and walls. In typical situations which trigger heave:

- the clay is fed with water, by leaking sewers, pipes, watering of gardens, etc.
- the ground is covered (especially in dry climates), so that evaporation is reduced or prevented
- mature trees are cut down, so that water demand is reduced.

Table 3.1 Clay shrinkage potential (Building Research Establishment, 1980)

Plasticity index: %	Clay fraction: %	Shrinkage potential
>35	>95	Very high
22–48	60–95	High
12–32	30–60	Medium
<18	<30	Low

The expansive potential of a clay soil may be assessed from the Atterberg Limits (Holtz and Gibbs, 1957; van der Merwe, 1964; Williams and Donaldson, 1980; Building Research Establishment, 1980). Figure 1.27 and Table 3.1 show some examples of how expansive clays are classified. The potential of some British clays assessed using Table 3.1 is shown in Table 3.2.

Similar (but if anything more dramatic) effects can be produced when anhydrite (calcium sulphate) comes in contact with water. The hydration of anhydrite produces gypsum (hydrated calcium sulphate), and this is accompanied by a large volume increase. Pressures of up to 70 000 kPa can be produced, and under extreme conditions violent upheaval of blocks of ground can occur.

Black shales are compressed clays that are rich in iron sulphides typically in the form of pyrite. When exposed the iron sulphide oxidizes, producing sulphuric acid and sulphates. Both of these products are extremely aggressive to concrete, and in addition the conversion of sulphides to sulphates produces a volume increase and hence heave.

Industrial by-products can make good fills or aggregates. These include slags from the production of iron and steel. Some types of slag such as those produced in open-hearth or electric-arc steel furnaces using basic

Table 3.2 Shrinkage potential of some common clays (Driscoll, 1984)

Clay type	Plasticity index: %	Clay fraction: %	Shrinkage potential
London Clay	28	65	Medium high
London Clay	52	60	Very high
Weald Clay	43	62	High
Kimmeridge Clay	53	67	High/very high
Boulder clay	32	–	Medium
Reading Beds	41	56	High
Gault	72	–	Very high
Gault	60	59	Very high
Lower Lias	68	69	Medium
Clayey silt	31	–	Low

fluxes are likely to exhibit instability (Crawford and Burn, 1969; Spanovich and Fewell, 1970). Heaves as much as 33% of the thickness of fill have been reported (Bailey and Reitz, 1970). Spontaneous heating (probably arising from hydration temperatures) in some iron and steel slag fills may also cause problems, as has been the case with many loose colliery discards.

Weathering during the life of a project

Weathering is normally considered to be a process which occurs over a relatively long period, but it has already been noted that some processes, such as the weathering of pyrite (typically found in black shales), can take place during the life of a structure, causing immense amounts of damage. Most hard clays and weak shales can be expected to undergo very significant amounts of weathering, softening and disintegration in the presence of water once they are exposed.

The oxidation of pyrite involves expansion. Ground movements associated with this however usually result from subsequent reactions between the oxidation products and construction materials or other rock, soil or groundwater components. Heaves as much as 300 mm have been reported in black shales. These rocks are the most common source of sedimentary pyrite and outcrop extensively worldwide. Problems attributed to the oxidation of pyrite have been reported in many countries including the USA (Dougherty and Barsotti, 1972), Canada (Quigley and Vogan, 1970; Penner *et al.*, 1973; Gillot *et al.*, 1974), Australia (Haldane *et al.*, 1970), the UK (Nixon, 1978), France (Millot, 1970) and Norway (Moum and Rosenqvist, 1959). The oxidation process and subsequent reactions are described by Taylor and Cripps (1984). The oxidation of pyrite and subsequent reactions generally involve volume increases; some of these reactions are shown in Table 3.3.

Table 3.3 Expansion in terms of crystalline solid due to mineral alteration (Taylor and Cripps, 1984)

Original mineral	Product mineral	Expansion: %	Reference
Pyrite FeS_2	Jarosite	115	Penner *et al.*, 1973
	Melanterite	536	Shamburger *et al.*, 1975
	Anhydrous ferrous sulphate	350	Fasiska *et al.*, 1974
Calcite	Gypsum	103	Penner *et al.*, 1973
	Gypsum	60	Shamburger *et al.*, 1975
Illite	Alunite	8	Shamburger *et al.*, 1975
	Jarosite	10	Shamburger *et al.*, 1975

Frost heave

A further mechanism of weathering occurs in the chalk, which is a soft high-porosity carbonate rock found in England and much of northern Europe. Frost heave occurs when the ground becomes frozen and water is drawn upwards into ice lenses.

In the UK frost will only penetrate about 0·5 m below ground level, even in the most severe winter. Therefore only those types of civil engineering structure with very shallow foundations (houses and roads) are likely to be affected by natural freezing. However, the ground underneath cold stores can be subject to serious heave if measures are not taken to provide adequate insulation.

Frost heave will only occur in fine grained but permeable ground, where capillary suction can draw water to the surface during the relatively short periods of cold weather. Chalk, oolitic (porous) limestone, and silty sand are the materials most likely to be affected, whilst clean gravels and clays will remain untouched. The presence of a water table within a metre or so of the ground surface will make matters worse.

Managing geotechnical risk

Sources

This section is based on the following sources:

- Clayton, C.R.I. (1998). Talking point. *Ground Eng.*, **31**(4), 15.
- Clayton, C.R.I. (2001). Managing geotechnical risk: time for a change? *Proc. ICE, Geotech. Eng.*, Paper 149. Pages 3–11.
- Hall, J.W., Cruickshank, I.C. and Godfrey, P.S. (2001). Software-supported risk management for the construction industry. *Proc. ICE, Civ. Eng.* Pages 42–48. Paper 12272.
- CIRIA (2001) RiskCom: Software tool for managing and communicating risks.

We gratefully acknowledge the permissions of Prof. C.R.I. Clayton, of Dr J.W. Hall and his co-authors, and of CIRIA (Construction Industry Research and Information Association – see also www.ciria.org.uk) to make use of these sources.

Overview

Most experienced geotechnical engineers and engineering geologists intuitively appreciate the risks associated with their sites. The passing on of this experience to students and younger colleagues, however, requires a more systematic method than the 'watch me and learn' approach so that learning is independent of time served. In addition, many sites can be very complex and no one manager can have a complete overview on a daily basis. Perhaps just as importantly in this litigious age of the blame culture, it can be argued that not to have evaluated risk in a systematic way (e.g. as embodied by a risk register) on a site where something ultimately goes wrong, is to be immediately disadvantaged in the eyes of the courts. Finally, of course, an appreciation of risk does concentrate the minds of all geotechnical practitioners to find a better way of evaluating sites and optimizing design processes than the 'routine' approaches of the past to solving the far from routine hazards presented by ground conditions.

Comment (Clayton, 1998)

In his 'Talking point' in the journal Ground Engineering, *Clayton (1998) points out that most engineers would happily accept that the standardization of the way we do things is not only a 'good thing' but is also essential to ensure that the client gets a quality product at a low price. But how realistic is this view when applied to geotechnical engineering?*

A large part of the financial risk to civil engineering construction and building is associated with geotechnical hazards. Managing these risks by appropriate site investigation and good design, and dealing with the inevitable surprises produced by unforeseen ground conditions during construction, must help increase the chances of delivering new infrastructure on time and to budget. To be more competitive in the global market for construction, however, requires that costs should be reduced significantly.

There are broadly two approaches that can be adopted to achieve reductions in cost that are associated with the geotechnical components of construction. The first involves standardizing our products, improving the management of the process, and driving costs down through competition. The second demands a radical rethink of what is done, the abandonment of many accepted procedures and approaches, and a paradigm leap to a new position based upon exploiting world-class research in the international geotechnical community in order to bring fresh thinking to the challenges of difficult ground conditions.

Standardization and the introduction of quality systems have brought many benefits. They are invaluable in providing a ready basis for competitive bidding, and can greatly help the flow of information between the increasing number of contributors to a project. But in geotechnical engineering there has been fierce competition and considerable standardisation for decades. Accompanying this is a general feeling that quality (for example of site investigations) has been compromised by prices that have reduced in real terms;

that too many geotechnical hazards are not properly dealt with during design; and that technical advance has been slow.

Perhaps some of the problem with site investigations arises from the fact that many clients commission this type of work principally because the construction professionals they employ would feel open to claims of negligence if the site investigation were omitted. How much better it would be if they regarded site investigation and geotechnical engineering as a means by which they could optimise use of their sites, while finding effective low-risk solutions for in-ground construction. For this to happen, every part of the geotechnical process needs to be rethought, with a view to achieving reductions in cost and in the time required for design and construction, while improving the quality and reliability of the product.

Because the ground is variable it will always be full of surprises. Geotechnical engineering can never be treated as routine, but there must be ways by which repetitive design issues can be dealt with, leaving the site investigation practitioners free to exercise their skill in identifying hazards and avoiding them where possible, while making projects easier and cheaper to build.

What is risk management?

A systematic approach to risk management

As pointed out by CIRIA (2001), everyone should be concerned with risk management, because risk and uncertainty with potentially damaging consequences are inherent in all construction projects. The success or failure of the project and the site investigation practitioner's business depends on the approach to risk.

Risk management is not new. Traditionally it has been applied instinctively with risks remaining implicit and managed by judgement informed by experience. The systematic approach makes risks explicit, formally describing them and making them easier to manage. In other words, systematic risk management is a management tool, which for best results requires practical experience and training in the use of the techniques. Once learnt, it supports the user in decision-making and informs instinctive judgment. In brief, systematic risk management helps to:

- identify, assess and rank risks by making risks explicit
- focus on the major risks from the project
- make informed decisions on the provision for adversity, e.g. mitigation measures
- minimize potential damage should the worst happen
- control the uncertain aspects of construction projects
- clarify and formalize the personnel roles in the risk management process

- identify the opportunities to enhance project performance (especially financial performance).

Although it is not possible to remove all uncertainties, systematic risk management improves the chances of the project being completed on time, within budget and to the required quality, with proper provision for safety and environmental issues.

What is risk?

Many people have attempted to provide a definitive statement of the meaning of risk, but there is still a great deal of debate regarding which definition should be used. CIRIA (2001) define risk as

> *the likelihood of an unwanted uncertain event, and its unwanted consequences for objectives.*

The measurement of risk

The likelihood, or more technically, the probability of an adverse event, is usually expressed in terms of the number of such events expected to occur during a defined duration. For instance, for a project lasting one year, if the event is expected once every ten years, it will have a probability of occurrence of 0·1. (Time is not the only measure of likelihood; for example, the likelihood of production defects in steel bars might be measured per kilometre of bar produced.)

The consequence of an adverse event, sometimes called damage, is often expressed in monetary terms. However, it is more appropriate in the case of fatalities or serious delays to use other performance measures. The risk rating, as the combination of likelihood and consequence (i.e. likelihood $\times$ consequence), will usually be expressed in pounds or euros or dollars per year in an annual budget. The methods work equally well for other units of measurement (such as € per activity or project), as long as the units are applied consistently throughout the assessment.

Consider carefully events which, though unlikely, are potentially catastrophic. The risk may appear insignificant, yet disaster can happen. If the consequences are unacceptable, then the risk must be avoided or at least mitigated, e.g. through insurance against fire or flood. One should not lose sight of the fact that risk can be measured in terms of positive impacts often referred to as 'opportunities'.

Hazard and risk

There is a distinction between hazard and risk. A hazard has the potential to do harm or cause a loss, but the degree of risk from the hazard depends also on the circumstances. For example, petrol is a hazardous substance but the risk from it depends on:

- its nature (it is inflammable)
- how it is used
- how it is controlled
- who is exposed
- what is being done.

In many hazardous situations, risks can be reduced to acceptable levels by the quality of risk management applied. For example, a large housing development was to be constructed close to an old fireworks factory. Arsenic, discovered only after work had begun, needed to be removed to meet stringent environmental standards. Great extra costs were incurred owing to the complexity of the clean-up, resulting in delays and changes to the main contract. The risk could have been reduced by a land contamination survey before the site was purchased.

What risks are there?

Consider the distinction between risks that affect projects and those that affect the contractor's business. Risks are specific to a project, interactive and sometimes cumulative: they all affect cost and benefit. A wide range of risks arising from a project may threaten the business of the 'contractor' (here the term 'contractor' is taken to mean the site investigation practitioner, the design consultant, or the construction firm carrying out the building work, i.e. the party or parties responsible for assessing, avoiding, mitigating and transferring risk). One should assess which of the strategic level risks:

- could have serious impact on the success of the project
- are increased by the project
- are reduced by your project
- create other opportunities.

Expecting the unexpected

It is impossible to identify all risks. To believe this has been done is counterproductive to risk management and dangerous. Always expect the unexpected. Systematic risk management will help anticipation of the unforeseen as well as minimize damage caused by identified problems.

Opportunities, risk and value

Risk and opportunity go hand in hand. For this reason, there is usually a commercial benefit, or 'added value' from risk control measures. For example, it may be decided to provide a hoist instead of a set of ladders on site to reduce the risk of people falling. The added value of this risk control measure may be that the hoist increases people's mobility and, as a result, their productivity. Consideration of potential opportunities arising from the risk control activities calls for little extra effort during the risk management process and can aid the decision-making by giving a more complete picture of likely outcomes. It is useful to note that the techniques used in the application of systematic risk management are similar to those used in the management process known as value management, outputs from each being closely linked and interdependent. Benefits, both in terms of time and cost savings and improved effectiveness, can be achieved by closely co-ordinating inputs to these two activities.

Ownership of risk

The term 'ownership of risk' has a variety of meanings which include:

- having a stake in the benefit or harm that may arise from the activity that leads to the risk
- responsibility for the risk
- accountability for the control of risk
- financial responsibility for the whole or part of the harm arising from the risk should it materialize.

When a risk materializes, the harm that occurs is seldom restricted to one organization or person. Some are direct 'stakeholders' and some indirect. Whereas the former may include the contractor's organization, team, etc., the latter may include the public and and the contractor's customers and clients, their insurance companies and professional advisors. Deciding who 'owns' risk is sometimes difficult. Risk management will help address this issue.

The true cost of risk

The true cost of risk can be much higher than is immediately apparent. Much of it can be indirect and uninsured. The uninsured cost of health and safety risk can be many times the direct costs on a construction site. The risk issues therefore can be much more complex than appears at first sight. It is helpful to keep the task of identifying the cost of risk simple and focus on:

- what is important
- actions which control risk.

Complacency is dangerous

Organizations experienced in executing construction projects naturally develop procedures that respond to the risks that they have encountered. They can be particularly vulnerable, however, to new risks such as those resulting from change or innovation. Change may also render some of their risk control practices obsolete.

Transfer and spreading of risk

If an attempt to transfer risk to others is made but the transfer of control of that risk is resisted, it will generally lead to an overall increase in the cost of risk. It can also mean paying not only for the transfer of the risk but also for the consequences of the risk if it materializes. Moreover, it may add to any legal costs incurred in sorting out the responsibility. This can escalate the consequences even further.

Some large companies that can afford to absorb losses actively seek high-risk/high-profit projects in areas where they are highly experienced and can gain a strong competitive advantage. Smaller companies may prefer to spread the risk to reduce its overall effect, at a lower return on their investment, because they cannot afford the higher risk should it occur. Good management of risks does not necessarily imply a reduction in risk. For example, if oil exploration wells cost £1 million and have a 1:10 chance of striking oil, a small oil company may prefer to take a 10% share in each of ten wells at £100 000 per well rather than wholly financing one. Thereby the company increases its probability of success but reduces the return on its capital if successful.

More harm than good

Some risk-reducing measures can do more harm than good. For example, the provision of marginally effective risk control measures may create a false sense of security which increases the risk. A costly risk management measure may prevent more cost effective measures being provided elsewhere. In addition, what may be a risk reduction measure for the project may cause increased risk to the contractor's business.

For example, the risks of road traffic accidents along a section of motorway roadworks may be reduced by speed limits, warning signs, speed cameras, and coning off motorway lanes so that there is a buffer between construction staff and passing traffic. However, the traffic disruption and delay that such measures may cause could lead to many motorists finding alternative routes, thereby increasing traffic load on less safe roads and the overall potential for accidents.

Risk to contractor's investment

When money is invested in a project, risks are accepted in order to achieve a desired benefit from the project. At the outset, uncertainties in the prediction of cost and benefit generate a range of likely outcomes, as shown in Fig. 3.4.

Risk management will help determine and control this range. The base case represents the best prediction which can be made in the early stages of project definition. Construction risks can increase the cost, delay completion and, consequently, delay payback and reduce the forecast benefit.

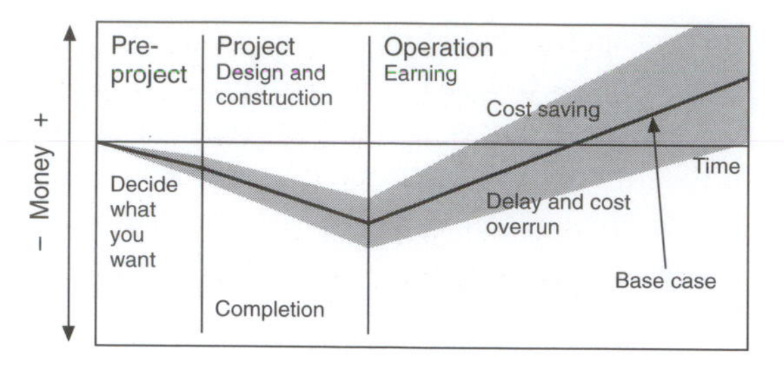

Fig. 3.4 Project cash diagram showing the escalating influence of cost risk over the life of the project, after CIRIA (2001)

Whilst rigorous risk assessment and management has its greatest impact during the pre-project period prior to project approval, like design, risk management is an iterative process and offers greater overall benefit when applied throughout the project life cycle. For this reason it is often better to encourage a 'whole team' approach bringing in clients, designers, contractors and subcontractors at the earliest possible stage.

A springboard for action

Good management of the risks faced by the contractor's business is essential to its success. Risks arise from uncertainty in any factor which affects the achievement of the project objectives. The greater the uncertainty, the more flexible the response and the greater the effort devoted to risk management must be. Systematic risk management is not an end in itself, but a process to help identify and focus on priorities, make informed decisions and take appropriate action. Identifying the major risks allows the procurement strategy to be effectively planned and monitored.

Why ground conditions are a risk

Overview

As pointed out by Clayton (2001), construction is a risky business. No construction project is risk-free. 'Risk can be managed, minimised, shared, transferred, or accepted. It cannot be ignored' (Latham, 1994). Subsurface projects present an enormous risk for the primary project 'stakeholders', that is, the owner and contractor. Realistically, not all risk for subsurface conditions can be entirely avoided or eliminated (Hatem, 1998). Risks occur as a result of vulnerability to hazards. As soil mechanics has become a mature part of engineering science, and engineers have come to regard it as a central part of the core skills obtained by them when undergraduates, geotechnical design has increasingly been seen as just another part of civil engineering design. But despite the advances in the under-pinning science, geotechnical design will

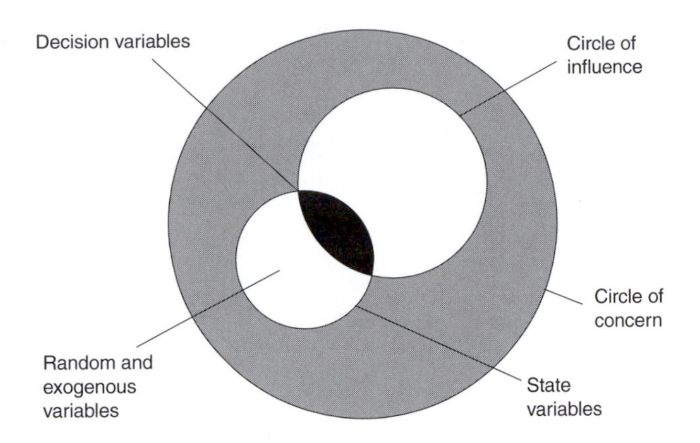

Fig. 3.5 State variables and decision variables, after Haimes (1998)

never be the same as, for example, structural design, for the following reasons.

- The properties and distribution of the ground and groundwater beneath a construction site are predetermined, and therefore largely outside of our control. In any endeavour there are matters of concern (e.g. the stability of slopes), and these are controlled by the state variables (e.g. strength parameters, slope angles, groundwater conditions). The circle of influence (Haimes, 1998) shown in Fig. 3.5 contains those things that the engineer can affect (in this example, slope angle and groundwater conditions). But in geotechnics the circle of influence does not cover all the state variables, and the designer must find suitable ways to live with those matters of concern (e.g. soil properties) that are beyond control.
- Ground and groundwater conditions can be highly variable, both from place to place and with depth (Fig. 3.6). For example, stiffness and strength can vary by 6 orders of magnitude, and permeability by 13. This is in sharp contrast with other construction materials, such as steel and concrete, which are man-made, to predetermined specifications, have a relatively predictable mean strength, and typically a standard deviation of less than 10% of the mean.
- Construction in the ground is normally carried out at the start of any project, so that delays at this stage will affect the later stages of construction.

The additional costs of unexpected ground conditions can be very large. In a study of ten (admittedly exceptional) highway projects, cost overruns averaged some 35% of the tendered sum (Tyrell *et al.*, 1983). Over half of these costs could be related to geotechnical problems, and 49% were attributed to just two factors: interpretation and planning of site

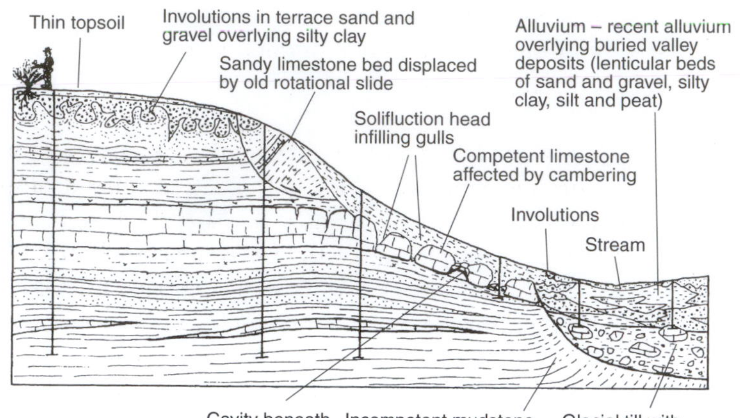

Fig. 3.6 An example of ground variability, after Fookes (1997)

investigations. In a later survey of 58 highway projects, eight showed that the out-turn costs of over half the projects exceeded the tender sum by 23% or more, with only a quarter of the projects having cost overruns of less than 10%. Given the likely profit margins of those involved in any construction project, such cost overruns could be disastrous.

Problems with the current approach

The traditional approach is based upon scientific method, set in a deterministic framework (Clayton, 2001). In best practice, engineers use a desk study to hypothesise about likely ground conditions, possible geotechnical problems and options for ground-related forms of construction. They then design as comprehensive a ground investigation as possible and attempt to determine the actual conditions, so that a complete design can be produced. The process is almost exactly as foretold by Terzaghi (1936).

> *The major part of the college training of civil engineers consists in the absorption of the laws and rules which apply to relatively simple and well-defined materials, such as steel or concrete. This type of education breeds the illusion that everything connected with engineering should and can be computed on the basis of a priori assumptions.... Engineers imagine that the future science of foundations will consist in carrying out the following program:*
>
> - *drill a hole into the ground*
> - *send the soil samples obtained from the hole through a laboratory with standard apparatus served by conscientious human automatons*

- *collect the figures*
- *introduce them into equations, and*
- *compute the result.*

There are inherent problems with the current approach. Firstly, since the ground conditions beneath a site are fixed, appropriate forms and methods of construction must be identified that are best suited to safe, cheap and environmentally sound construction. Risks to construction will be related not only to ground hazards, but also (because of vulnerability) to the form of construction that is selected by the designer.

Secondly, it may be uneconomical, however much expertise, time and money is used, to obtain either a complete picture of the ground conditions beneath a site, or the precise properties of the ground from place to place. Ground investigations are never likely to investigate more than a tiny fraction of the ground and probably less than one part in a million of the soil that will affect subsequent construction. In addition, the time between project conception and final detailed design of all its components can be very long. Increasingly, design (and especially geotechnical design) of major components such as piles or retaining structures is being subcontracted to the constructor, or to specialist subcontractors. While the final design remains undecided, it cannot be possible to plan a completely effective ground investigation.

Even where it is possible (e.g. on a site with very uniform conditions) to obtain a complete picture of the ground, predictions of behaviour made during design can be expected to be approximate at best. Although the under-pinning engineering sciences (soil and rock mechanics) are now well developed, the accuracy of many routine ground-related design calculations remains very poor. There are many examples in the literature. Figure 3.7 shows the results of a recent pile prediction competition. Figure 3.8 shows how inaccurate predictions of foundation settlements based upon standard penetration test data can be, even using the best methods available.

There are numerous ways in which the ground can cause problems for construction, for example due to chemical attack, heave, subsidence, groundwater flow, slope failure, excessive foundation settlement, and so on. Because of the considerable range of risks the ground can pose, it is relatively easy for an inexperienced or non-specialist designer, perhaps using routine procedures, to fail to recognize a critical mechanism of damage or failure that may threaten either the financial viability or health and safety of a project. If a mechanism of damage (a limit state) is not foreseen then it cannot be designed for, and it is often for this reason that ground-related problems occur. They will have a disproportionate effect on the cost and completion of any project, since problems

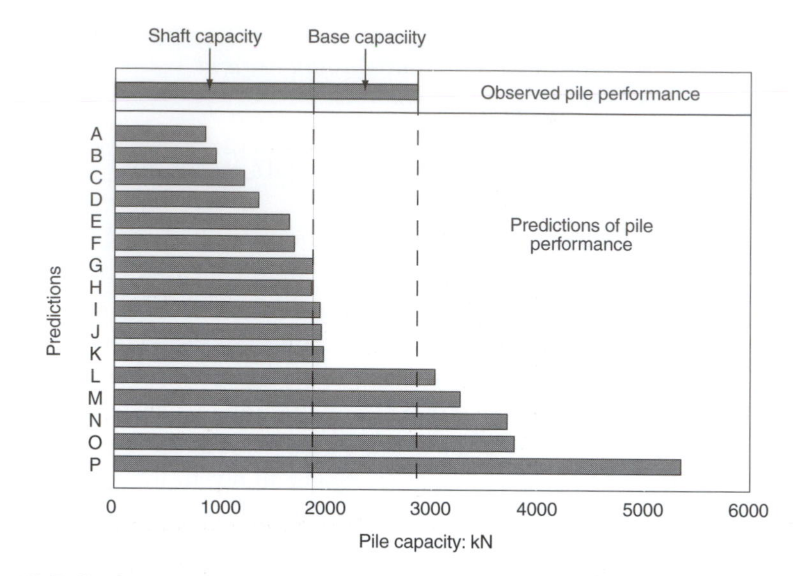

Fig. 3.7 Comparison of predicted and observed pile capacities, after Wheeler (1999)

occurring at an early stage of construction will often lead not only to the additional costs of putting things right but also to irrecoverable delays, which are themselves costly.

The evidence shows that even when traditional procurement methods were in use, with carefully selected consultants and contractors carrying out the work, cost overruns could be very high for certain types of project (Fig. 3.9). Note that with the traditional levels of expenditure on ground investigation (typically less than 1%), 13 cost overruns on

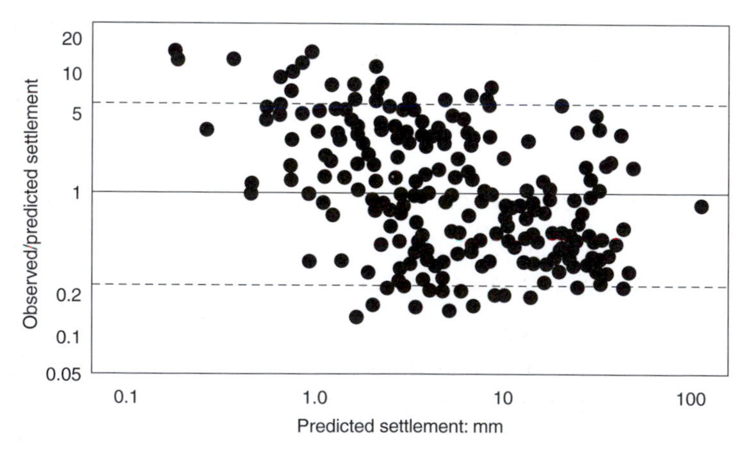

Fig. 3.8 Predicted and observed settlements for foundations on sand, after Clayton et al. *(1988)*

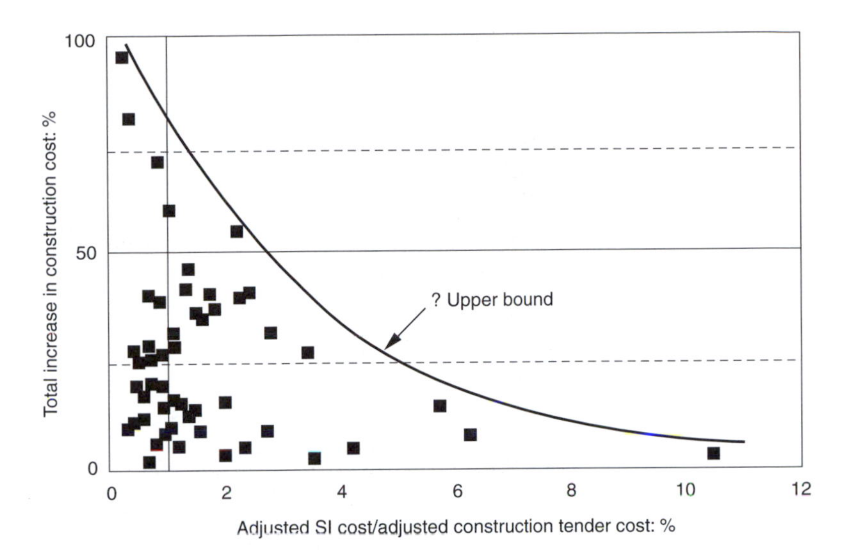

Fig. 3.9 Increase in construction cost as a function of expenditure on site investigation, after Mott MacDonald and Soil Mechanics Ltd (1994)

highway projects were found to be as much as 100%, while expenditure on site investigation of 6% of construction cost appeared to be necessary to guarantee an overspend of less than 10%. With other projects, however, it has been possible to get away with minimal geotechnical skill. As Brierley (1998) has said:

> *To be frank, most ground conditions are not that complex, most projects do not have large amounts of underground work as compared to the overall completed project components, and most contractors are sufficiently familiar with ground conditions in their chosen area of activity to avoid major disasters. Hence, most participants in construction are able to obtain what they want from the process without putting a lot of thought into how best to handle the sub-surface aspect of the work.*

Eliminating uncertainty – new methods of working

As Clayton (2001) has noted, the evidence from recent studies suggests that in current circumstances:

- there will never be sufficient time or money available to investigate with sufficient thoroughness the properties of all the ground to be affected by construction
- the form of ground-related construction is unlikely to be known in its entirety at the time that ground investigation would traditionally be carried out.

This being the case, a new strategy is required. Those involved 'in construction' should therefore:

- accept that ground conditions will always be, to a greater or lesser extent, uncertain
- introduce geotechnical factors into risk management systems
- identify geotechnical hazards in the early stages of project planning
- decide whether a 'routine' or 'sophisticated' geotechnical approach is most suitable for the project
- emphasize appropriate design techniques (e.g. systematic design, conceptual design, robust detailed design)
- allow for a fragmented, fast and multi-staged ground investigation process at the planning stages.

Improving the design process

Effective geotechnical design is an essential part of the risk management process (Clayton, 2001). In the first half of the 20th century, when analytical techniques were poorly developed, engineers had little to rely on other than their understanding of underlying principles and some simplistic empirical or semi-empirical methods of calculation. As analysis has become more sophisticated and texts on all aspects of geotechnical engineering have made information more accessible to the practising engineer, there has been increasing emphasis on analysis and numerical modelling, and decreasing attention to good design principles. University courses need to respond to this changing environment by enhanced teaching of geotechnical design and by developing the engineering judgement of undergraduates.

In other spheres of engineering, systematic design, based on engineering principles, has developed (Pahl and Beitz, 1996). Increasing competitive pressure in manufacturing has led to a need to design more rapidly and with more certainty, and to produce new goods that meet perceived needs more effectively. Systematic design (Fig. 3.10) provides a framework within which innovation can be used, while maximizing certainty of outcome. The first stage requires the designer to determine as simply and precisely as possible the fundamental requirements of the client. Most clients have little or no engineering knowledge, few preconceived ideas about how their needs can be met, and no idea of, or interest in, geotechnical matters. Expressing the needs of the project in the most general way leads to a design specification against which the final design can be checked while not prejudging the technical solution.

In order to fulfil the design specification it will be necessary to understand the functions and sub-functions of the various parts of the proposed project, and to satisfy their requirements. The design specification might

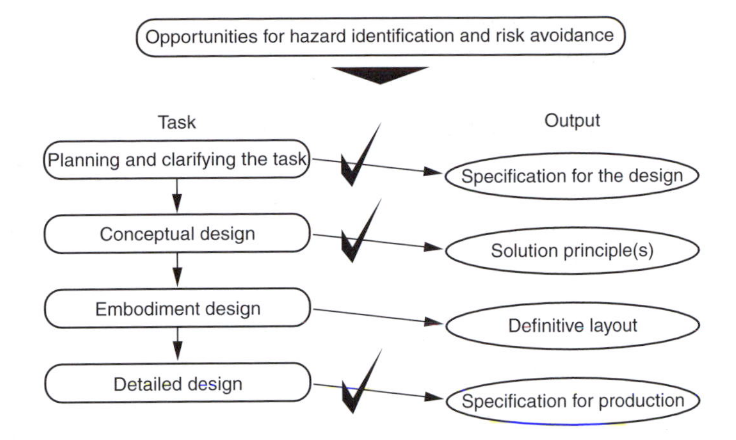

Fig. 3.10 Elements of systematic engineering design, after Pahl and Beitz (1996)

call for 10 000 m^2 of usable office space, with under-office parking. The geotechnical designer would typically have to consider a wide range of issues, including vertical support for building loads, lateral support for basement excavations, de-watering, pavement design for access roads, and slope design for temporary works and for permanent landscaping. The conceptual design process needs to consciously breakdown the project into these functions and sub-functions, and to find the best combination of solutions bearing in mind fitness for purpose, likely cost, uncertainty, and environmental and health and safety impact. In a well managed design process, a number of design variants will be developed for each function or sub-function, and will be objectively judged against the design specification.

Conceptual design is an important counterbalance to the bottom-up process of analysis (literally 'a detailed examination of the elements or structure of something' in the *Oxford English Dictionary*). Conlon (1989) reports what must be a common feeling among most experienced geotechnical engineers:

> *I have made my greatest engineering contributions not by solving difficult problems, but by avoiding them.*

Another useful top-down systems approach considers the effects of different design components on each other, in an attempt not only to identify critical interactions but also to assess which are likely to be most important (Hudson, 1992). Consider the example of a design for an oil pipeline, to be laid across the bed of the North Sea in up to 400 m of water. A number of design subcontracts were let, to assess the stability of soil slopes over which the pipe was to pass, the likelihood of rockfalls from adjacent sub-sea cliffs impacting on the pipe, and the suitable

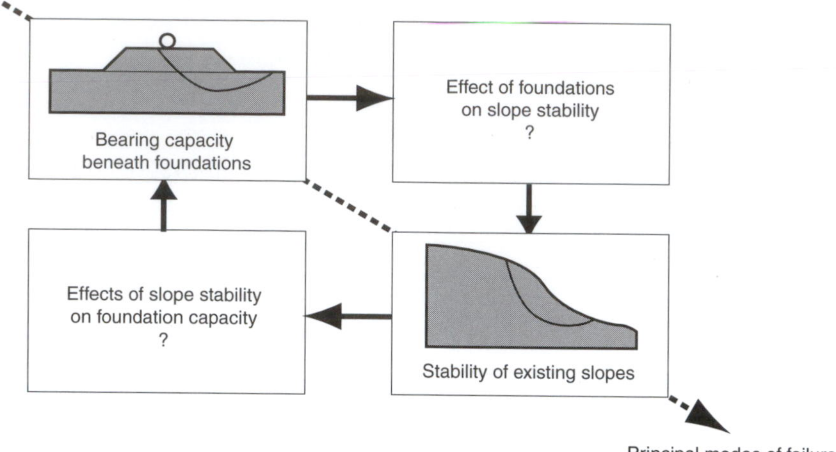

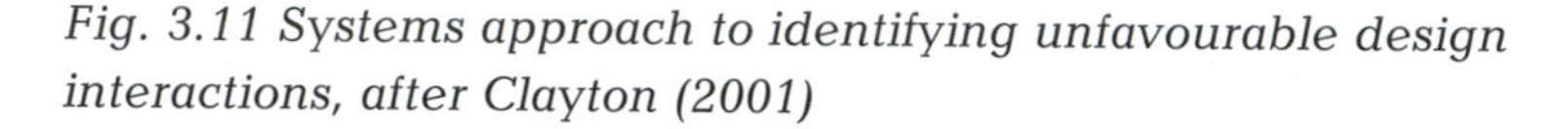
Fig. 3.11 Systems approach to identifying unfavourable design interactions, after Clayton (2001)

geometry for small gravel embankments that were to be placed at intervals along the sea bed in order to support the pipe. The design produced excellent assessments of all these, but apparently failed to identify and design for the critical interactions between the slopes and the bearing capacity of the gravel foundations shown in Fig. 3.11.

An important phase of geotechnical design comes during detailing. Design details such as the connection of anchors to a sheet pile wall, or the joints used in sprayed concrete lined tunnels, can often be vitally important. Dam designers have long understood the importance of detailing, and have used a 'belt and braces' approach (sometimes termed 'defence in depth') where specific details (e.g. the stability of a critical slope or the performance of a cut-off/filter system) have been identified as critically important.

Making geotechnical investigation more effective

In the context of risk management, the success of geotechnical investigations should be judged in the light of what can be foreseen, rather than what is actually found (Clayton, 2001). Given the advances in soil mechanics, geotechnical engineering and engineering geology in the UK over the past 50 years, and the large amount of infrastructure development that has taken place during that period, there should be few surprises left in terms of the possible behaviour of the ground during construction.

The latest edition of the British Standards Institution's *Code of Practice for Site Investigations* (BS 5930: 1999) has introduced a welcome change of emphasis by dividing ground investigations into three phases:

- desk study and site reconnaissance
- detailed investigation for design
- construction review.

However, in the context of risk management, and especially in the light of more widely dispersed geotechnical design, further changes can be expected. For example:

- For effective hazard identification a search of existing information on the site needs to be carried out very early, perhaps before the client's final decision to invest. The interpretation of these data requires skill and judgement, suggesting that one or more geotechnical experts should be involved (albeit briefly) during the formation of the development concept.
- It is increasingly likely that more than one phase of ground investigation will be required as different elements of geotechnical design will be undertaken at different times. Evidence from design-and-build contracts suggests that it is already common practice to carry out at least two phases of investigation when there are significant amounts of contractor-designed temporary or permanent geotechnical works.
- The speed of ground investigations (from instruction to provision of data) carried out for routine projects will need to increase, in order to serve a fast-track subcontracted industry. This will favour in situ testing methods (cone penetration test, standard penetration test, geophysics) and fast laboratory testing (undrained triaxial testing, bender element testing). The location and type of investigation should be carefully planned in order to provide data for design while testing any hypotheses developed from existing geological and geotechnical data.
- Ground conditions should be observed and recorded during construction, when more ground is likely to be exposed than at any other stage of the work. The construction review should compare what is exposed with what was foreseen at the hazard identification stage, and designs should be sufficiently flexible to accept any necessary changes.
- In the relatively few cases where cost benefits can be recognized, the use of extensive investigations, sophisticated sampling, in situ and laboratory testing will be justified, even though there are likely to be significant delays to a project's programme as a result. Recent experience suggests that dams, tunnels and deep basements for inner-city buildings are some of the types of construction that may benefit from a more sophisticated approach. It is doubtful if this traditional, deterministic, approach will be of value for highway projects.
- In the future it is unlikely that many clients will be willing to accept responsibility for 'unforeseen ground conditions'. Designs should be

sufficiently robust to accept the full range of expected ground conditions. This may mean that monitoring is required, or that the observational method (Nicholson *et al.*, 1999) is used.

It seems likely that large, time-consuming ground investigations will become rare and that if the client is to be best served, the emphasis should move to a more effective use of geotechnical engineering skill and experience. This should take the form of hazard and risk identification and effective geotechnical design, coupled in most projects with fast low-cost boring, penetrometer and laboratory testing.

Implementing a geotechnical risk system

Based upon the Highways Agency's *Value for Money Manual* (1996), the following procedure seems to be appropriate for geotechnical risk analysis (Clayton, 2001):

- creation of a 'team' of geotechnical experts (for a simple building job this might in fact comprise a single geotechnical engineer, but for large construction projects a minimum of three experienced geotechnical engineers/geologists would be required)
- collection of published and unpublished data on the ground conditions at the site
- identification of the likely range of forms of construction that might be used
- brainstorming to identify hazards and risks to different forms of construction, based upon existing information
- use of group experience to rank risks according to impact
- establishment of a risk register (see below)
- association of the various risks with the different phases of the project
- estimate costs and variances, and probability (based upon group experience).

For risk analysis to be effective it is important that (i) a thorough search and review of existing ground-related information is carried out and (ii) sufficient experience, both geotechnical and geological, is brought to bear in identifying hazards, risks and consequences.

The risk register

As pointed out by Hall *et al.* (2001), systematic risk management aims to add value to the construction process by:

- encouraging ingenious identification of hazards and opportunities
- using estimates of risk as a basis for decision-making
- controlling uncertainty
- minimizing the potential damages should the worst happen

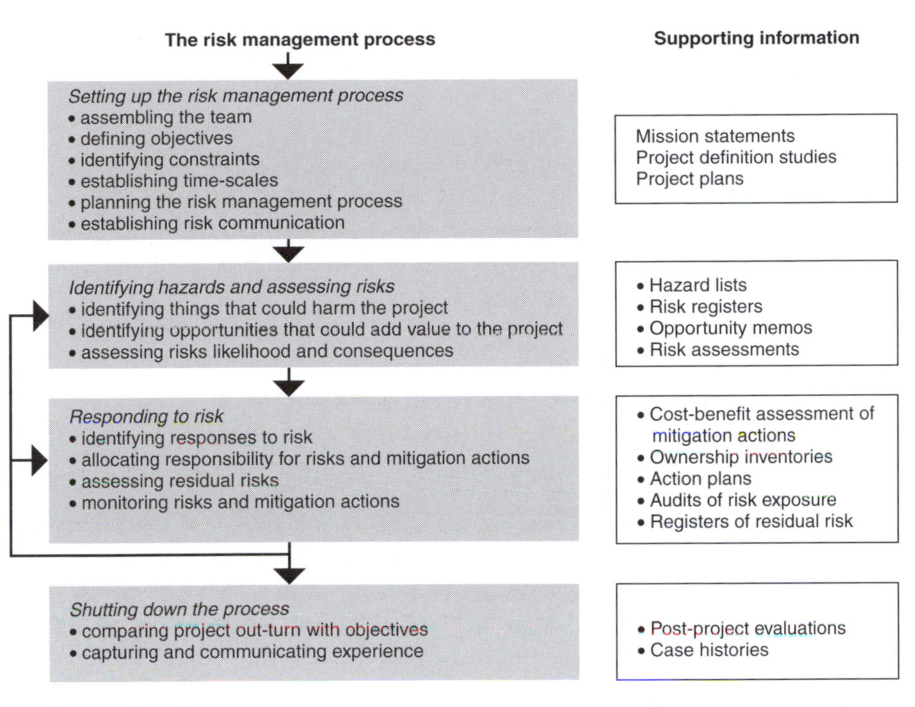

Fig. 3.12 Steps in the risk management process, together with supporting information, after Hall et al. *(2001)*

- communicating risks between project participants
- clearly establishing ownership of risks and risk mitigation actions.

Recent initiatives have endeavoured to formalize the risk management process (Godfrey, 1996; Simon *et al.*, 1997; RAMP, 1998) and, although views of the risk management process differ somewhat, the main steps are now widely recognized (Fig. 3.12).

The cyclic risk management process corresponds closely to the process of successful project management in general. This said, it is important to recognize that systematic risk management could improve decision-making by 'informing judgement, but it cannot replace it' (Engineering Council, 1993).

Figure 3.12 illustrates the information repositories that support the risk management process by:

- recording the hazards that have been identified and the results of risk assessments
- providing an overview of the risk portfolio
- informing project players about risk ownership and their responsibilities for risk management

- monitoring the progress of risk management actions and the status of residual risks.

The steps in software-supported risk management

The risk management software developed for the construction industry follows well recognized steps in the risk management process (Fig. 3.13). A benefit of using software is that large amounts of best practice guidance can be embedded in the tool, with links to the guidance from each stage in the process. The aim of the tool is to add value to participants in the construction process, including clients, by:

- recording risks and risk management actions
- focusing attention on the most important risks

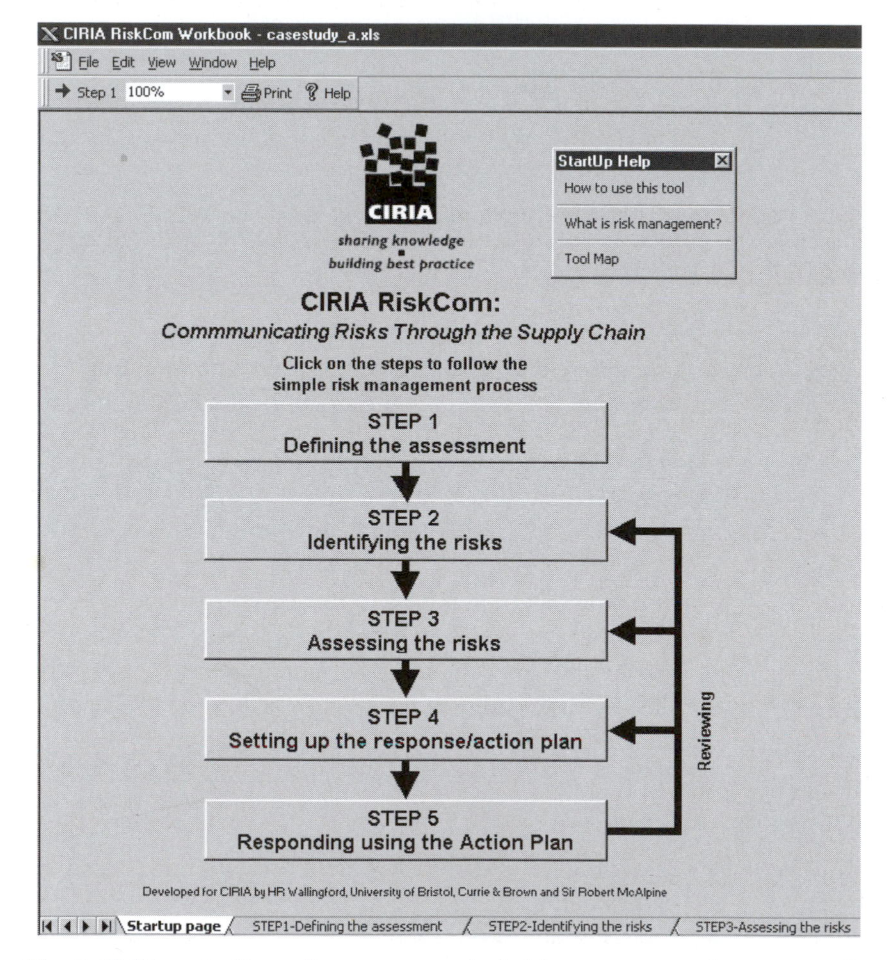

Fig. 3.13 Steps in the software-supported risk management process, after Hall et al. *(2001)*

- clarifying risk ownership and responsibilities
- providing a common format for risk communication throughout the supply chain
- providing a convenient and traceable mechanism for re-visiting risk assessments as a project proceeds
- disseminating best practice through a comprehensive knowledge base and case studies.

Step 1 – Setting up the risk management process

A potential danger of software to support risk registers and other information-handling aspects of risk management is that these activities can become the focus of risk management, while the focus should rightly be on adding value to the business process (Blockley and Godfrey, 2000). The first stage in the risk management process is therefore to define the process being assessed and identify how it contributes to success in business terms. The window shown in Fig. 3.14 helps to formalize this. Many of the items of information required at this first stage may be obvious to the risk manager. However, it serves essential purposes to

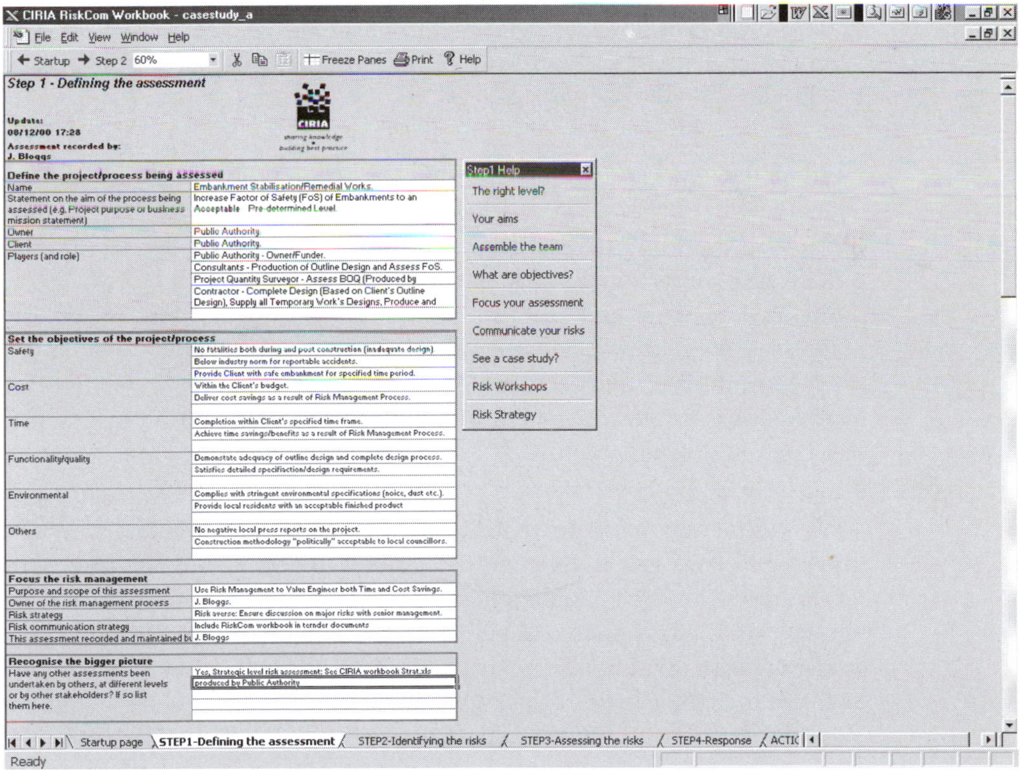

Fig. 3.14 Process attributes and objectives, after Hall et al. *(2001)*

- make the objectives of the process explicit and gain collective commitment to those objectives
- set a time scale for achieving the objectives
- identify players in the process and ensure that they are aware of their responsibilities
- identify the business context in which the process is being enacted and the linkages to the business objectives of the players
- establish a system for regular feedback between project members.

Naming the process. The starting point for risk management is to name the process being managed. The name should embody the purpose of the process, for example 'widening a highway' or, at a more detailed level, 'tendering for the highway surfacing contract'. The process will be clarified by clearly stating its attributes as follows.

- The *process owner* is the person with overall responsibility for the process achieving its objectives.
- The *players* are the other people who make a contribution to the process achieving its objectives. Any player may be a risk owner.
- The *objectives* are the measurable criteria against which the success of the process will be judged. The most common objectives will relate to time, cost, quality, safety and the environment, but there may also be project-specific objectives.

Setting the process in the business context. All of the players in the construction process need to address construction in the context of their overall business objectives. Risk management may be applied to strategic issues at organizational level, at a tactical level to groups of projects, to individual projects and to tasks within those projects (Fig. 3.15). The risk assessment should address issues relevant to the level at which it is being conducted and cross-reference related risk assessments that address risks at other levels to identify the linkages and leverages that are the key to adding value.

Planning the risk management. Responsibility for risk management and the scope of the risk assessment is clarified from the start by listing key attributes of the risk management process (Fig. 3.12) and identifying related risk assessments. Users are encouraged to address their risk strategy explicitly. They may wish to note, for example, that their strategy is to accept some classes of risk through self-insurance. The risk communication strategy informs the collection and dissemination of risk information and should involve all of the project players with responsibility for risk as well as wider stakeholders.

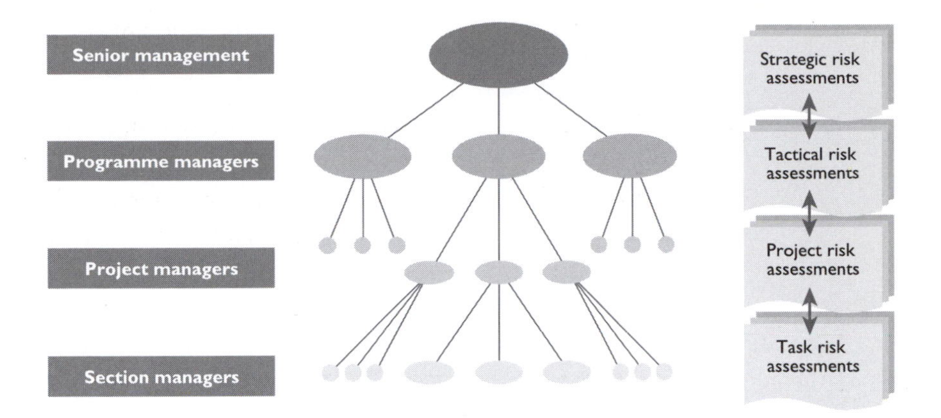

Fig. 3.15 Appropriate levels of risk management, after Hall et al. *(2001)*

Step 2 – Identifying the risks

Conventional terminology has been adopted: a hazard is a potential cause of harm or loss, for example 'adverse weather'; a risk is an uncertain event that, should it occur, will have an effect on achieving objectives, for example 'earthworks delayed by adverse weather'. Hazard analysis and risk identification is a highly creative stage in the risk management process. While a simple software application cannot replace this creative process, it can support it in important respects as follows.

- Software provides a convenient mechanism for recording risks when they are identified (Fig. 3.16).
- Searchable prompt lists support the creative process of risk identification.

Guidance on best practice on risk identification, for example, how to run a risk workshop can be embedded in the software.

The converse to a hazard is an opportunity, which is a potential way to add value. During the creative process of hazard analysis, opportunities to add value may also be identified, which can also be recorded in a software tool. While risks are identified formally, opportunities are recorded on a free format note pad that can be re-visited for more formal development later in the value management process (Fig. 3.16).

Step 3 – Assessing the risks

The aim of assessing risks is to focus management actions on the risks that matter. Risk assessment involves evaluating the likelihood of a specific risk materializing and consequences in terms of impact on objectives. Together, likelihood and consequence provide a measure of the significance of the risk. At the simplest level, risks are assessed based on the

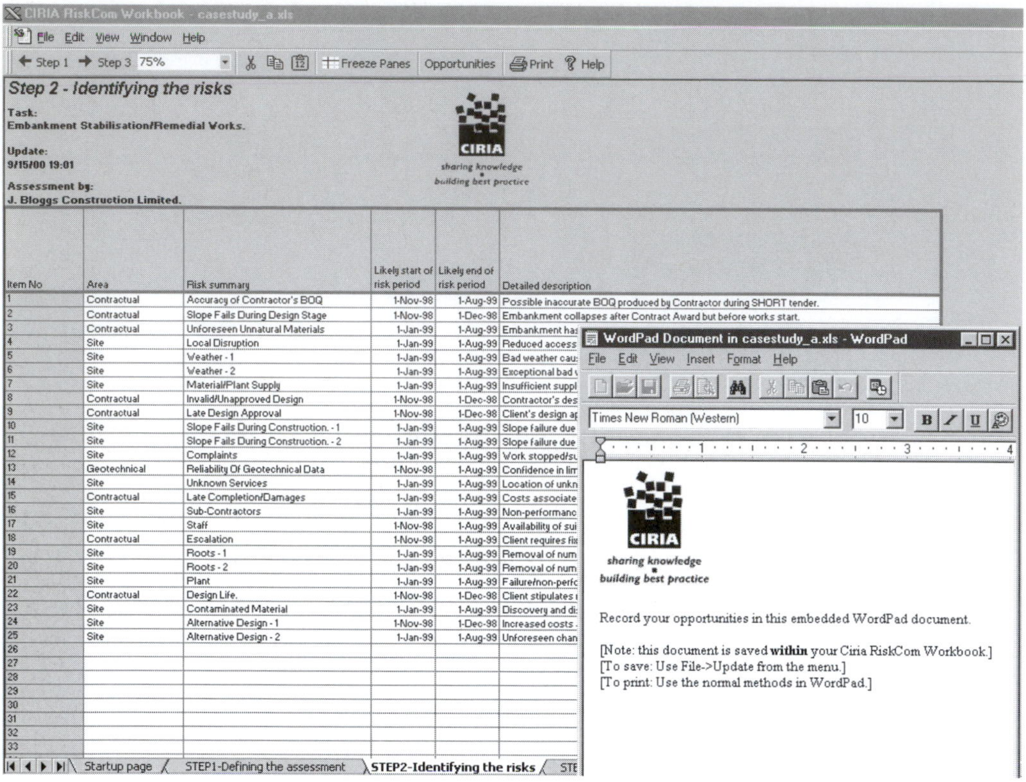

CIRIA RiskCom Workbook - casestudy_a.xls

File Edit View Window Help

Step 1 Step 3 75% Freeze Panes Opportunities Print Help

Step 2 - Identifying the risks

Task:
Embankment Stabilisation/Remedial Works.

Update:
9/15/00 19:01

Assessment by:
J. Bloggs Construction Limited.

CIRIA sharing knowledge • building best practice

Item No	Area	Risk summary	Likely start of risk period	Likely end of risk period	Detailed description
1	Contractual	Accuracy of Contractor's BOQ	1-Nov-98	1-Aug-99	Possible inaccurate BOQ produced by Contractor during SHORT tender.
2	Contractual	Slope Fails During Design Stage	1-Nov-98	1-Dec-98	Embankment collapses after Contract Award but before works start.
3	Contractual	Unforeseen Unnatural Materials	1-Jan-99	1-Aug-99	Embankment has
4	Site	Local Disruption	1-Jan-99	1-Aug-99	Reduced access
5	Site	Weather - 1	1-Jan-99	1-Aug-99	Bad weather cau
6	Site	Weather - 2	1-Jan-99	1-Aug-99	Exceptional bad
7	Site	Material/Plant Supply	1-Jan-99	1-Aug-99	Insufficient suppl
8	Contractual	Invalid/Unapproved Design	1-Nov-98	1-Dec-98	Contractor's des
9	Contractual	Late Design Approval	1-Nov-98	1-Dec-98	Client's design ap
10	Site	Slope Fails During Construction. - 1	1-Jan-99	1-Aug-99	Slope failure due
11	Site	Slope Fails During Construction. - 2	1-Jan-99	1-Aug-99	Slope failure due
12	Site	Complaints	1-Jan-99	1-Aug-99	Work stopped/su
13	Geotechnical	Reliability Of Geotechnical Data	1-Nov-98	1-Aug-99	Confidence in lim
14	Site	Unknown Services	1-Jan-99	1-Aug-99	Location of unkn
15	Contractual	Late Completion/Damages	1-Jan-99	1-Aug-99	Costs associate
16	Site	Sub-Contractors	1-Jan-99	1-Aug-99	Non-performanc
17	Site	Staff	1-Nov-98	1-Aug-99	Availability of sui
18	Contractual	Escalation	1-Nov-98	1-Aug-99	Client requires fix
19	Site	Roots - 1	1-Jan-99	1-Aug-99	Removal of num
20	Site	Roots - 2	1-Jan-99	1-Aug-99	Removal of num
21	Site	Plant	1-Jan-99	1-Aug-99	Failure/non-perfc
22	Contractual	Design Life.	1-Nov-98	1-Dec-98	Client stipulates
23	Site	Contaminated Material	1-Jan-99	1-Aug-99	Discovery and di
24	Site	Alternative Design - 1	1-Nov-98	1-Dec-98	Increased costs .
25	Site	Alternative Design - 2	1-Jan-99	1-Aug-99	Unforeseen chan
26					
27					
28					
29					
30					
31					
32					
33					

Startup page / STEP1-Defining the assessment / **STEP2-Identifying the risks** / STE

WordPad Document in casestudy_a.xls - WordPad

File Edit View Insert Format Help

Times New Roman (Western) 10

CIRIA sharing knowledge • building best practice

Record your opportunities in this embedded WordPad document.

[Note: this document is saved **within** your Ciria RiskCom Workbook.]
[To save: Use File->Update from the menu.]
[To print: Use the normal methods in WordPad.]

Fig. 3.16 Risk summaries in the software-supported risk register, after Hall et al. *(2001)*

judgement of the experts involved, individually or preferably through collective discussion. A scale of likelihood enables different risks to be compared without needing fully quantitative analysis. Consequences are evaluated against the objectives of the project (Fig. 3.17). Multiplying the judgements of likelihood and consequence provides a risk rating, which can be used to rank risks. In more complex situations, quantitative risk assessment can be used to inform the estimates of likelihood and consequences, taking account of interactions between different risks.

Step 4 – Setting the response/action plan

Risk analysis forms the basis for rational decision-making about responses to hazards. It is only through timely action that value can be added. Risk management software can support the development of effective risk control strategies by:

- forcing systematic consideration of each risk that has been identified
- providing guidance on generic risk control strategies

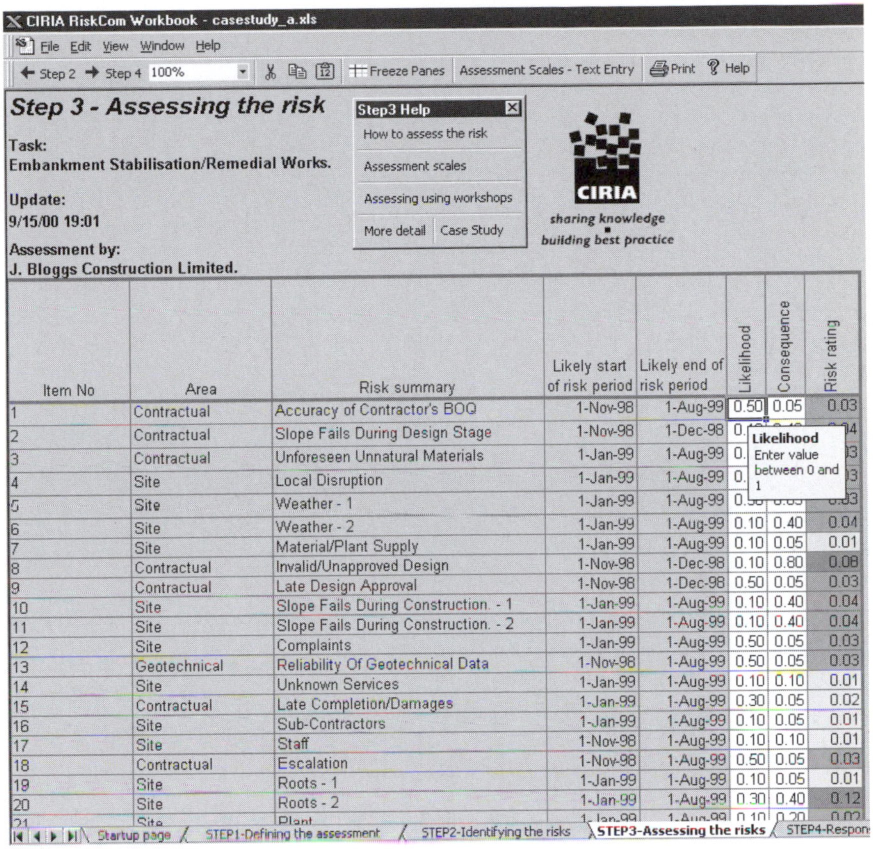

CIRIA RiskCom Workbook - casestudy_a.xls

File Edit View Window Help

Step 2 Step 4 100% Freeze Panes Assessment Scales - Text Entry Print Help

Step 3 - Assessing the risk

Task:
Embankment Stabilisation/Remedial Works.

Update:
9/15/00 19:01

Assessment by:
J. Bloggs Construction Limited.

Step3 Help
How to assess the risk
Assessment scales
Assessing using workshops
More detail | Case Study

CIRIA
sharing knowledge
building best practice

Item No	Area	Risk summary	Likely start of risk period	Likely end of risk period	Likelihood	Consequence	Risk rating
1	Contractual	Accuracy of Contractor's BOQ	1-Nov-98	1-Aug-99	0.50	0.05	0.03
2	Contractual	Slope Fails During Design Stage	1-Nov-98	1-Dec-98	[illegible]	[illegible]	[illegible]
3	Contractual	Unforeseen Unnatural Materials	1-Jan-99	1-Aug-99	[illegible]	[illegible]	[illegible]
4	Site	Local Disruption	1-Jan-99	1-Aug-99	[illegible]	[illegible]	[illegible]
5	Site	Weather - 1	1-Jan-99	1-Aug-99	[illegible]	[illegible]	[illegible]
6	Site	Weather - 2	1-Jan-99	1-Aug-99	0.10	0.40	0.04
7	Site	Material/Plant Supply	1-Jan-99	1-Aug-99	0.10	0.05	0.01
8	Contractual	Invalid/Unapproved Design	1-Nov-98	1-Dec-98	0.10	0.80	0.08
9	Contractual	Late Design Approval	1-Nov-98	1-Dec-98	0.50	0.05	0.03
10	Site	Slope Fails During Construction. - 1	1-Jan-99	1-Aug-99	0.10	0.40	0.04
11	Site	Slope Fails During Construction. - 2	1-Jan-99	1-Aug-99	0.10	0.40	0.04
12	Site	Complaints	1-Jan-99	1-Aug-99	0.50	0.05	0.03
13	Geotechnical	Reliability Of Geotechnical Data	1-Nov-98	1-Aug-99	0.50	0.05	0.03
14	Site	Unknown Services	1-Jan-99	1-Aug-99	0.10	0.10	0.01
15	Contractual	Late Completion/Damages	1-Jan-99	1-Aug-99	0.30	0.05	0.02
16	Site	Sub-Contractors	1-Jan-99	1-Aug-99	0.10	0.05	0.01
17	Site	Staff	1-Nov-98	1-Aug-99	0.10	0.10	0.01
18	Contractual	Escalation	1-Nov-98	1-Aug-99	0.50	0.05	0.03
19	Site	Roots - 1	1-Jan-99	1-Aug-99	0.10	0.05	0.01
20	Site	Roots - 2	1-Jan-99	1-Aug-99	0.30	0.40	0.12
21	Site	Plant	1-Jan-99	1-Aug-99	0.10	0.20	0.02

Likelihood
Enter value between 0 and 1

Startup page / STEP1-Defining the assessment / STEP2-Identifying the risks / **STEP3-Assessing the risks** / STEP4-Respon

Fig. 3.17 Risk assessment in the software-supported risk register, after Hall et al. *(2001)*

- concisely stating risk control actions, with links to more detailed software-based risk control statements and analysis of costs and benefits
- clearly identifying ownership of risks
- identifying those risks where it is agreed to be more cost effective to manage the outcome than to invest in reduction
- monitoring the dates when risks are expected to become live
- building confidence by demonstrating the application of good management
- enabling feedback and interaction between project team members.

Risk control strategies may reduce risks or change their ownership. Seldom will they remove risks altogether, so it is important to identify and analyse residual risks, clarifying their ownership (Fig. 3.18). By comparing the original risk with the residual risk, the potential value of risk management can be demonstrated. That value is only realized if the risk management is implemented as planned.

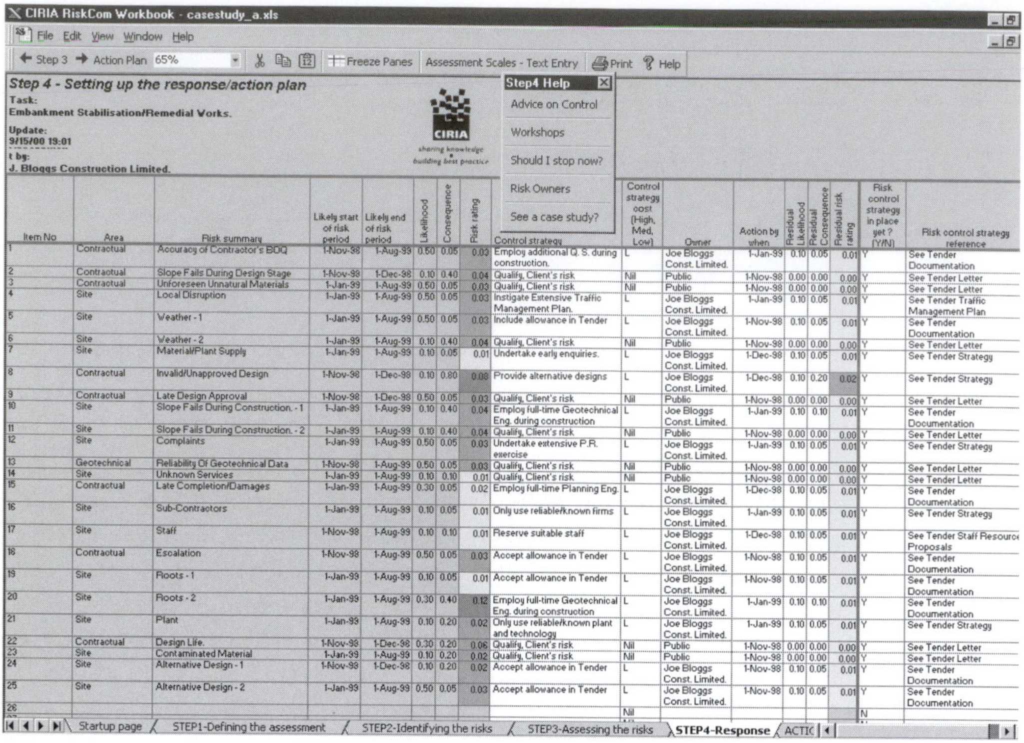

CIRIA RiskCom Workbook - casestudy_a.xls

File Edit View Window Help

Step 3 Action Plan 65% Freeze Panes Assessment Scales - Text Entry Print Help

Step4 Help: Advice on Control / Workshops / Should I stop now? / Risk Owners / See a case study?

Step 4 - Setting up the response/action plan

Task:
Embankment Stabilisation/Remedial Works.

Update:
9/15/00 19:01

t by:
J. Bloggs Construction Limited.

CIRIA sharing knowledge building best practice

Item No	Area	Risk summary	Likely start of risk period	Likely end of risk period	Likelihood	Consequence	Risk rating	Control strategy	Control strategy cost (High, Med, Low)	Owner	Action by when	Residual Likelihood	Residual Consequence	Residual risk rating	Risk control strategy in place yet ? (Y/N)	Risk control strategy reference
1	Contractual	Accuracy of Contractor's BOQ	1-Nov-98	1-Aug-99	0.50	0.05	0.03	Employ additional Q. S. during construction.	L	Joe Bloggs Const. Limited.	1-Jan-99	0.10	0.05	0.01	Y	See Tender Documentation
2	Contractual	Slope Fails During Design Stage	1-Nov-98	1-Dec-98	0.10	0.40	0.04	Qualify, Client's risk	Nil	Public	1-Nov-98	0.00	0.00	0.00	Y	See Tender Letter
3	Contractual	Unforeseen Unnatural Materials	1-Jan-99	1-Aug-99	0.50	0.05	0.03	Qualify, Client's risk	Nil	Public	1-Nov-98	0.00	0.00	0.00	Y	See Tender Letter
4	Site	Local Disruption	1-Jan-99	1-Aug-99	0.50	0.05	0.03	Instigate Extensive Traffic Management Plan.	L	Joe Bloggs Const. Limited.	1-Jan-99	0.10	0.05	0.01	Y	See Tender Traffic Management Plan
5	Site	Weather - 1	1-Jan-99	1-Aug-99	0.50	0.05	0.03	Include allowance in Tender	L	Joe Bloggs Const. Limited.	1-Nov-98	0.10	0.05	0.01	Y	See Tender Documentation
6	Site	Weather - 2	1-Jan-99	1-Aug-99	0.10	0.40	0.04	Qualify, Client's risk	Nil	Public	1-Nov-98	0.00	0.00	0.00	Y	See Tender Letter
7	Site	Material/Plant Supply	1-Jan-99	1-Aug-99	0.10	0.05	0.01	Undertake early enquiries.	L	Joe Bloggs Const. Limited.	1-Dec-98	0.10	0.05	0.01	Y	See Tender Strategy
8	Contractual	Invalid/Unapproved Design	1-Nov-98	1-Dec-98	0.10	0.80	0.08	Provide alternative designs	L	Joe Bloggs Const. Limited.	1-Dec-98	0.10	0.20	0.02	Y	See Tender Strategy
9	Contractual	Late Design Approval	1-Nov-98	1-Dec-98	0.50	0.05	0.03	Qualify, Client's risk	Nil	Public	1-Nov-98	0.00	0.00	0.00	Y	See Tender Letter
10	Site	Slope Fails During Construction. - 1	1-Jan-99	1-Aug-99	0.10	0.40	0.04	Employ full-time Geotechnical Eng. during construction	L	Joe Bloggs Const. Limited.	1-Jan-99	0.10	0.10	0.01	Y	See Tender Documentation
11	Site	Slope Fails During Construction. - 2	1-Jan-99	1-Aug-99	0.10	0.40	0.04	Qualify, Client's risk	Nil	Public	1-Nov-98	0.00	0.00	0.00	Y	See Tender Letter
12	Site	Complaints	1-Jan-99	1-Aug-99	0.50	0.05	0.03	Undertake extensive P.R. exercise	L	Joe Bloggs Const. Limited.	1-Jan-99	0.10	0.05	0.01	Y	See Tender Strategy
13	Geotechnical	Reliability Of Geotechnical Data	1-Nov-98	1-Aug-99	0.50	0.05	0.03	Qualify, Client's risk	Nil	Public	1-Nov-98	0.00	0.00	0.00	Y	See Tender Letter
14	Site	Unknown Services	1-Jan-99	1-Aug-99	0.10	0.10	0.01	Qualify, Client's risk	Nil	Public	1-Nov-98	0.00	0.00	0.00	Y	See Tender Letter
15	Contractual	Late Completion/Damages	1-Jan-99	1-Aug-99	0.30	0.05	0.02	Employ full-time Planning Eng.	L	Joe Bloggs Const. Limited.	1-Dec-98	0.10	0.05	0.01	Y	See Tender Documentation
16	Site	Sub-Contractors	1-Jan-99	1-Aug-99	0.10	0.05	0.01	Only use reliable/known firms	L	Joe Bloggs Const. Limited.	1-Jan-99	0.10	0.05	0.01	Y	See Tender Strategy
17	Site	Staff	1-Nov-98	1-Aug-99	0.10	0.10	0.01	Reserve suitable staff	L	Joe Bloggs Const. Limited.	1-Dec-98	0.10	0.05	0.01	Y	See Tender Staff Resource Proposals
18	Contractual	Escalation	1-Nov-98	1-Aug-99	0.50	0.05	0.03	Accept allowance in Tender	L	Joe Bloggs Const. Limited.	1-Nov-98	0.10	0.05	0.01	Y	See Tender Documentation
19	Site	Roots - 1	1-Jan-99	1-Aug-99	0.10	0.05	0.01	Accept allowance in Tender	L	Joe Bloggs Const. Limited.	1-Nov-98	0.10	0.05	0.01	Y	See Tender Documentation
20	Site	Roots - 2	1-Jan-99	1-Aug-99	0.30	0.40	0.12	Employ full-time Geotechnical Eng. during construction	L	Joe Bloggs Const. Limited.	1-Jan-99	0.10	0.10	0.01	Y	See Tender Documentation
21	Site	Plant	1-Jan-99	1-Aug-99	0.10	0.20	0.02	Only use reliable/known plant and technology	L	Joe Bloggs Const. Limited.	1-Jan-99	0.10	0.05	0.01	Y	See Tender Strategy
22	Contractual	Design Life.	1-Nov-98	1-Dec-98	0.30	0.20	0.06	Qualify, Client's risk	Nil	Public	1-Nov-98	0.00	0.00	0.00	Y	See Tender Letter
23	Site	Contaminated Material	1-Jan-99	1-Aug-99	0.10	0.20	0.02	Qualify, Client's risk	Nil	Public	1-Nov-98	0.00	0.00	0.00	Y	See Tender Letter
24	Site	Alternative Design - 1	1-Nov-98	1-Dec-98	0.10	0.20	0.02	Accept allowance in Tender	L	Joe Bloggs Const. Limited.	1-Nov-98	0.10	0.05	0.01	Y	See Tender Documentation
25	Site	Alternative Design - 2	1-Jan-99	1-Aug-99	0.50	0.05	0.03	Accept allowance in Tender	L	Joe Bloggs Const. Limited.	1-Nov-98	0.10	0.05	0.01	Y	See Tender Documentation
26									Nil						N	

Startup page / STEP1-Defining the assessment / STEP2-Identifying the risks / STEP3-Assessing the risks / **STEP4-Response** / ACTIC

Fig. 3.18 Residual risk analysis in the software-supported risk register, after Hall et al. *(2001)*

Step 5 – Responding using the action plan

Risk management software provides outputs in a range of formats including a summary of top risks, a list of most imminent risks and risks sorted according to owner. These summaries can be circulated electronically or as hard copies. The aim is for the risk registers to become a standard format for communicating risk between project partners, leading to a more collaborative approach to risk management throughout the supply chain.

At the start of the project the client could undertake the initial risk assessment and then communicate it to the designer by forwarding the completed risk register. The designer could then expand on it during the design process so that it can be passed to contractors at tender, and so on through the supply chain. Similarly, tenderers can be asked to submit assessments of the risks identified when preparing their tenders, together with outline plans to manage those risks.

CHAPTER FOUR

Parameter determination: classic and modern methods

Overview: key terminology, parameters and test types

Typically, soil comprises a skeleton of soil grains in frictional contact with each other forming an open-packed structure (loose/soft) or close-packed structure (dense/hard). The soil particles may be microscopic in the case of *clays* (which may range in hardness from *soft* to *stiff*), just visible in the case of *silts*, and clearly visible in the case of *sands* (which may range in *density* from *loose* to *dense*, and in *particle size* from *fine* to *coarse*) and the larger particle sized *gravels*. The distribution of particle size is called *grading*. The *soil skeleton*, which can also be *cemented*, forms an interstitial system of connecting spaces or *pores*. The pores in the soil will usually contain some moisture even in *unsaturated soils*. The flow of pore water can be restricted by the small size of the pores and degree of saturation thus giving rise to low *permeability* k, particularly in clays. During construction, for saturated soils the change in load or *total stress* σ is shared between the soil structure and the *pore pressure* u. The time dependent flow of water in soil under applied load is referred to as *consolidation* (pore water flowing out of a loaded zone) or *swelling* (pore water flowing into an unloaded zone) and is the means whereby total stress change is transferred from pore pressure to structural loading of the soil skeleton as measured by *effective stress* $\sigma' = \sigma - u$, the parameter that uniquely controls all deformation in soils. It is this time dependency that gives saturated clays their unique behaviour whereby they have a *short term* or *undrained strength* s_u that is different from the *long term* or *drained strength* s_d. This is why soil supported structures (e.g. foundations) and soil structures (e.g. embankments and cuttings) have *short term stability* and *long term stability*, e.g. why Victorian-era railway cuttings in England failed half a century after construction. The maximum capacity of the soil skeleton to support load is called the *shear strength* because soil fails in shear. This strength depends on the frictional nature of the interparticle contact and is measured by the coefficient of friction or *angle of shearing resistance* ϕ', and by the constant, *cohesion* c', with respect to effective stress as designated by the prime notation thus $'$. The deformability of the soil skeleton is measured by elastic theory deformation moduli such as *Young's modulus E*, *Poisson's ratio* ν and *Shear modulus G*. Because of formation history such as deposition by wind or water, soil in situ possesses *fabric* or geometric orientation of particles that gives rise to *anisotropy*, i.e. different properties in different directions.

Soil can be geologically loaded to a *maximum past pressure* or *preconsolidation pressure*. This pre-load constitutes a *yield point*. At stresses less than yield the soil behaves elastically, i.e. the strains are nearly recoverable. At stresses more than yield the soil behaves plastically, i.e. the strains are not recoverable and the mathematical *theory of plasticity* is sometimes used to describe the post-yield soil behaviour, e.g. in finding the *bearing capacity* of footings and piles. *Stress distributions*, however, can be described generally using the mathematical *theory of elasticity* that is also used for the prediction of vertical movement such as *settlement* (downward) or *heave* (upwards).

Soil properties can be studied and parameters measured in a variety of tests. The most common and most useful test is the *triaxial test*, which is carried out in the laboratory so that test conditions can be carefully controlled. Other important laboratory test equipment are the *resonant column apparatus*, which measures maximum shear modulus, the *hollow cylinder apparatus*, which is an element test and can apply rotation of *principal stresses* (e.g. as pertain under moving wheel loads), and the *ring shear apparatus*, which measures *residual shear strength* on an established shear surface (e.g. such as may control stability on natural slopes that have previously slipped).

Soil may be characterized by plasticity index tests that give rise to a range of indices including the *liquid limit*, *plastic limit*, *plasticity index* and *liquidity index*. These indices have been correlated empirically with soil parameters such as undrained Young's modulus.

Field tests include penetration tests such as the *standard penetration test* (SPT) (split spoon hammered into the ground) and the *cone penetration test* (CPT) (cone pushed into the ground by hydraulic means), which require empirical correlation with soil parameters. Other field testing equipment include the *pressuremeter*, which expands a cylindrical casing against the sides of a borehole (the Camkometer – short for 'Cambridge *k*-zero meter' – is a self-boring pressuremeter) and the *dilatometer*, which expands a spade-shaped diaphragm after the dilatometer has been pushed into the ground.

The widespread availability of commercial finite element stress analysis software has concentrated attention on measuring soil parameters, particularly ground stiffness. This has led increasingly to the use of seismic test apparatus to measure *shear wave velocity*. *Up-hole*, *down-hole* and *cross-hole* methods use boreholes. The *seismic cone* penetration test uses a hammer at the surface to produce vibrations detected by a receiver in the cone. Using a hammer or a frequency-controlled vibrator at the ground surface generates *surface waves*. These include *Rayleigh waves* that travel parallel to the ground surface to a depth of about one wave length thus testing the soil in the mass (i.e. including the effects of fissuring and jointing) in a non-invasive way. The resulting ground vibrations are detected by an array of vertically polarized sensors or *geophones*. From surface wave tests, shear wave velocity is correlated with *wavelength* and these data can be interpreted to give *stiffness–depth profiles*. Shear wave velocity measurements can be used to *characterize* soils as well as to provide useful data for estimating sampling disturbance.

It is important to make the distinction between a *soil property* and a *soil parameter*. A soil property is independent of test type and can be used to characterize soils (e.g. shear wave velocity). A soil parameter is dependent on test type (e.g. undrained strength) but is useful for design purposes, particularly when correlated with field performance of full-scale works.

Milestones in research: the past 30 years

Stress path dependency of soil moduli

In the UK during the 1960s and 1970s the major discovery was that soil, unlike other engineering material, has properties that are dependent on loading pattern or 'stress path'. Deformation moduli are not unique material constants but vary according to the actual stress path. This understanding led Bishop and Wesley (1975) to develop a new kind of hydraulic triaxial apparatus in which the stress paths encountered in practice could be followed. Complex loadings could be applied that simulated real geological, construction and in-service conditions where vertical and horizontal stresses both change at the same time. As shown by Simons (1971) in Table 4.1 for a circular foundation in London Clay at Bradwell, conventional methods overestimated the stress path method of settlement calculation by 240%!

Table 4.1 Example of settlement calculations for a circular foundation in London Clay at Bradwell, after Simons (1971)

Settlement type	Settlement: mm	
	Conventional method	Stress path method
A = Immediate	32·7	17·2
B = Consolidation	109·0	40·9
Total settlement = A + B	141·7	58·1

Table 4.2 Measurements of soil stiffness, after Jardine et al. *(1984)*

Soil type	Undrained shear strength: kPa	Ratio of moduli, $E_{u(0\cdot1\%)}/E_{u(0\cdot01\%)}$	Error: %
North Sea clay	122	0·185	540
Ham River sand	1085	0·518	193
London Clay	123	0·371	270
Upper Chalk	1350	0·723	138

Small strain stiffness of soil controlling ground movements

During the 1980s researchers and consultants were puzzled by the huge difference between ground stiffness measured in the lab by triaxial tests and that back-analysed from field observations using finite element modelling of real structures like foundations, retaining walls and tunnels. Jardine *et al.* (1984) realized that conventional methods of strain measurement in the triaxial test included bedding errors that exaggerated the amount of measured deformation giving artificially low stiffness values. They used local strain transducers inside the triaxial cell to measure soil deformations directly on the test specimen. As shown in Table 4.2, these were not small errors of a few per cent – some over 500% were measured!

Seismic means for measuring operational ground stiffness

During the late 1980s and early 1990s, soil stiffness was being measured in the laboratory using small strain dynamic resonant column apparatus. Investigators were struck by the similarity of these *dynamic* moduli to *static* moduli back-analysed from movements around real static structures

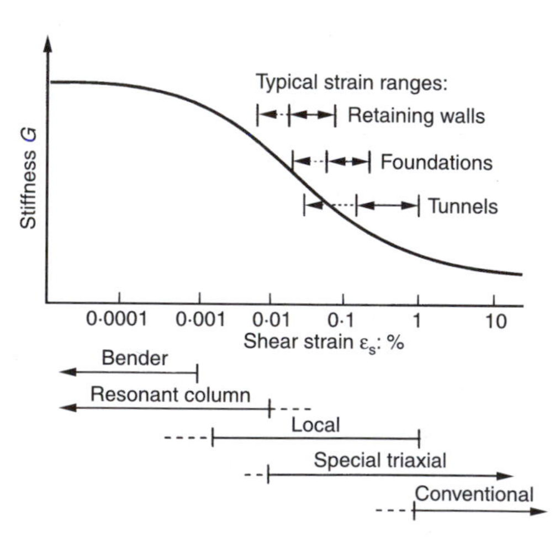

Fig. 4.1 Typical variation of stiffness with strain for most soils

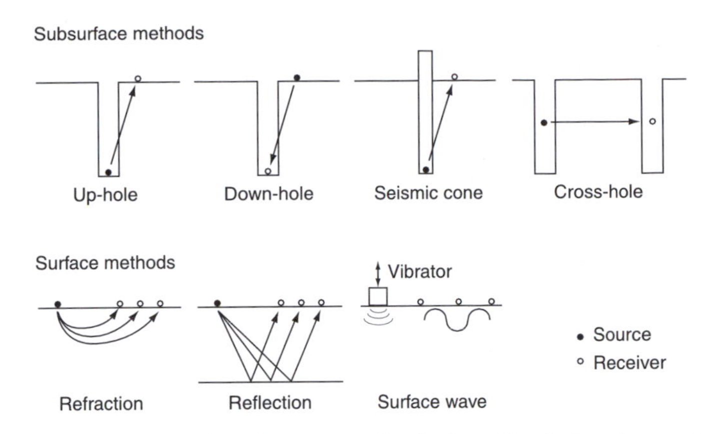

Fig. 4.2 Subsurface and surface geophysical methods for determining stiffness

like retaining walls and excavations. They then realized the differences in moduli measured in the past between static tests like the conventional triaxial, on the one hand, and dynamic tests like the resonant column, on the other, were related to strain level, i.e. one test measured small strain moduli and the other large strain moduli, not that one test was 'dynamic' and the other 'static' (see Fig. 4.1).

Somewhat unexpectedly the seismic-like resonant column dynamic test measured stiffnesses close to the field static operational values – but this is because they both present small strain behaviour. This encouraged researchers to look again at seismic methods for measuring soil and rock stiffnesses in situ. This has given rise to the commercial development of a range of seismic tests in the field including the seismic cone penetration test (SCPT), cross-hole and down-hole shear wave velocity measurement, and the surface wave (Rayleigh wave) methods of SASW (Spectral Analysis of Surface Waves), which uses a hammer as the seismic source, and CSW (Continuous Surface Wave), which uses a frequency-controlled vibrator as the seismic source (Fig. 4.2).

Introduction to key laboratory and field methods

For a comprehensive review of test equipment and methods for parameter determination in the laboratory and field, we refer the reader to the book *Site investigation* by Clayton *et al.* (1995b). In contrast, and in order to provide the necessary focus required by this short book, we set ourselves the task of selecting and describing only three laboratory and six field methods for parameter determination. We chose those test methods that we believe are the most useful because they have some or all of the following attributes:

- is historically important to the establishment of geotechnical knowledge and hence giving connections to well established precedent and empirical correlation (e.g. the shear vane test)
- is widely used in the present day (e.g. cone penetration test, consolidation test)
- affords the maximum amount of data for the minimum cost
- benefits from value added by up-to-date technology (e.g. quick undrained triaxial compression test plus bender elements testing for small strain stiffness evaluation)
- represents probably the best way to determine design parameters in many cases (e.g. seismic methods for stiffness–depth profiling).

For the laboratory tests we chose the triaxial, consolidation and ring shear tests. For the field tests we chose the shear vane, plate loading test, the standard penetration test, the cone penetration test, and the pressuremeter test as well as describing the seismic methods that we believe are the future of site characterization. Below we explain why we made these selections. There are, of course, other tests and test apparatus that can also play an important role, particularly in specialized site investigations. Some of these are mentioned in the case studies given elsewhere in this book.

Laboratory tests

Review of laboratory test types

There are many test types widely used in commercial and research laboratories. Most of these are summarized in our Short Course Notes: Laboratory Test Types that follow.

Sampling disturbance

Sources

This section is based on the following sources.

- Hight, D.W. (1996). Moderator's report on Session 3: drilling, boring, sampling and description. *Advances in Site Investigation Practice*. Thomas Telford, London.
- Clayton, C.R.I., Siddique, A. and Hopper, R.J. (1998). Effects of sampler design on tube sampling disturbance – numerical and analytical investigations. *Géotechnique*, **48**(6), 847–867.
- Clayton, C.R.I. and Siddique, A. (1999). Tube sampling disturbance – forgotten truths and new perspectives. *Proc. Inst. Civ. Eng. Geotech. Eng.*, **137**, July, 127–135.
- Hight, D.W. (2000). Sampling methods: evaluation of disturbance and new practical techniques for high quality sampling in soils. Keynote Lecture, *Proc. 7th Nat. Cong. of the Portuguese Geotech. Soc.*, Porto.

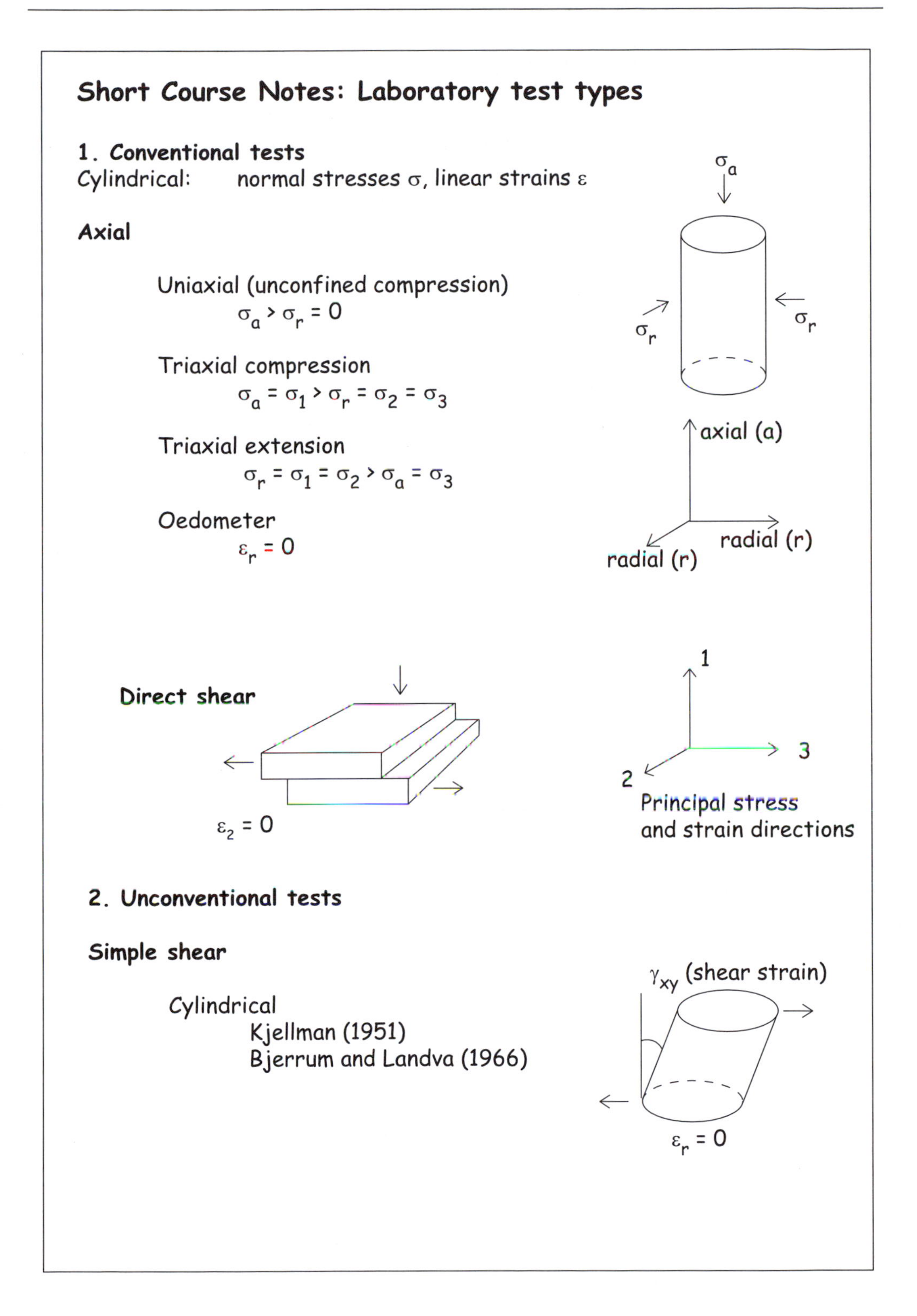

Short Course Notes: Laboratory test types
1. Conventional tests
Cylindrical: normal stresses σ, linear strains ε
Axial
Uniaxial (unconfined compression)
$\sigma_a > \sigma_r = 0$
Triaxial compression
$\sigma_a = \sigma_1 > \sigma_r = \sigma_2 = \sigma_3$
Triaxial extension
$\sigma_r = \sigma_1 = \sigma_2 > \sigma_a = \sigma_3$
Oedometer
$\varepsilon_r = 0$
σ_a
σ_r
σ_r
axial (a)
radial (r)
radial (r)
Direct shear
$\varepsilon_2 = 0$
1
3
2
Principal stress and strain directions
2. Unconventional tests
Simple shear
Cylindrical
Kjellman (1951)
Bjerrum and Landva (1966)
γ_{xy} (shear strain)
$\varepsilon_r = 0$

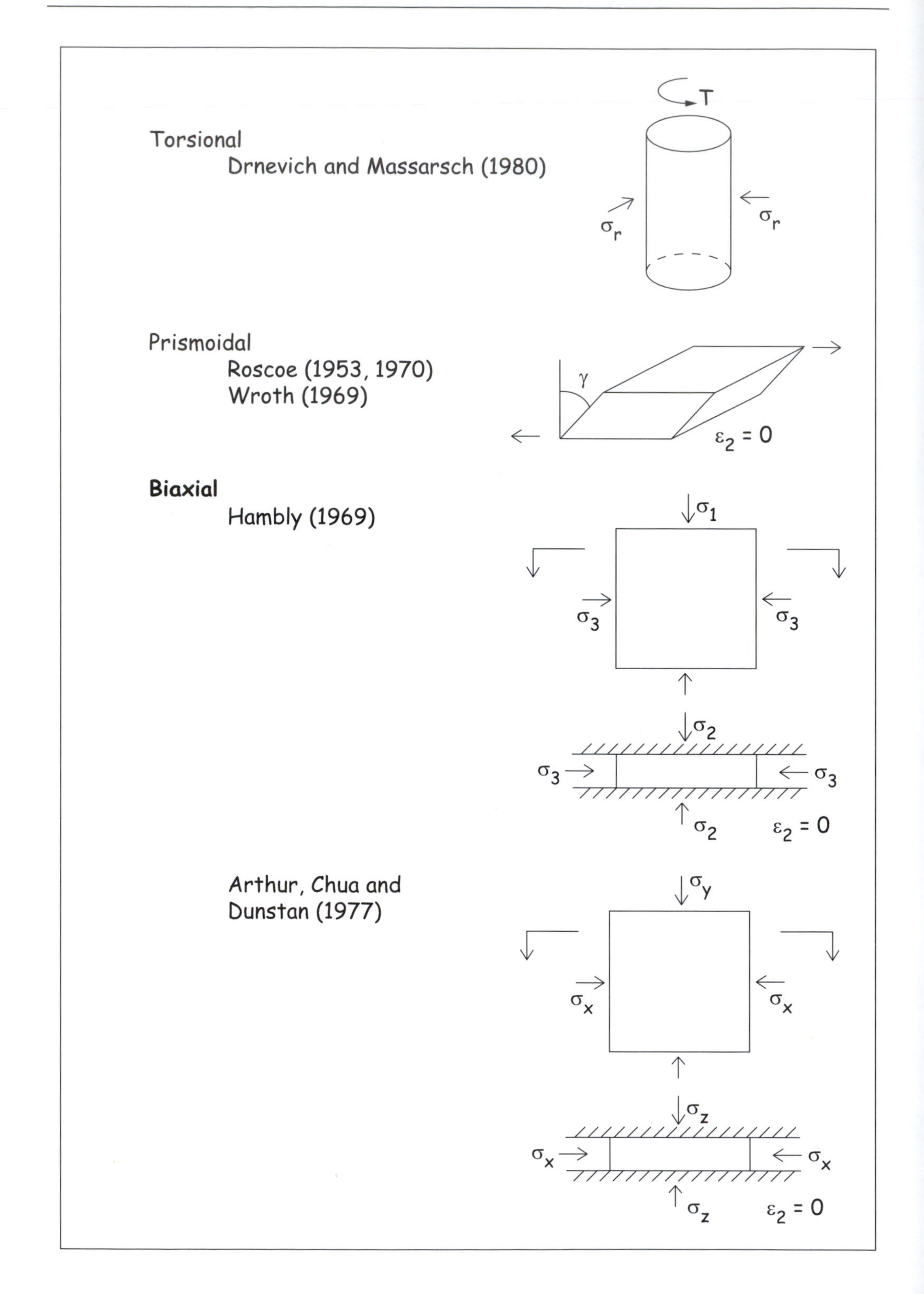
Torsional
Drnevich and Massarsch (1980)
T
σr
σr
Prismoidal
Roscoe (1953, 1970)
Wroth (1969)
γ
ε2 = 0
Biaxial
Hambly (1969)
σ1
σ3
σ3
σ2
σ3
σ3
σ2
ε2 = 0
Arthur, Chua and
Dunstan (1977)
σy
σx
σx
σz
σx
σx
σz
ε2 = 0

True triaxial

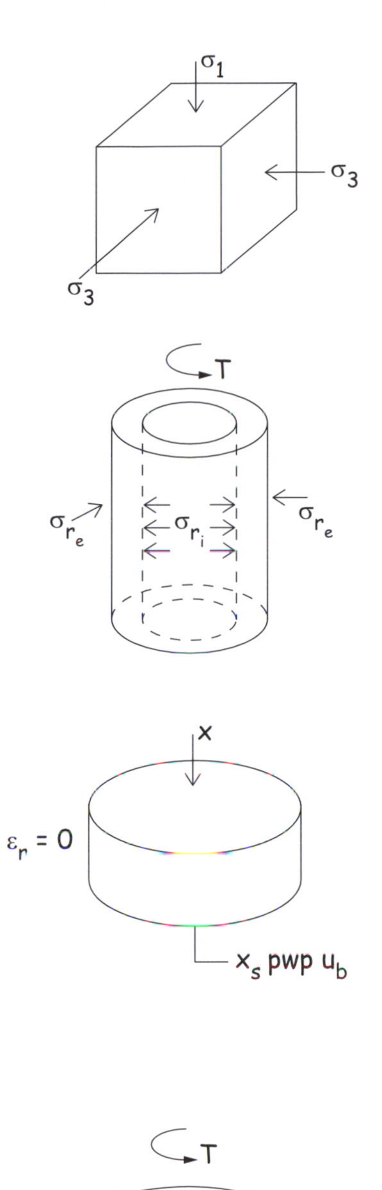

Cuboidal

Scott and Ko (1969)
Shibata and Kanube (1965)
Pearce (1971)
Arthur and Menzies (1972)
Green (1971)
Lade and Duncan (1973)
Mitchell (1973)

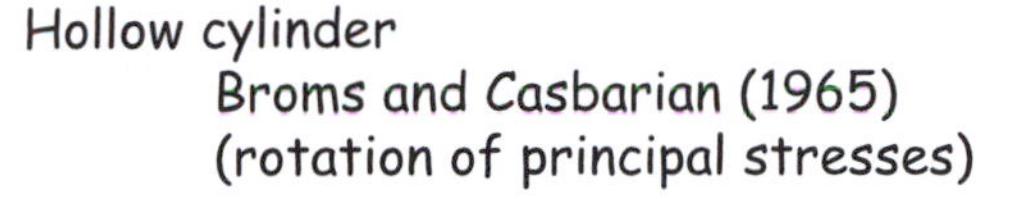

Hollow cylinder

Broms and Casbarian (1965)
(rotation of principal stresses)

Oedometer
(Consolidation test)

$\varepsilon_r = 0$

Smith and Harvey (1969)
Wissa, Christian and Davis (1971)

u_b = const

Lowe, Jonas and Obrician (1969)

σ_a, fluid pressure, u_b applied/measured

Rowe and Barden (1966)
Bishop, Green and Skinner (1973)

Ring shear

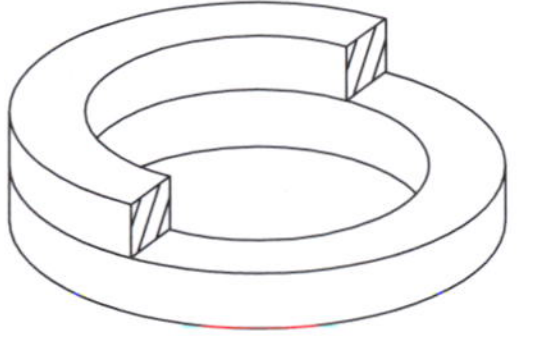

Bishop, Green, Garga,
Andresen and Brown (1971)

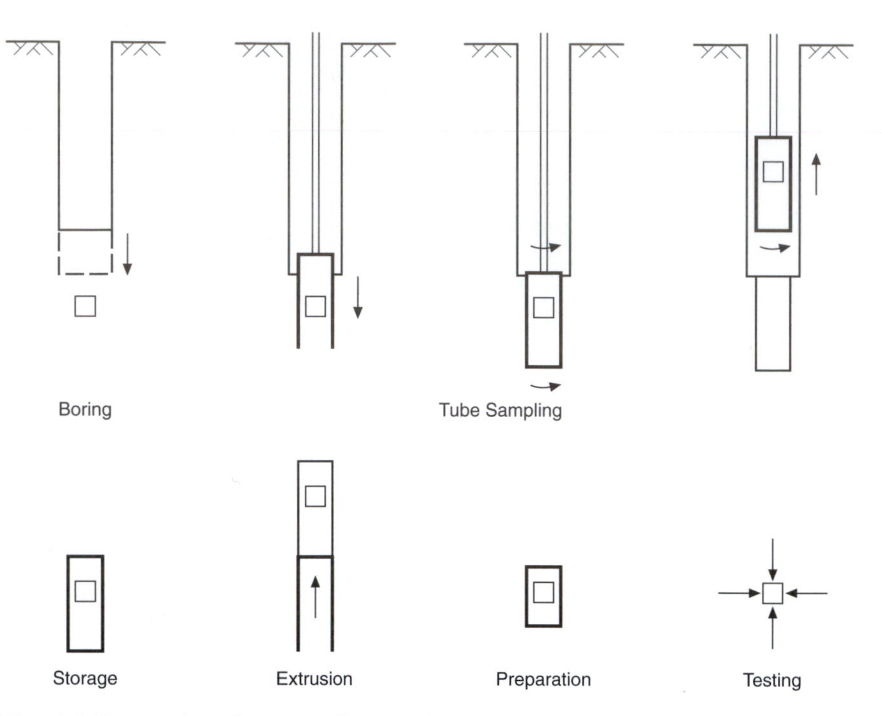

Fig. 4.3 Stages in tube sampling and preparing soil specimen for laboratory test (Hight, 2000)

- Leroueil, S. (2001). Some fundamental aspects of soft clay behaviour and practical implications. *Proc. 3rd Int. Conf. On Soft Soil Engineering, Hong Kong.* Lee *et al.* (eds). Balkema, Rotterdam, pp. 37–53.

We gratefully acknowledge permissions to make verbatim extracts from these sources from Dr D.W. Hight, Prof. C.R.I. Clayton and Prof. S. Leroueil.

Causes of sampling disturbance

Hight (2000) has pointed out that the most common method of obtaining a sample is by forming a borehole and pushing or driving a tube into the ground, involving the stages illustrated in Fig. 4.3. Disturbances can occur at each of the stages shown in Fig. 4.3, but we will concentrate on those caused by penetration of the sampling tube.

Sampling effects in soft clays

Hight (2000) gives typical results of a conventional soft clay site investigation, involving sampling and laboratory testing (Fig. 4.4). These are from unconsolidated undrained (UU) triaxial compression (TC) tests on samples of Singapore marine clay, the majority of which were taken with a thick-walled open drive sampler. There is a large scatter in the

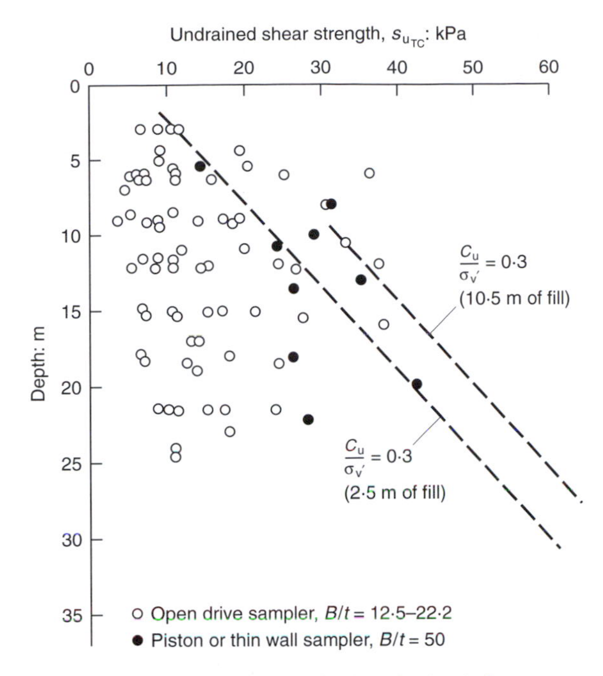

Fig. 4.4 Results of unconsolidated undrained triaxial compression tests on thick-walled and thin-walled tube samples of Singapore marine clay (Hight, 2000)

data, most of which falls below the best estimate of in situ shear strength in compression. Samples taken with a thin-walled piston sampler lead to higher strengths. The effect of sampler type is illustrated graphically in Fig. 4.5, which compares the results of unconfined compression (UC) tests on samples of Ariake clay, taken from a depth of 10 m, with:

- a Sherbrooke down-hole block sampler (Lefebvre and Poulin, 1979), which carves blocks approximately 250 mm in diameter and 350 mm high
- a Laval sampling system (La Rochelle *et al.*, 1981), which uses a 660 mm long, 200 mm diameter piston-less tube, with no inside clearance, a 5° cutting edge and 4 mm wall thickness
- a Japanese standard piston sampler, which uses a 1 m long, 75 mm internal diameter tube, with no inside clearance, a 6° cutting edge and 1·5 mm wall thickness
- a Norwegian 54 mm diameter fixed piston sampler (NG154, Andersen and Kolstad, 1979), which has a 54 mm internal diameter tube, 768 mm long, and up to 13 mm wall thickness
- the so-called ELE fixed piston sampler, which has a 101 mm internal diameter tube, only 500 mm long, with no inside clearance, a 30° cutting edge, and 1·7 mm wall thickness

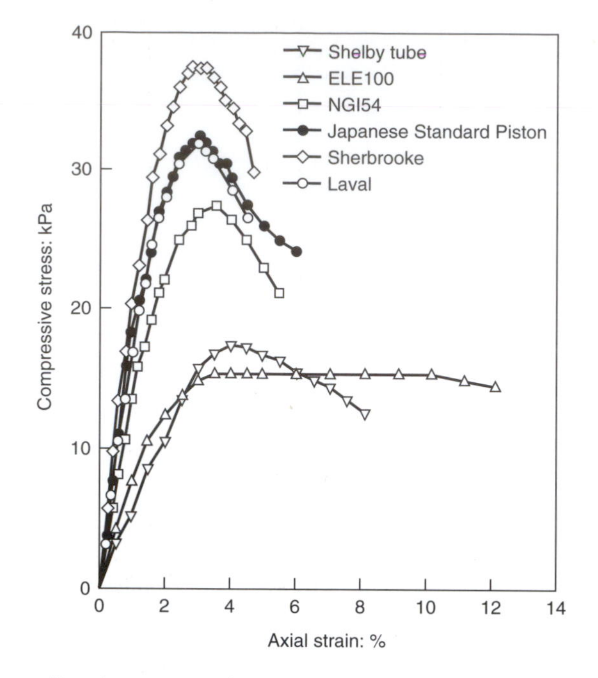

Fig. 4.5 Unconfined compression tests on Ariake clay (data from Tanaka and Tanaka, 1999), after Hight (2000)

- a piston-less Shelby tube, 72 mm internal diameter, 1·65 mm wall thickness and 610 mm long.

The advantage of both UC and UU tests is that they show the full imprint of sampling effects, and the UC data in Fig. 4.4 confirm that block samples taken by rotary methods can be of higher quality than even the best tube samples.

Strain path method applied to tube sampling

Hight (2000) points out that the most significant advance in the past two decades in understanding disturbance by tube sampling has been the introduction of the strain path method by Professor Baligh and his co-workers (Baligh, 1985; Baligh *et al.*, 1987) and its application to the deep penetration of a sampling tube into the ground. Clayton *et al.* (1998) extended the work of Baligh *et al.* (1987) to examine the effects of different details of sampling tube geometry. Clayton and Siddique (1999) considered four types of tube sampler, which have been widely used in the UK, plus a fifth experimental sampler. The dimensions in Fig. 4.6 were obtained by measuring actual cutting shoes. The samplers used were as follows.

- *Sampler 1.* A simple cutting shoe used in conjunction with the conventional U100 (described in BS 5930 as the 'general purpose sampler'

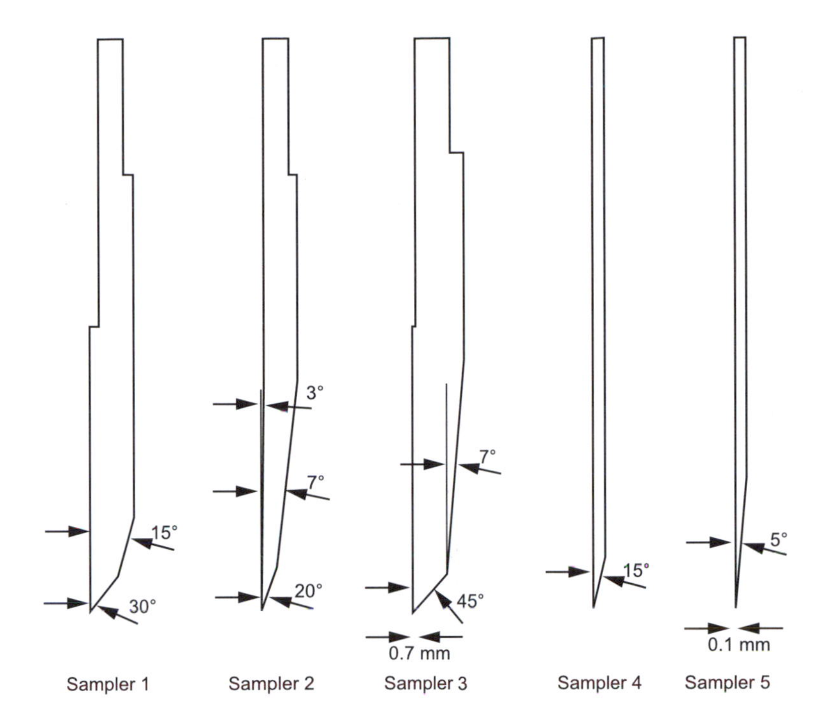

Fig. 4.6 Cutting shoe geometries analysed using the strain path method (Clayton and Siddique, 1999)

(British Standards Institution, 1999)). The shoe is relatively blunt by ISSMFE standards, and has an area ratio of 27%. Inside clearance is obtained by stepping out from the cutting shoe to the screw-on metal sampler tube above it. This sampler is hammered into the ground.

- *Sampler 2.* A more sophisticated version of Sampler 1. Although superficially similar, and having an almost identical area ratio, this sampler has an inside taper angle which removes the sudden step between the cutting shoe and sampler tube that exists in Sampler 1. In addition, the external cutting-edge taper angles are much smaller than for Sampler 1.
- *Sampler 3.* A version of the U100 cutting shoe which is used in the UK in combination with plastic liners. The area ratio is much higher, at nearly 48%. A small stepped inside clearance exists between the shoe and the liner.
- *Sampler 4.* The thin-walled sampler tube described by Harrison (1991), variations of which have been widely used in higher quality ground investigations in stiff clays, in order to provide samples for small-strain stiffness and stress path tests. The sampler has a 15° cutting-edge taper, and is pushed (rather than hammered) into the ground.

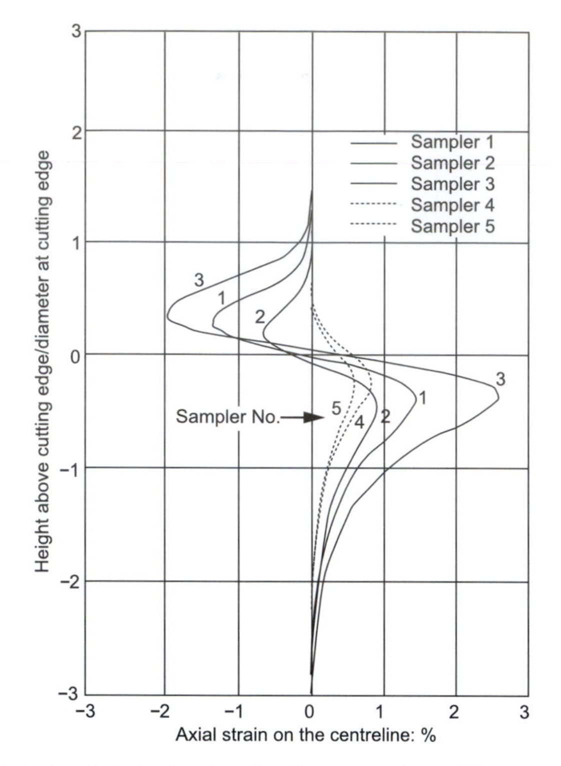

Fig. 4.7 Predicted axial strains for the five samplers (Clayton and Siddique, 1999)

- *Sampler 5.* This has been included for comparison. It is an experimental sampler used by Hight (2002) to sample the soft clay at Bothkennar in Scotland. As far as is known, it has not been used in stiffer clays, but there appears little reason why it should not be so used in the future. This tube has a sharper cutting edge than Sampler 4, but for practical reasons has a 0·1 mm wide flat at its extremity. Like Sampler 4, it is pushed into the ground.

The predicted axial strains on the centreline are shown in Fig. 4.7. Over the past decade much testing has been carried out in the UK using local strain stiffness measurement. Research tests carried out on very high quality samples have revealed that natural (i.e. bonded, rather than disturbed or remoulded) clays fail at strains of the order of 0·75–2% (Schjetne, 1971; Heymann, 1998). By comparison with the figures already cited, it can be seen that the strains imposed by Samplers 1 and 3 (which have peaks in compression of 1·5% and 2·6% respectively) will certainly cause very significant damage to a cohesive soil. Sampler 2 performs considerably better (peak axial compressive strain 0·93%) because of its superior cutting-edge design, but is still no match for

Sampler 4, the pushed thin-wall tube described by Harrison (peak axial strain 0·85%) which also has the advantage of being pushed rather than driven. The analysis of Sampler 5 suggests, however, that Sampler 4 could be further improved by reducing its cutting-edge taper angle. Sampler 5, with a peak axial compressive strain of only 0·61%, could deliver significantly higher quality tube samples, provided the other factors leading to disturbance are controlled, and the device proved practical for use in firm to stiff clays.

Strains at the periphery of a sample

Hight (2000) points out that the effects of strains at the periphery when sampling clay are to create a zone of destructured or remoulded soil in which excess pore pressures are generated which are positive in normally-consolidated or lightly over-consolidated clays. The distribution of pore pressures in the sample immediately after sampling is illustrated in Fig. 4.8(a). Shear-induced pore pressures are positive in both the central and peripheral zones, being higher in the peripheral zone.

With time, there is pore pressure equilibration and water content redistribution. In the soft clay, the outer zone consolidates and the water content reduces; the central part swells and the water content increases. Overall, the effects of the peripheral strains superimposed on those of the centreline strains are to reduce further the effective stress in the soft clay. Figure 4.8(b) shows measured water contents across the diameter of tube samples of soft clay which are higher in the centre. In summary, the predicted effects of the combined tube sampling strains on samples of soft clay are:

- a reduction in mean effective stress
- damage to structure, manifest as shrinking of the soil's bounding surface, and
- an increase in water content in the centre of the sample.

Sampling effects in stiff clays

Hight (2000) gives the results of a recent investigation involving the sampling and laboratory testing of London Clay (Fig. 4.9(a)). London Clay is a stiff to hard plastic over-consolidated and fissured clay. The samples were taken by either driving in thick-walled sampling tubes (U100) or pushing in thin-walled Shelby tubes and were tested in UU triaxial compression. As with conventional soft clay investigations, the scatter, particularly at depth, is typical, and is actually greater than that shown by standard penetration test (SPT) data (Fig. 4.9(b)).

As with soft clays, the effects of tube sampling in over-consolidated clays are to:

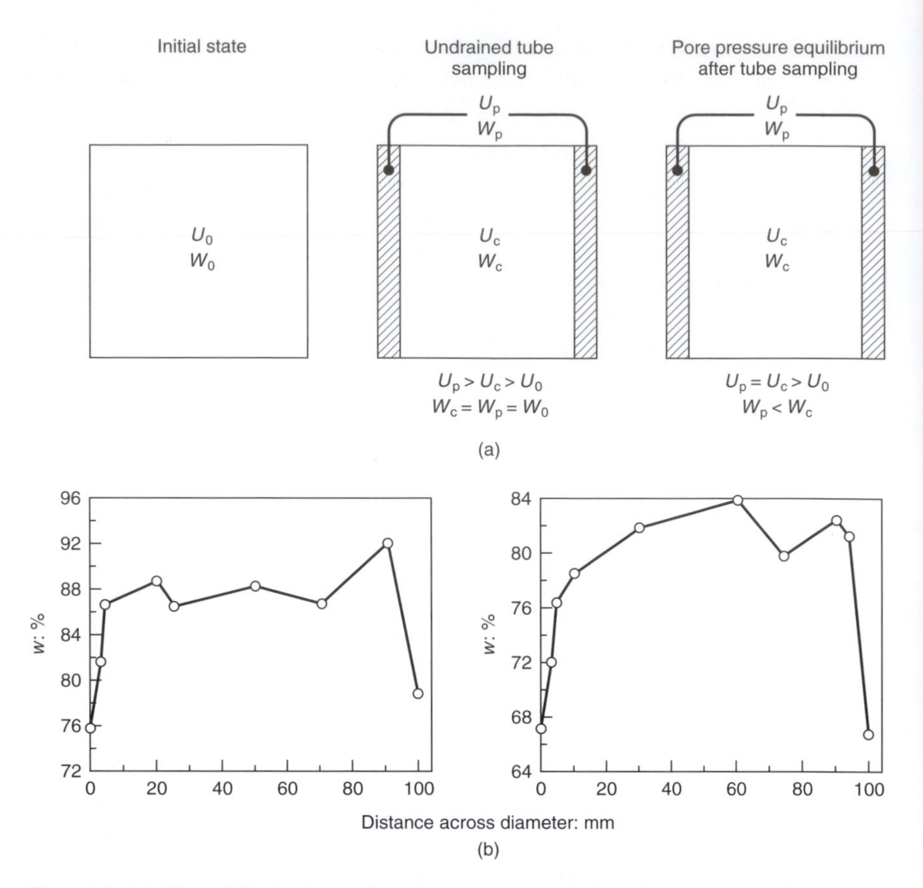

Fig. 4.8 (a) Simplified view of pore pressure and water content changes after tube sampling of normally consolidated or lightly over-consolidated clay (Hight, 2000). (b) Measured water content distributions across diameter of tube samples of soft clay (Bjerrum, 1972)

- change the mean effective stress in the sample
- cause water content redistribution
- damage the structure, i.e. the effects of ageing, cementing, etc.
- modify the fabric.

There are important differences, however, as illustrated in Figs 4.10 and 4.11. The effects of centreline tube sampling strains in London Clay are to produce a small increase in pore pressure (Georgiannou and Hight, 1994). The effects of the intense shearing around the periphery of the sample are to produce negative excess pore pressures in this zone (Vaughan *et al.*, 1993). As these pore pressures equilibrate, there is water content redistribution with water being drawn into the outer peripheral zone from the central zone (Fig. 4.10(a)). Evidence to support this is presented in Fig. 4.10(b).

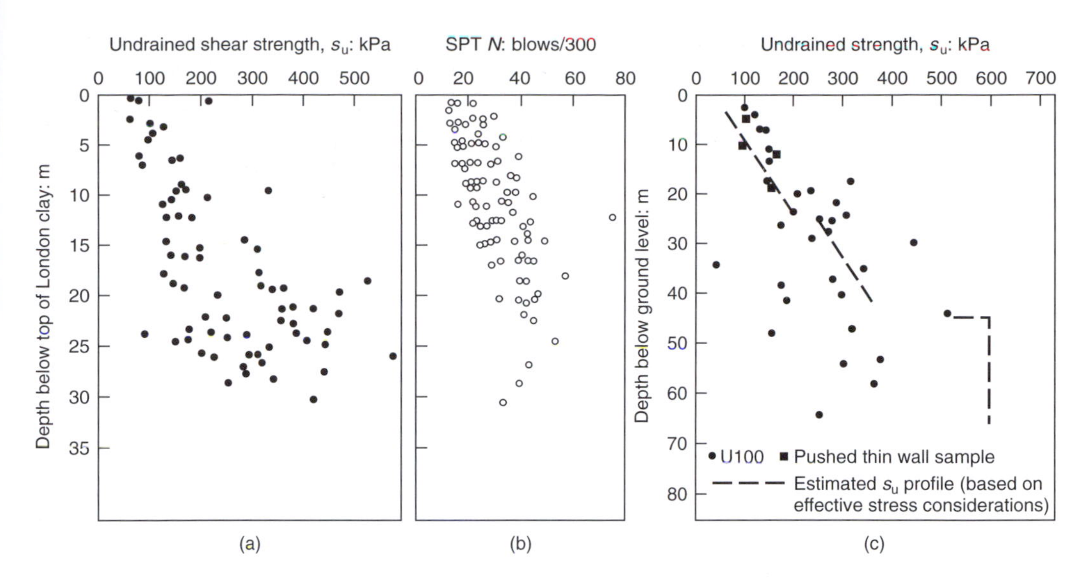

Fig. 4.9 Results of a conventional site investigation in London Clay. (a) s_u from UU triaxial compression tests on 100 mm diameter U100 samples. (b) SPT N values. (c) Best estimate of strength profile (from Hight and Jardine, 1993)

Surprisingly, strengths measured in UUTC (unconsolidated-undrained triaxial compression) tests on thick-walled and thin-walled tube samples are sometimes similar (Fig. 4.11(b)), despite the differences in initial effective stress. This suggests a compensating effect of different levels of damage in the two types of sample.

The damage to structure can be seen as a lowering of the failure envelope (shrinking of the boundary surface) as shown in Fig. 4.12. The damage is progressive and the amount by which the failure envelope is lowered increases with increasing levels of disturbance, i.e. from blocks to rotary cores, to thin-walled tubes, to thick-walled tubes. At the same time, there is a change in mean effective stress in the samples: an increase in tube samples, which is higher in thick-walled than thin-walled tubes; a reduction in rotary cores. The net effect, because of the particular effective stress paths followed by these materials in undrained shear, is that similar strengths may be measured in UUTC tests on these different sample types, despite their different levels of damage.

Hight and Jardine (1993), combining data from UU, CIU (consolidated isotropically-undrained) and CAU (consolidated anisotropically-undrained) tests on rotary cored samples of London Clay from different depths, have shown that the clays, in fact, exhibit a family of local bounding surfaces/failure envelopes which expand with depth, reflecting increasing lithification and cementing (Fig. 4.12). From this they are able

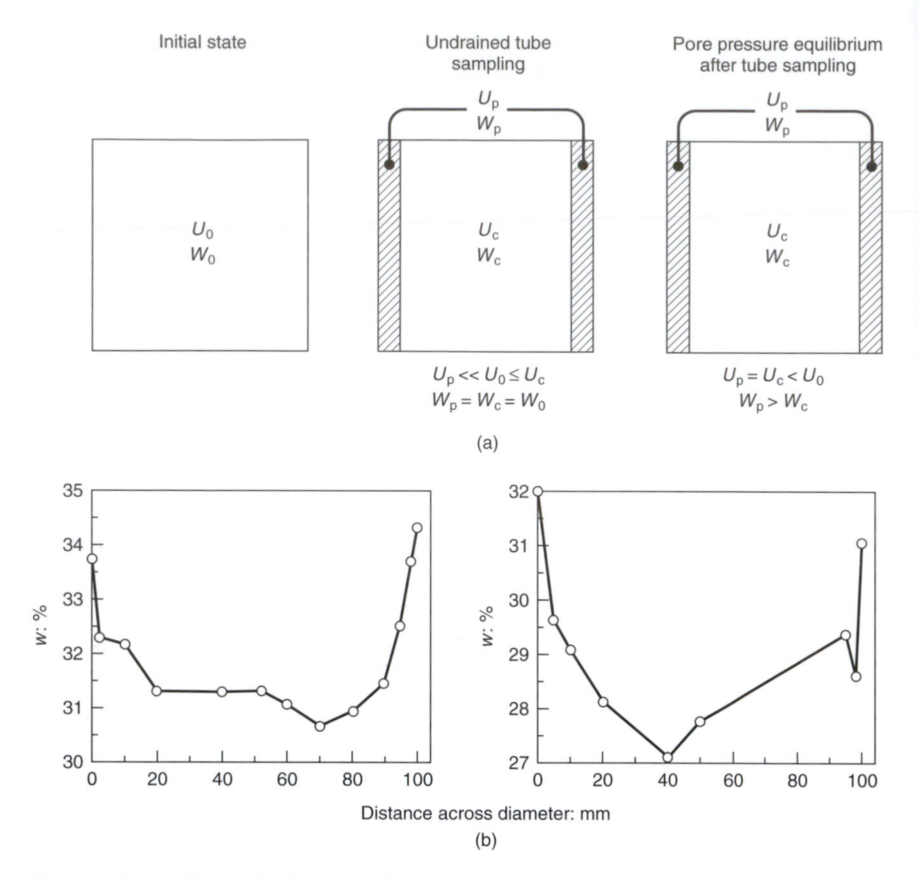

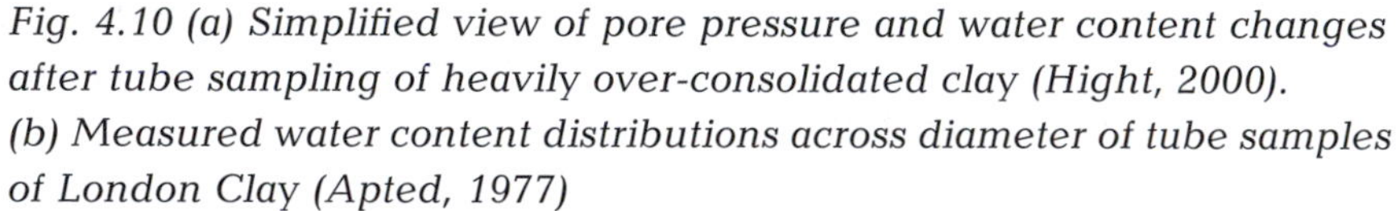
Fig. 4.10 (a) Simplified view of pore pressure and water content changes after tube sampling of heavily over-consolidated clay (Hight, 2000).
(b) Measured water content distributions across diameter of tube samples of London Clay (Apted, 1977)

to estimate the in situ strength in triaxial compression. Such an estimate has been superimposed on a typical UUTC profile in Fig. 4.9(c).

Sampling in sands

Hight (2000) points out that tube sampling in sands will almost certainly be drained, so that volumetric strains as well as shear strains will occur. It is highly probable that the magnitude of the shear strains will be sufficient to destructure the sand, particularly as yield strains in granular materials are so low. The volumetric strains caused by shear will depend on the initial density of the sand being sampled. Thus, initially loose sand will contract and increase in density, while initially dense sand will dilate and reduce in density. The increase in density of initially loose sand is unconservative and potentially dangerous, since it is in these materials that the problems of collapse and flow can occur.

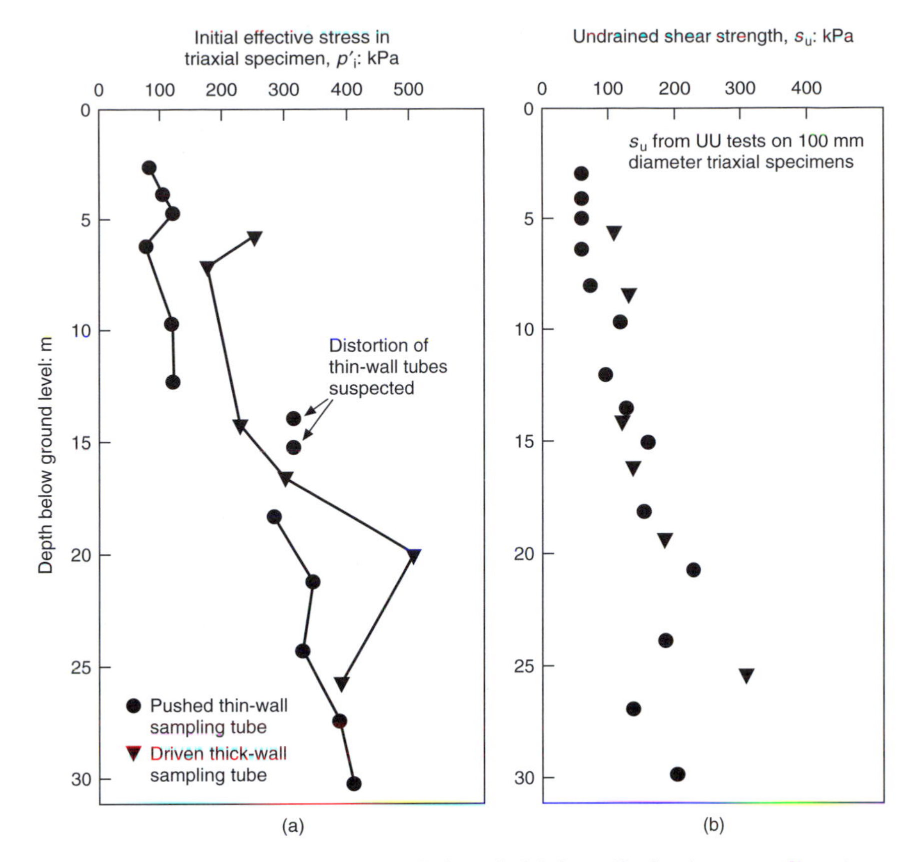

Fig. 4.11 Comparison of thin-walled and thick-walled tube sampling in London Clay. (a) Initial effective stress, p_i'. *(b) Undrained shear strength* s_u *(from Hight, 1996)*

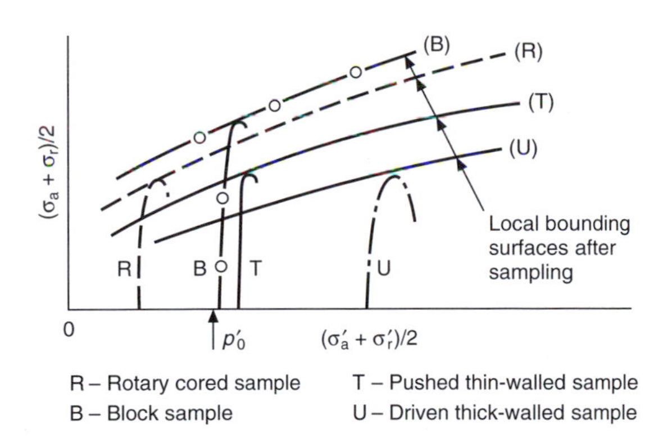

Fig. 4.12 Effects of different types of sampling in stiff plastic clay (Hight, 2000)

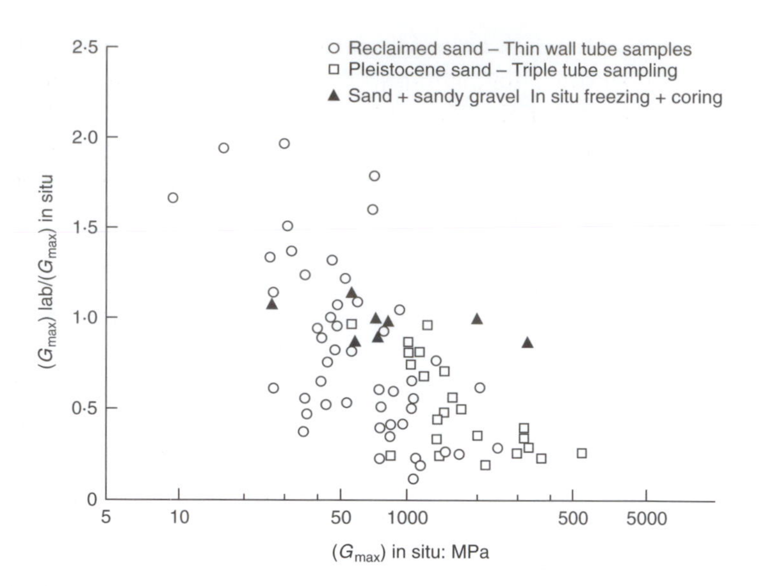

Fig. 4.13 Assessing sample disturbance through comparisons of G_{max} *from in situ and laboratory measurements in sand (Hight, 2000)*

It would appear that truly undisturbed samples of sand can only be obtained if the in situ sand structure is fixed in some way before sampling. The most common method of doing this is by freezing the ground in situ before carving out large-diameter samples by rotary coring (Yoshimi *et al.*, 1978, 1994; Davila *et al.*, 1992).

The effects of sampling in sands and the benefits of coring after in situ freezing are summarized in Fig. 4.13. This graph presents a comparison of field and laboratory measurements of G_{max} (G_0, the modulus at zero strain, is equivalent to G_{max}, the maximum shear modulus) in two types of sand sampled by conventional coring and by coring after in situ freezing. The data are presented as the ratio of G_{max} measured in the laboratory to G_{max} measured in situ versus G_{max} measured in situ. It can be seen that coring after freezing leads to $(G_{max})_{lab}/(G_{max})_{in\,situ} \approx 1$ for a wide range of $(G_{max})_{in\,situ}$ values, suggesting that damage and changes in density have been avoided.

Assessment of sample quality

Hight (2000) reviews various methods for assessing sample quality, i.e. the level of sample disturbance, as discussed below.

Fabric inspection

Although of major importance for assessing some of the potential effects of sampling, in particular the risk of there having been water content redistribution between adjacent sand and clay layers, fabric inspection is not sufficient to determine the likely level of disturbance. Only gross distortion, for example in the distorted peripheral zone, can be seen, whereas only relatively small strains cause yield and damage to a bonded structure. In addition, strain histories in the central zone of the sample involve unloading so that the maximum imposed strains cannot be deduced.

Measurement of initial effective stress, p_i'

Measurement of initial effective stress, p_i', in samples taken from the ground and set up in the laboratory, has been advocated for many years, e.g. Ladd and Lambe (1963). Comparison of p_i' with the value of p' after perfect sampling, p_{ps}', provides an indicator of level of disturbance. Comparing mean effective stress in situ and after sampling of Bothkennar clay (Fig. 4.14) shows how measurements of p_i' indicate a difference in quality between Laval and conventional UK piston samples. However, the measurement of p_i' alone is not sufficient, as it cannot indicate the amount of destructuring that has occurred.

Measurement of strains during reconsolidation

In reconsolidating samples to in situ stresses, the strains will depend on both the reduction in effective stress that has occurred as a result of

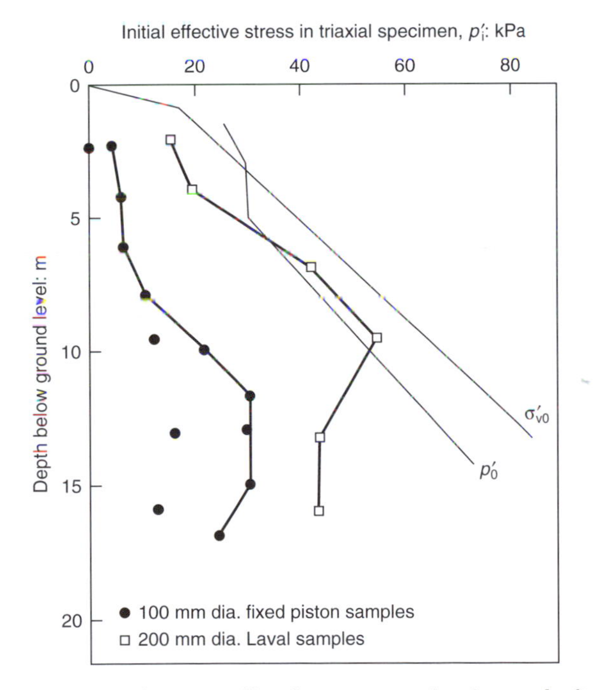

Fig. 4.14 Comparison of mean effective stresses in situ and after sampling of Bothkennar clay

Table 4.3 Proposed criteria to evaluate sample disturbance (Lunne et al., *1997)*

Over-consolidation ratio	Ratio of voids ratio change to original voids ratio, $\Delta e/e_0$			
	Very good to excellent [a]	Good to fair [a]	Poor [a]	Very poor [a]
1–2	<0·04	0·04–0·07	0·07–0·14	>0·14
2–4	<0·03	0·03–0·05	0·05–0·10	>0·10

[a] The description refers to samples used for measurement of mechanical properties.

sampling and the amount of destructuring. The absolute value of the strains will depend on the reconsolidation path followed and the soil compressibility. To take account of the latter, Lunne *et al.* (1997) have proposed expressing the volume strains in terms of $\Delta e/e_0$, where Δe is the change in void ratio and e_0 the initial void ratio. They proposed criteria for evaluating sample disturbance as shown in Table 4.3.

Comparison of field and laboratory measurements of shear wave velocity/dynamic shear modulus

The relationships illustrated in Fig. 4.15 set out the basis for adopting comparisons of shear wave velocity V_s or small-strain shear modulus G_{max} in the laboratory and field as a measure of mechanical disturbance for the case of sands. In sand, V_s depends on stress state and void ratio (Fig. 4.15(a)), degree of cementing (Fig. 4.15(b)), age (Fig. 4.15(c)) and contact distribution. Figure 4.15(c) illustrates how a disturbance, in this case a twist, such as might be applied during rotary coring, changes V_s. The effects of ageing are removed, and, since the void ratio actually reduces, the reduction in V_s must be related to a change in contact distribution. A similar argument applies to clays in which the development of structure at constant stress is associated with an increase in V_s while destructuring during sampling results in a reduction in V_s.

For the comparisons to be valid, the laboratory samples must be representative, in terms of fabric, discontinuities, etc. They must be restored to their in situ stress state, because of the dependence of V_s on stress state. Because of the dependence of V_s on void ratio, allowances must be made for changes in void ratio during reconsolidation to in situ stresses. Measurements of V_s should also be made with shear wave propagation in the same direction as in the field, with the same plane of polarization and at a similar frequency.

An example of a comparison between in situ and laboratory shear wave velocity measurements in Bothkennar clay is presented in Fig. 4.16. In situ measurements of V_s were made using the seismic cone. Laboratory

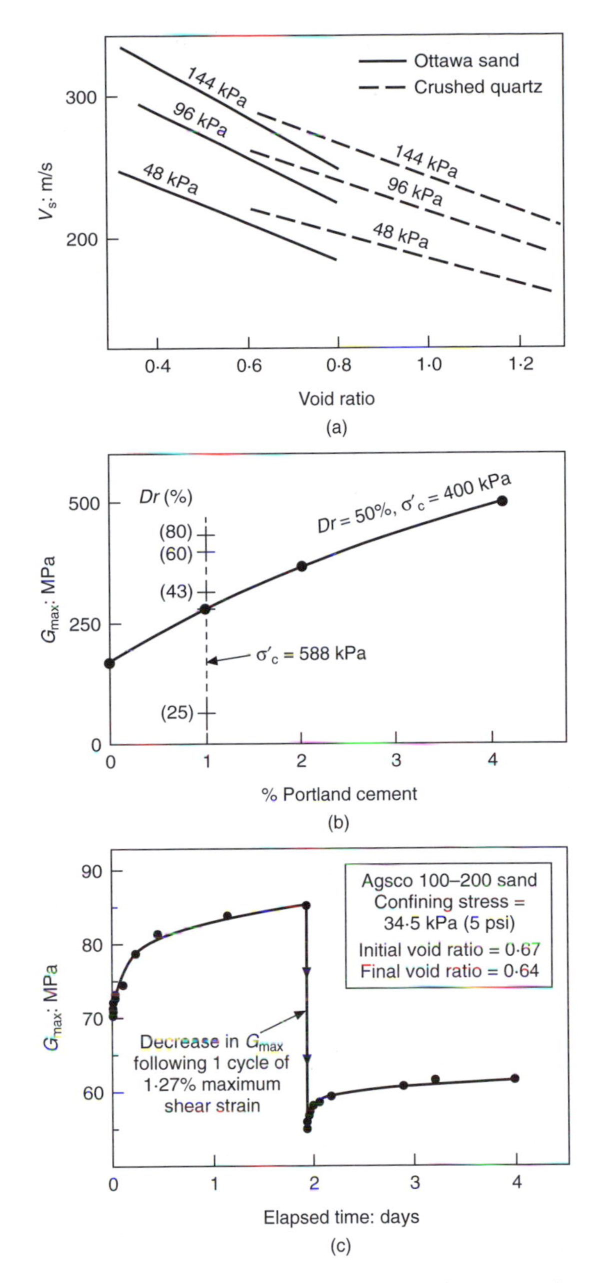

Fig. 4.15 Background to the use of V_s *(or* G_{max}*) for assessing sample disturbance in sand. (a) Dependency of* V_s *on void ratio and stress level. (b) Dependency of* G_{max} *on degree of cementing. (c) Sensitivity of* G_{max} *to age and disturbance (Hight, 2000)*

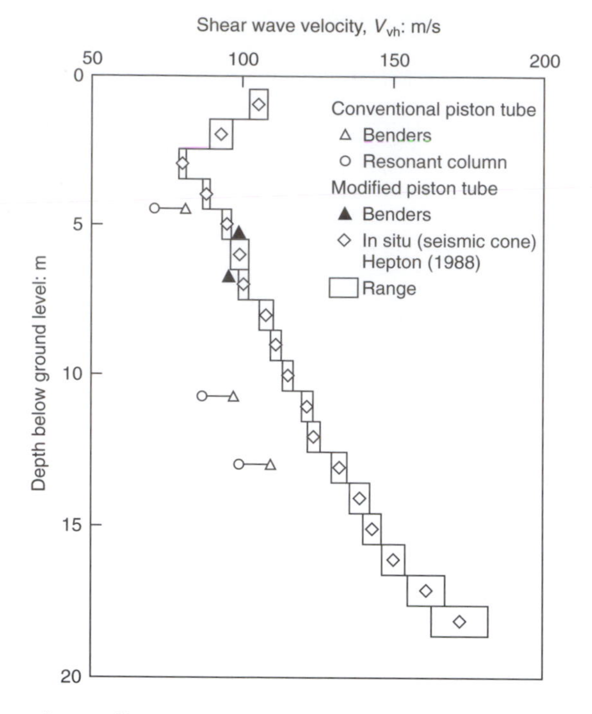

Fig. 4.16 Sample quality assessment based on a comparison of laboratory and in situ shear wave velocities for Bothkennar clay (Hight, 2000)

measurements were made using bender elements, or the resonant column apparatus, on piston samples taken with either the conventional tube (30° cutting edge) or the modified tube (5° cutting edge) and then reconsolidated to in situ stresses. The improved performance of the modified tube compared with the conventional tube, which has been described above, is confirmed by the shear wave velocity measurements. There is reasonable agreement between the in situ values and those measured on samples taken with the modified tube; values measured on samples taken with the conventional tube are lower because of structural damage.

Use of high-quality samples to provide design parameters

Hight (2000) warns that there are dangers in adopting these methods when taking high-quality samples of soft clay, stiff clay and sands. There are inherent dangers in introducing these methods if they are to be used in conjunction with empirical or semi-empirical design procedures which have been developed on the basis of poorer quality samples and which ignore anisotropy of soil properties.

Hight (2000) illustrates this point by means of a hypothetical embankment constructed on the Bothkennar soft clay, a structured soft clay

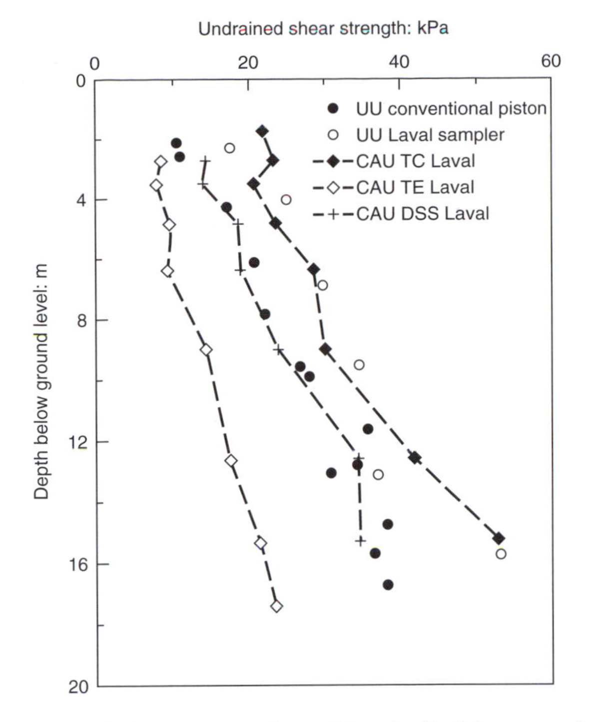

Fig. 4.17 Undrained shear strength profiles in Bothkennar clay (Hight, 2000)

from the UK National Soft Clay Research Site in Scotland. Undrained strength profiles at the Bothkennar soft clay site are shown in Fig. 4.17 as follows:

- the profile based on UUTC tests on conventional piston samples
- the profile based on UUTC tests on Laval samples
- the profiles based on CAU tests in triaxial compression (TC), triaxial extension (TE) and simple shear DSS on Laval samples – these profiles have been adjusted to allow for rate effects and disturbance and represent the best estimates of these particular strengths.

Note how the improved sample quality (Laval versus conventional piston) leads to a higher strength profile, based on UUTC tests, and how the UUTC profile for piston samples corresponds approximately to the best estimate of simple shear strength, and the UUTC profile for Laval samples corresponds approximately to the best estimate of in situ shear strength in triaxial compression.

As illustrated in Fig. 4.18, the strength available around the critical potential failure surface a–b–c beneath an embankment constructed on this clay will vary by an amount reflecting the level of anisotropy in the

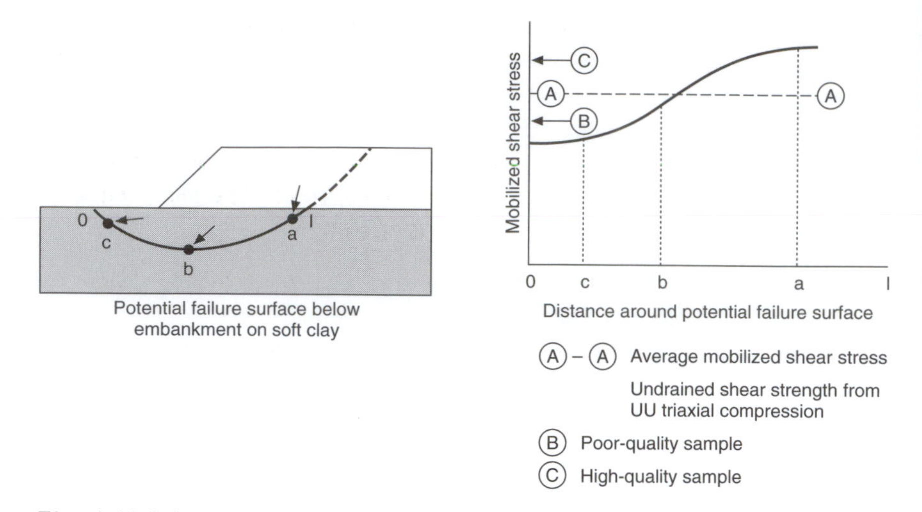

Fig. 4.18 Inherent dangers in improving sample quality while neglecting anisotropy (Hight, 2000)

soil. The average mobilized strength is indicated as A–A. Experience has shown that this is often close to the average strength measured in simple shear which in this case is similar to the strength measured in UU tests on poor-quality samples! A safe design results, therefore, when UUTC data are combined with a conventional factor of safety. An improvement in sample quality will lead to a higher UUTC strength profile (see Fig. 4.17), and if the same design procedure is adopted, i.e. using UUTC strengths to represent an average mobilized strength, an unsafe design may result, unless the factor of safety is modified.

Clayton and Siddique (1999) make a similar point. They reflect that although the basic mechanisms of sample disturbance have been understood for at least 50 years, routine practice in the UK and elsewhere has often been poor. Provided it is borne in mind that these practices, combined with relatively simplistic approaches to laboratory testing and analysis, are part of a semi-empirical design approach, then there seems little objection to their continued use for routine, low-cost construction projects. Larger and more demanding forms of construction will, however, provide an opportunity to use sophisticated analytical techniques to reduce construction costs. The best quality samples are required for this approach.

Disturbance by groundworks and other means

Leroueil (2001) points out that soil microstructure, carefully preserved by the sampling methods described above, must also be preserved during field works, if the determined design parameters are to be realistic in practice. The following case studies given by Leroueil (2001) illustrate this point.

Case study: Bang Bo test excavation, Thailand
Phien-Wej and Chavalitjiraphan (1991) describe a test excavation in Bang Bo district, 80 km south-east of Bangkok. The test excavation had four side slopes and a depth of 4 m as follows:

- a non-treated 1 : 4 slope that was expected to be the most critical and to fail in short term conditions
- a 1 : 3 slope treated with soil-cement piles that was not expected to fail
- a 1 : 4 slope treated with sand compaction piles that was not expected to fail
- a 1 : 6 non-treated slope that was expected to fail in long term conditions.

Surprisingly, the 1 : 4 non-treated slope turned out to be the most stable one and failure could only be induced after steepening the slope to 1 : 3. On the other hand, the three other slopes failed before completion of excavation to full depth! Phien-Wej and Chavalitjiraphan (1991) attributed this behaviour to disturbance of the soil by heavy equipment used prior to excavation and during soil improvement, and to the fact that the non-treated 1 : 6 slope was initially used as an access path to the test area. This hypothesis was confirmed by pore pressures in excess of the hydrostatic pressure observed at the start of excavation in all slope areas, except the 1 : 4 slope.

Case study: compensation grouting in Singapore marine clay
Shirlaw *et al.* (1999) describe an assessment of compensation grouting in Singapore marine clay in relation to tunnel projects. From the observations presented in the paper, heave at the surface was in the order of 20 mm at the end of the grouting period (Trial 4). Excess pore pressures had been generated and progressively dissipated over a period of about 100 days. At that time, nine of the thirty monitoring points showed an absolute settlement or no heave at all. It is thought that the clay had been partly destructured by grouting and reconsolidated under essentially the same overburden stress at a void ratio smaller than the initial one.

Leroueil (2001) concludes that application of dead loads, use of heavy equipment and generation of vibrations or deformations have to be considered carefully if preservation of microstructure of soft clay is important. Destructuring of clayey soils can also be obtained by swelling associated with a reduction of effective stress in soils containing swelling minerals (Leroueil and Vaughan, 1990) and by freezing (Graham and

Au, 1985; Leroueil *et al.*, 1991). Lunne *et al.* (2001) also shows how clays taken at large depths under water can be destructured by gas ex-solution.

Triaxial test

Historical perspective

As pointed out by Skempton (1955) in his Inaugural Lecture as Professor of Soil Mechanics at Imperial College, the triaxial apparatus

> *was first used by von Karman (1911) for tests on marble, but during the past 25 years [since 1930] many improvements have been made, and it now constitutes an indispensable piece of equipment in geotechnical laboratories. With the triaxial apparatus Terzaghi (1932) proved that $\phi = 0$ in undrained tests on a saturated clay and Rendulic (1937) observed the pore pressures set up during shear. Modern triaxial testing technique for clays originated principally at Harvard (Casagrande, 1941), whilst pore pressure measurements have been developed at MIT (Taylor, 1944), US Bureau of Reclamation (Holtz, 1947) and Imperial College (Bishop and Eldin, 1950).*

The triaxial test has retained its pre-eminence to the present day as the most advanced soil testing laboratory apparatus in everyday use. This is not only because of its versatility (see below), but also because test conditions can be carefully controlled to study their effect on soil behaviour (e.g. drainage conditions, testing rate, degree of saturation, etc.).

Drawbacks

Because of sampling disturbance and the small size of test specimens, the soil tested might not be representative of the soil mass. This is because the sample will be disturbed to some extent by removal from the ground. In addition, the soil might include joints and fissures on a scale larger than in the sample itself. The triaxial test specimen prepared from the sample will be of intact soil only and thus not represent the soil in the mass. In addition, the stress system is one of radial symmetry. The cell pressure is thus always equal to two of the principal stresses (see Short Course Notes: Triaxial Test). To study the influence of principal stress rotation (such as can occur under an embankment or a moving wheel load) and the influence of the intermediate principal stress on soil behaviour requires the use of a hollow cylinder apparatus.

Triaxial apparatus description

In the triaxial test, a right cylindrical specimen of soil is sheathed in a rubber membrane and placed inside a triaxial cell filled with pressurized water (Fig. 4.19(a)). Drainage is provided between the pores of the soil and the outside of the cell via a porous stone interface and a pore water duct.

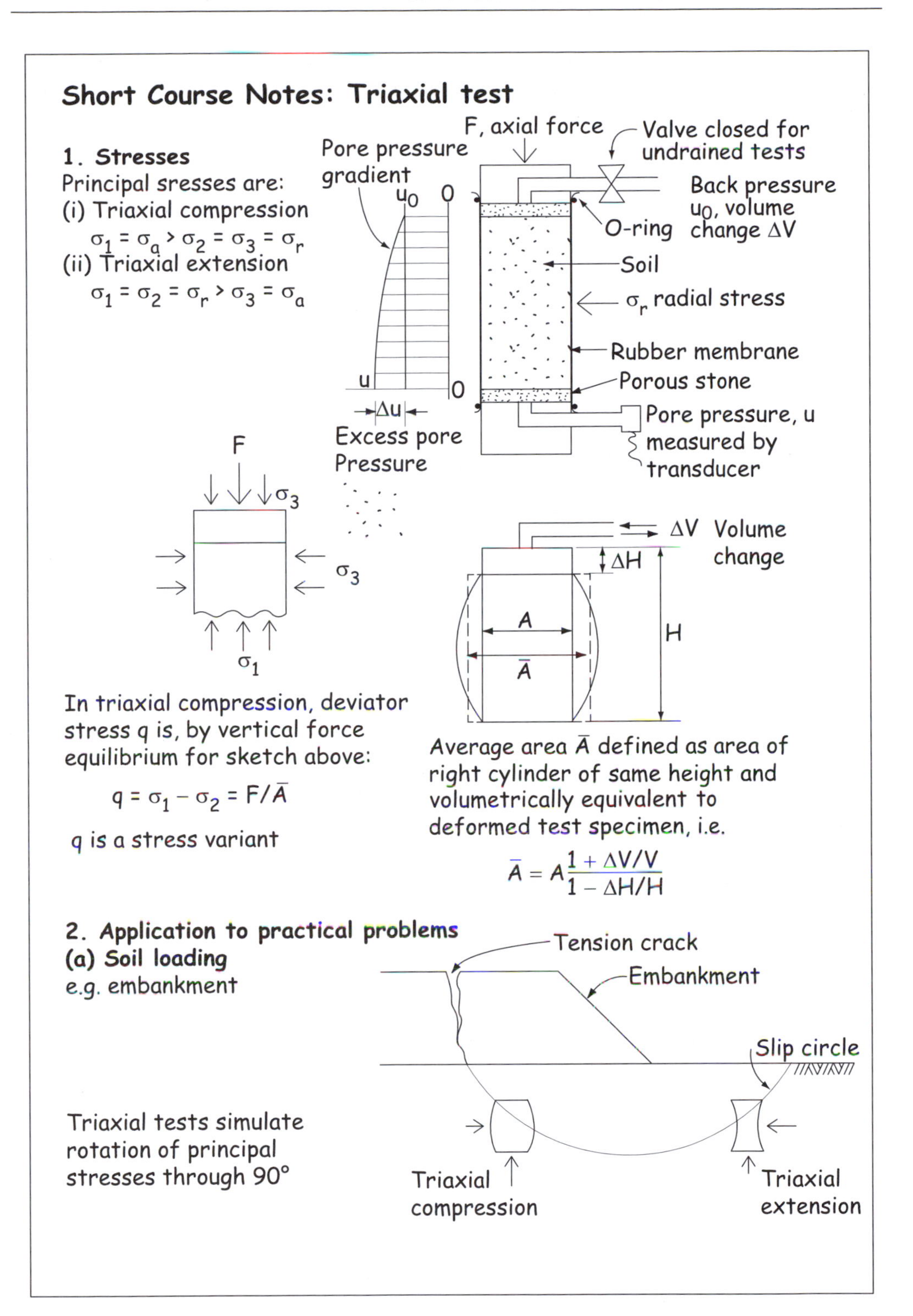
Short Course Notes: Triaxial test
1. Stresses
Principal sresses are:
(i) Triaxial compression
$\sigma_1 = \sigma_a > \sigma_2 = \sigma_3 = \sigma_r$
(ii) Triaxial extension
$\sigma_1 = \sigma_2 = \sigma_r > \sigma_3 = \sigma_a$
F, axial force
Pore pressure gradient
u_0
0
u
0
Δu
Excess pore Pressure
Valve closed for undrained tests
Back pressure u_0, volume change ΔV
O-ring
Soil
σ_r radial stress
Rubber membrane
Porous stone
Pore pressure, u measured by transducer
F
σ_3
σ_3
σ_1
ΔV Volume change
ΔH
H
A
$\bar{A}$
In triaxial compression, deviator stress q is, by vertical force equilibrium for sketch above:
$q = \sigma_1 - \sigma_2 = F/\bar{A}$
q is a stress variant
Average area $\bar{A}$ defined as area of right cylinder of same height and volumetrically equivalent to deformed test specimen, i.e.
$\bar{A} = A\frac{1 + \Delta V/V}{1 - \Delta H/H}$
2. Application to practical problems
(a) Soil loading
e.g. embankment
Tension crack
Embankment
Slip circle
Triaxial tests simulate rotation of principal stresses through 90°
Triaxial compression
Triaxial extension

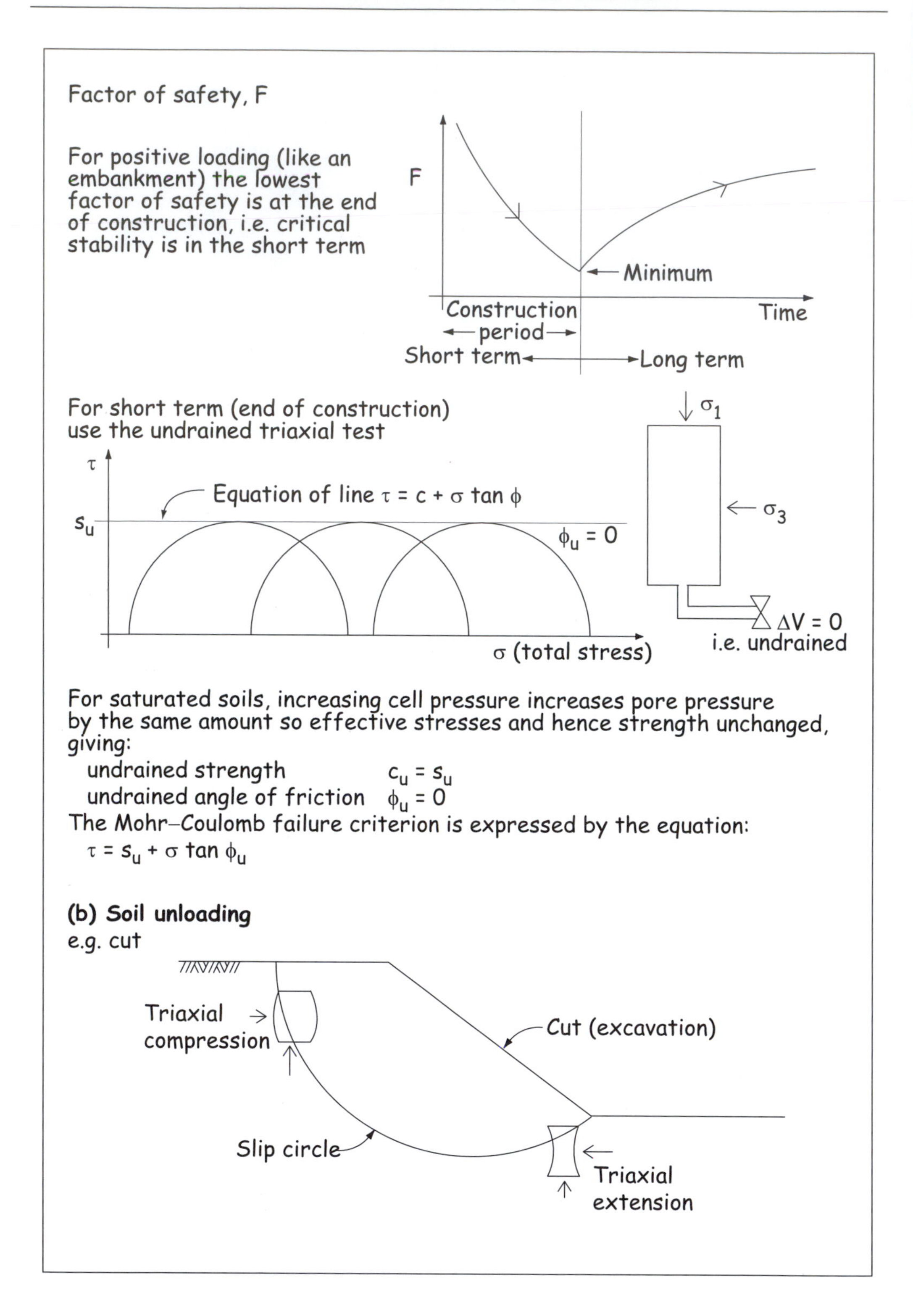
Factor of safety, F
For positive loading (like an embankment) the lowest factor of safety is at the end of construction, i.e. critical stability is in the short term
F
Minimum
Construction period
Time
Short term
Long term
For short term (end of construction) use the undrained triaxial test
τ
Equation of line τ = c + σ tan ϕ
s_u
ϕ_u = 0
σ (total stress)
σ_1
σ_3
ΔV = 0
i.e. undrained
For saturated soils, increasing cell pressure increases pore pressure by the same amount so effective stresses and hence strength unchanged, giving:
undrained strength c_u = s_u
undrained angle of friction ϕ_u = 0
The Mohr–Coulomb failure criterion is expressed by the equation:
τ = s_u + σ tan ϕ_u
(b) Soil unloading
e.g. cut
Triaxial compression
Cut (excavation)
Slip circle
Triaxial extension

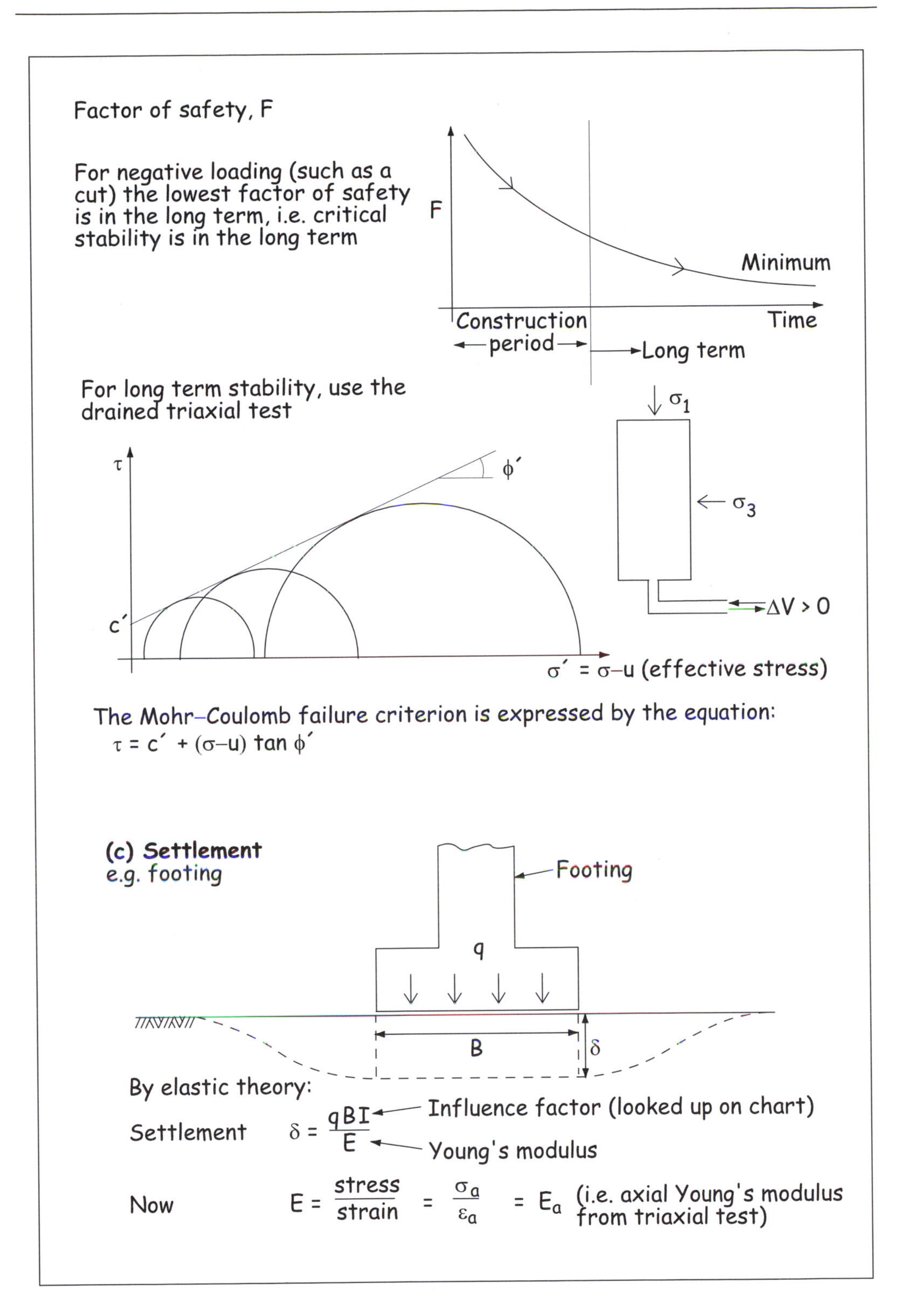
Factor of safety, F
For negative loading (such as a cut) the lowest factor of safety is in the long term, i.e. critical stability is in the long term
F
Minimum
Construction period
Time
Long term
For long term stability, use the drained triaxial test
τ
φ´
c´
σ´ = σ–u (effective stress)
σ1
σ3
ΔV > 0
The Mohr–Coulomb failure criterion is expressed by the equation:
τ = c´ + (σ–u) tan φ´
(c) Settlement
e.g. footing
Footing
q
B
δ
By elastic theory:
Settlement δ = qBI / E
Influence factor (looked up on chart)
Young's modulus
Now E = stress / strain = σa / εa = Ea (i.e. axial Young's modulus from triaxial test)

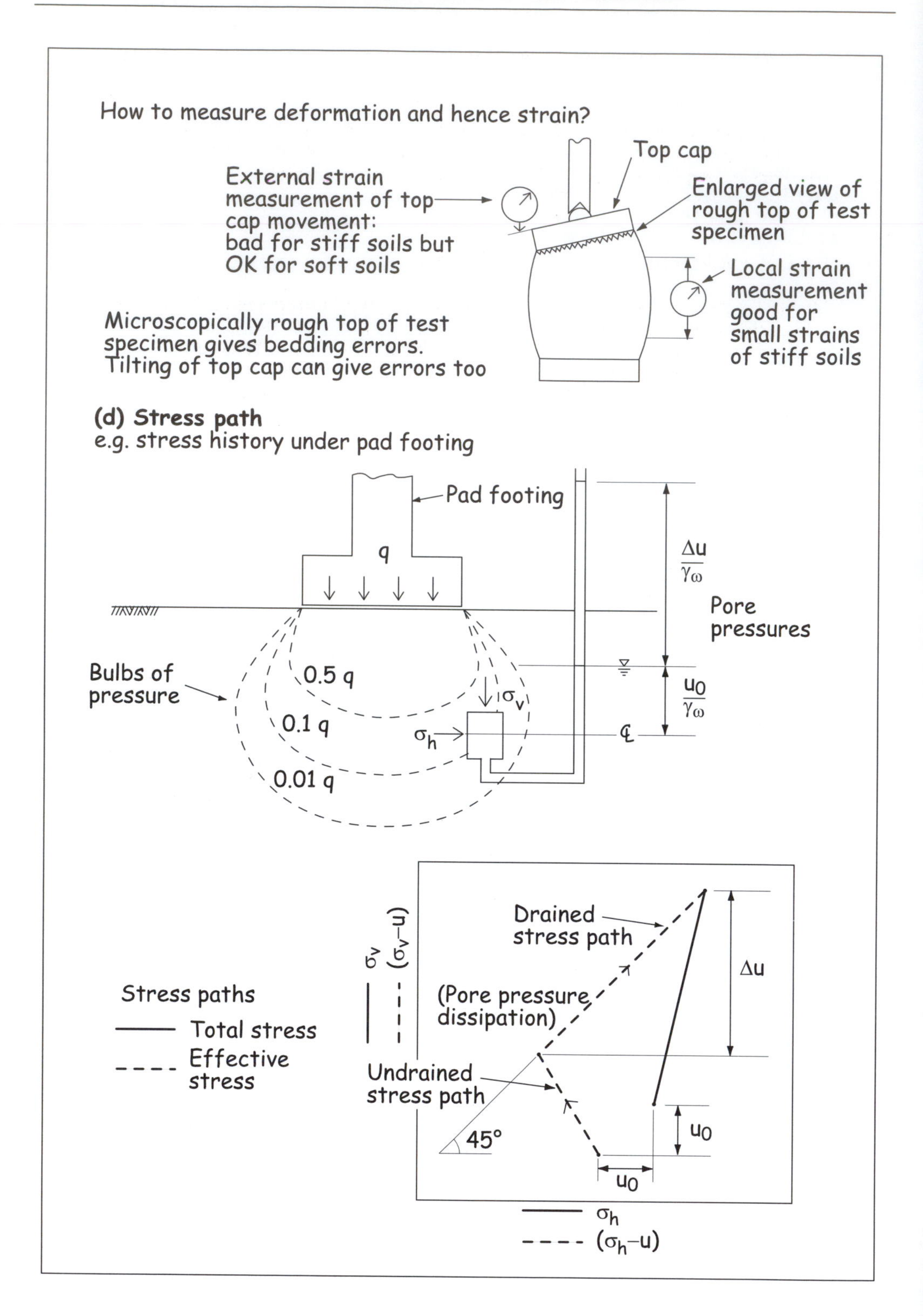
How to measure deformation and hence strain?
External strain measurement of top cap movement: bad for stiff soils but OK for soft soils
Top cap
Enlarged view of rough top of test specimen
Local strain measurement good for small strains of stiff soils
Microscopically rough top of test specimen gives bedding errors. Tilting of top cap can give errors too
(d) Stress path
e.g. stress history under pad footing
Pad footing
q
Δu/γω
Pore pressures
u0/γω
Bulbs of pressure
0.5 q
0.1 q
0.01 q
σv
σh
℄
Stress paths
Total stress
Effective stress
σv
(σv–u)
Drained stress path
(Pore pressure dissipation)
Undrained stress path
Δu
u0
45°
u0
σh
(σh–u)

For more information on stress path settlement analysis and for our "Fifteen commandments of triaxial testing", see **Short course in foundation engineering**, by Simons and Menzies (2000)

3. The cyclic and dynamic triaxial test

Normally these tests are carried out in the undrained state to study the build-up of pore pressure with number of cycles and the consequent reduction in stiffness and strength. Cyclic tests are low-frequency (<0.1 Hz), dynamic tests are high frequency (>1 Hz). For example, most earthquake frequency spectra are in the 0.1–10 Hz range.

Typical dynamic triaxial results are as follows.

(a) Strain control

Input: Test control by sinusoidal variation of deformation

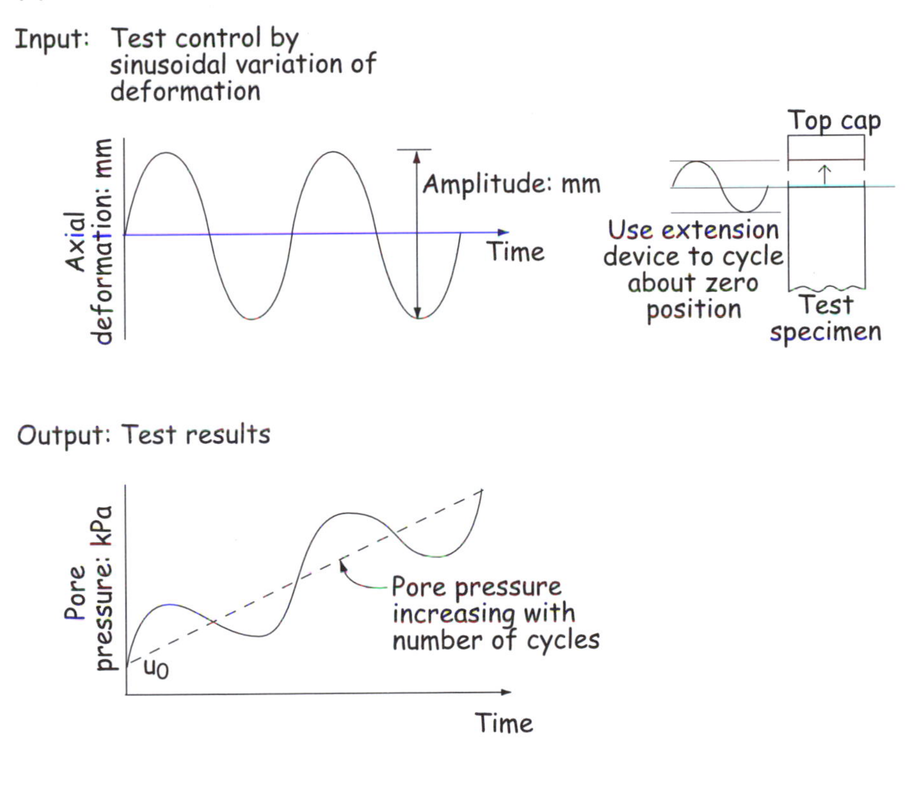

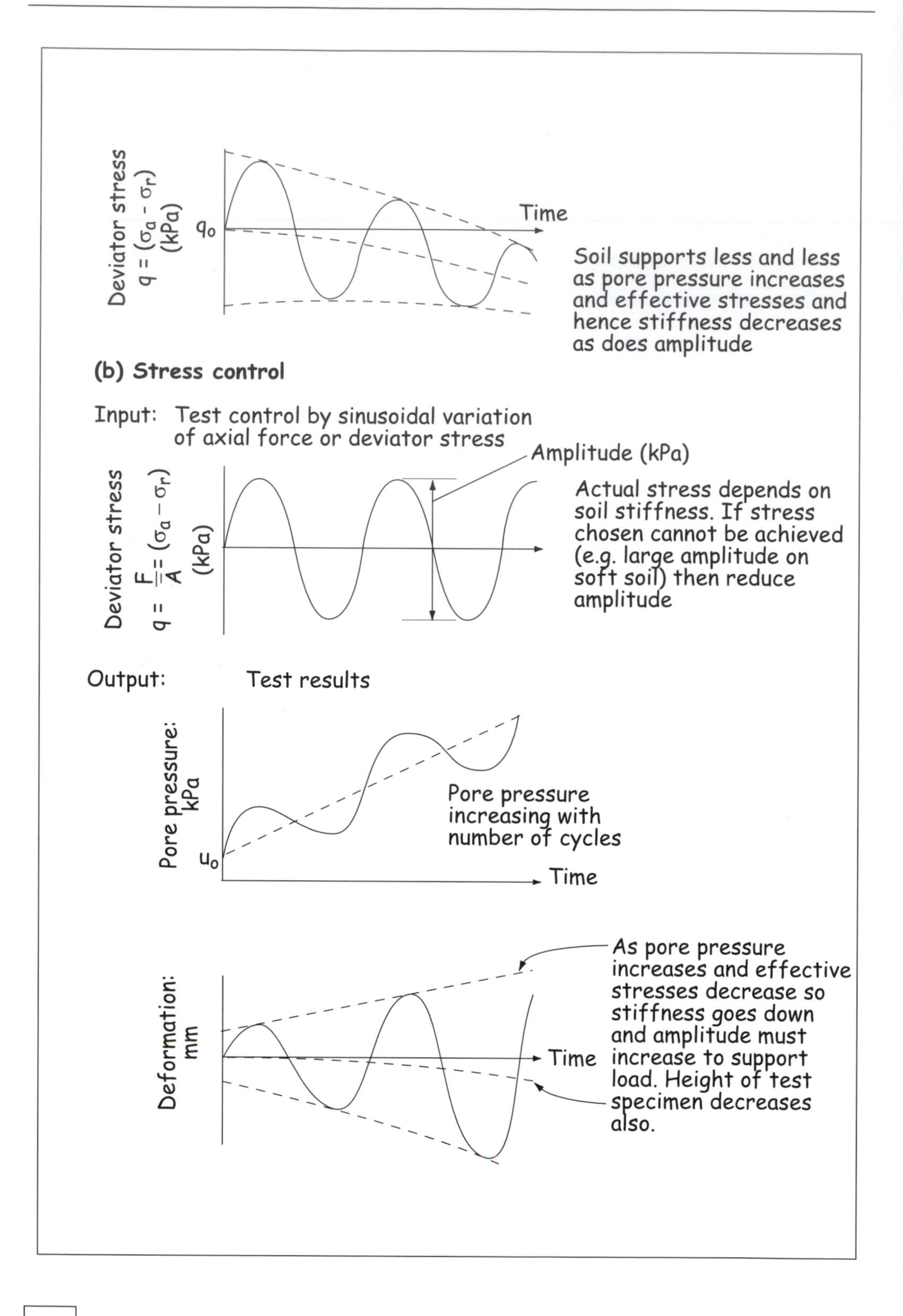
Deviator stress
$q = (\sigma_a - \sigma_r)$
(kPa)
q_o
Time
Soil supports less and less as pore pressure increases and effective stresses and hence stiffness decreases as does amplitude
(b) Stress control
Input: Test control by sinusoidal variation of axial force or deviator stress
Amplitude (kPa)
Deviator stress
$q = \frac{F}{A} = (\sigma_a - \sigma_r)$
(kPa)
Actual stress depends on soil stiffness. If stress chosen cannot be achieved (e.g. large amplitude on soft soil) then reduce amplitude
Output: Test results
Pore pressure: kPa
u_o
Pore pressure increasing with number of cycles
Time
Deformation: mm
Time
As pore pressure increases and effective stresses decrease so stiffness goes down and amplitude must increase to support load. Height of test specimen decreases also.

4. Bender elements

Bender elements are piezoelectric transducers embedded into the top cap and base pedestal of the triaxial apparatus. They are also embedded in the soil test specimen. By energizing the transmitter bender with a single sinusoidal pulse, a shear wave is propagated and received at the opposite receiver bender. The shear wave velocity V_s is thus found. From the density of the soil ρ, the shear modulus can be found by $G_0 = \rho V_s^2$. Shear modulus G is related to Young's modulus E and Poisson's ratio ν by $G = E/2\ (1 + \nu)$.

ρ is soil density

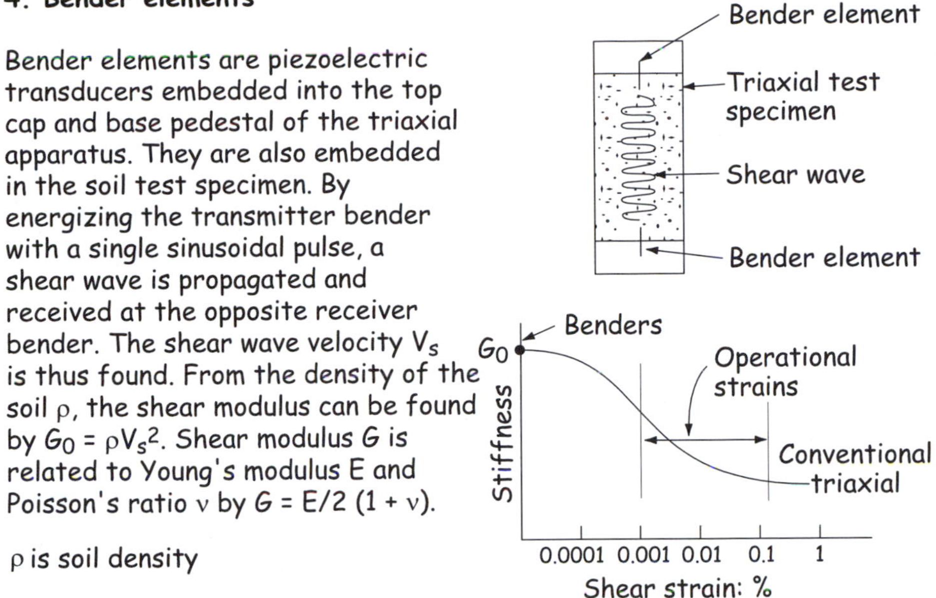

Table: Examples of stiffness degradation for different geomaterials (after Heymann, 1998)

Material	$E_{0.01}/E_0$	$E_{0.1}/E_0$	$E_{1.0}/E_0$
Intact chalk	0.87-0.93	0.42	failed
London Clay	0.83-0.97	0.35-0.58	0.11-0.20
Bothkennar clay	0.75-0.81	0.36-0.55	0.11-0.21

From the table above, approximate conservative reduction factors may be inferred to give notional values of operational stiffness E_{op} as:

$E_{op} \approx 0.5\ E_0$ for soft clays, and

$E_{op} \approx 0.85\ E_0$ for stiff clays and soft rocks

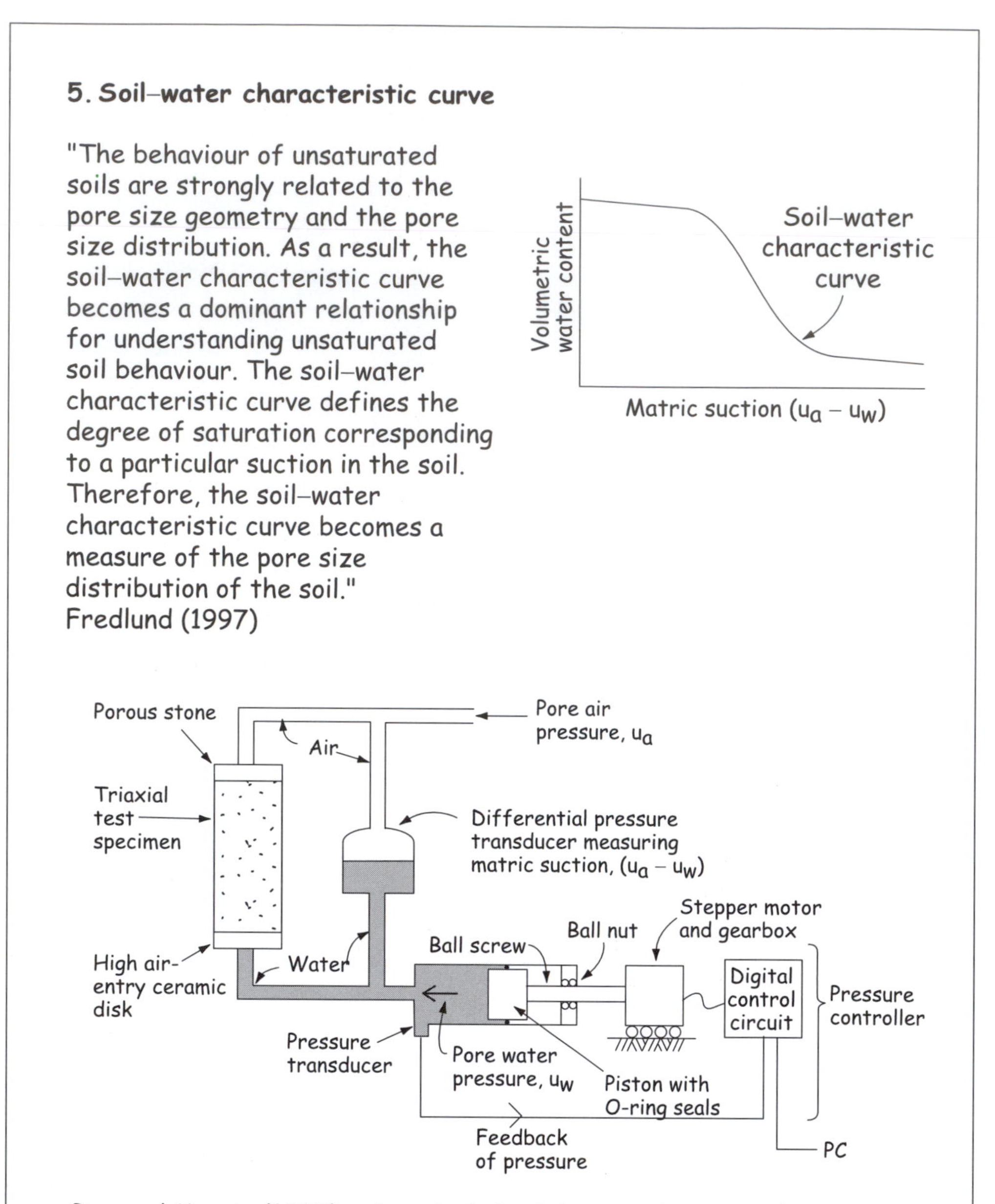

Ray and Morris (1995) automated obtaining matric suction/saturation curves. The procedure is continuous, yielding the entire curve. As shown above, a GDS Instruments Ltd pressure controller was put under computer control and set to withdraw water from the bottom of the sample at a slow constant rate and the differential pressure transducer measured the corresponding matric solution.

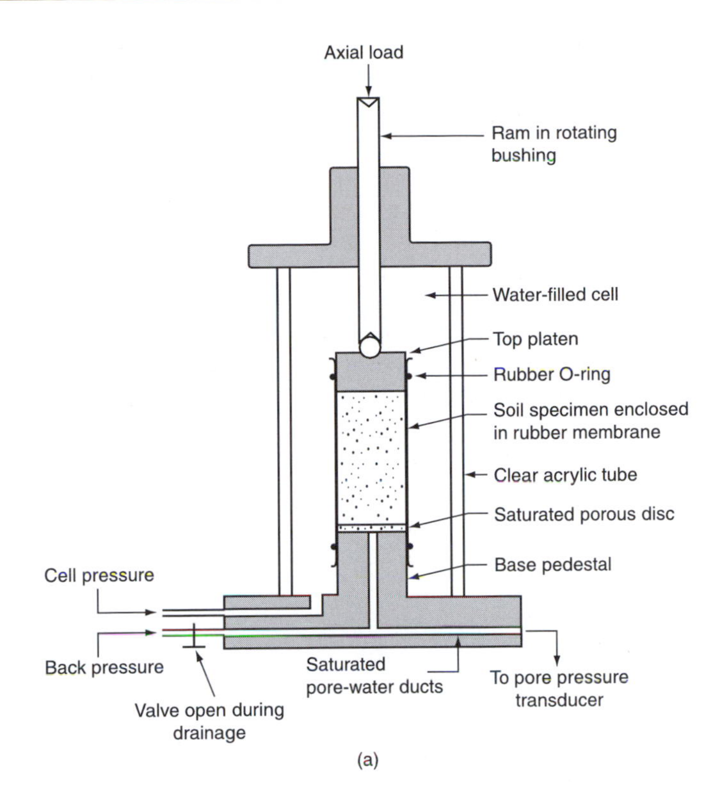

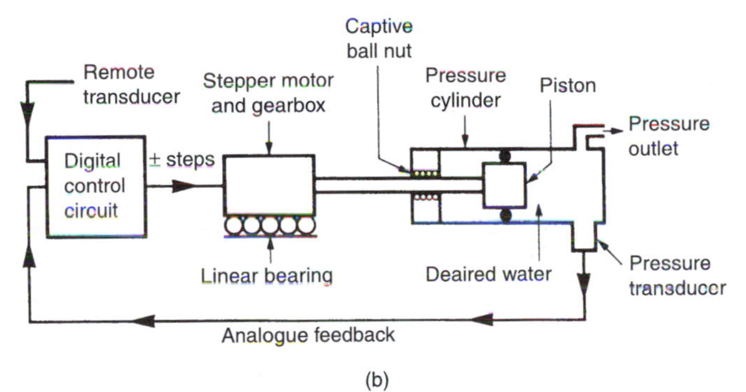

Fig. 4.19 (a) Typical set-up for triaxial compression tests. (b) Closed loop control screw pump schematic diagram (permission of GDS Instruments Ltd)

Through this duct, pore pressure can be measured, back-pressure applied and pore water volume change measured during consolidation and swelling. Axial stress is applied with a piston that passes through the top or base of the cell. The piston or loading ram is usually actuated by an electric motor and gearbox turning a screw or sometimes by hydraulic means.

LIVERPOOL JOHN MOORES UNIVERSITY
LEARNING SERVICES

Cell pressure and back-pressure are normally applied by means of compressed air–water bladder devices or by a closed-loop control screw pump of the GDS type shown schematically in Fig. 4.19(b). As well as not requiring a compressed air supply, the microprocessor-controlled screw pump has the added advantage that it provides measurement of volume change by counting the steps to the stepping motor and can

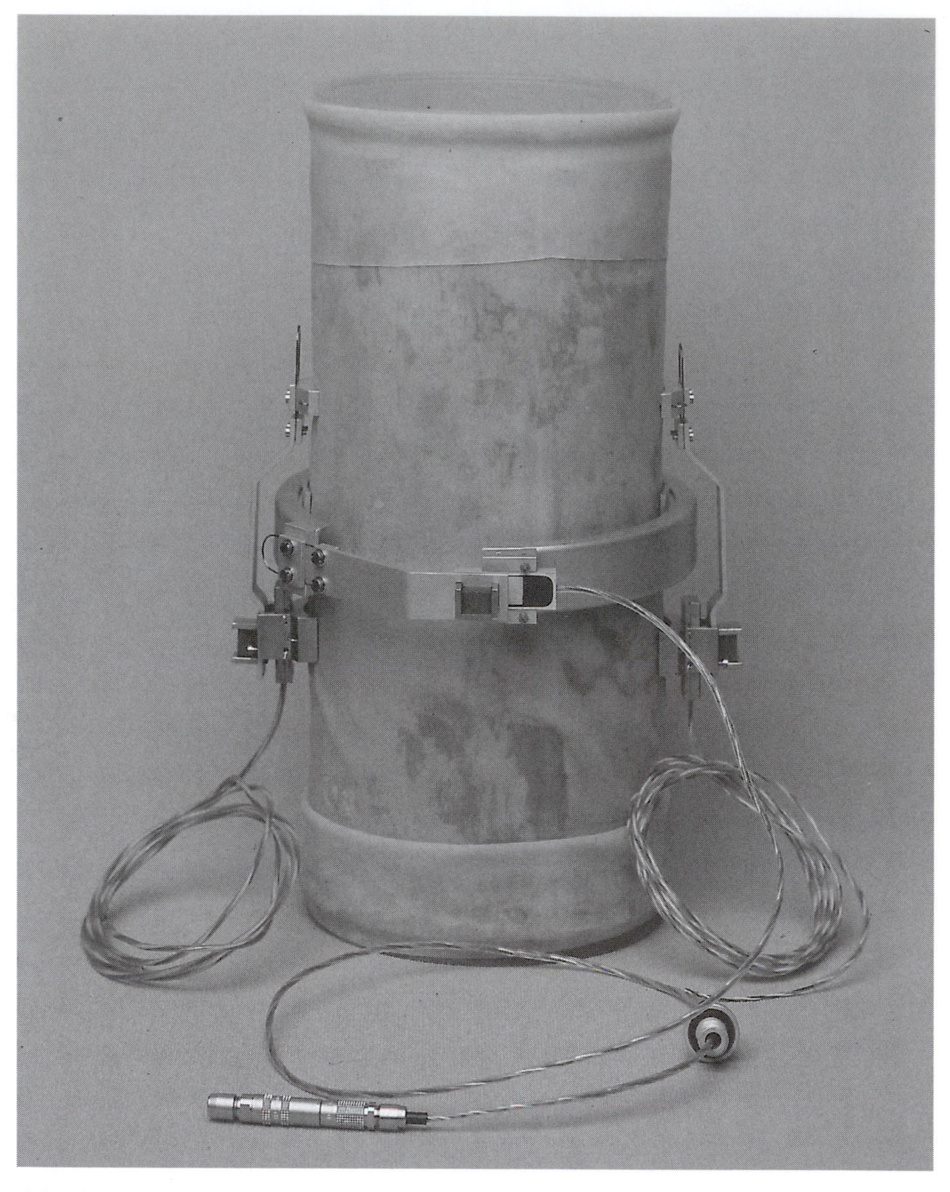

Fig. 4.20 Axial and radial Hall effect local strain transducers fitted to a triaxial test specimen (permission of GDS Instruments Ltd)

resolve volume change to 1 mm^3. In addition, the device has a computer interface. When the loading frame has a computer interface as well, computer automation of the triaxial test is enabled whereby complex loadings can be applied which follow real geological, construction and in-service conditions where vertical and horizontal stresses both change at the same time. This is important because soil is a loading-dependent material, i.e. the deformation moduli are not unique material constants but are related to the actual loading pattern or stress path.

Test interpretation

For an outline of test interpretation, see Short Course Notes: Triaxial Test. For a comprehensive treatment of triaxial test interpretation and for our 'Fifteen Commandments of Triaxial Testing' see *A Short Course in Foundation Engineering* Second Edition (Simons and Menzies, 2000).

Modern adaptations

In more advanced triaxial apparatus, local strain measurement can be carried out by means of local strain transducers attached directly to the test specimen (Fig. 4.20). This is important because the axial deformation measured does not include bedding errors between the test specimen and the top cap and so soil stiffness is measured correctly. This is particularly important for stiff clays (where bedding errors are a large proportion of measured strains), but not so important for soft clays (where bedding errors are a small proportion of measured strains). Another means for indirectly measuring soil stiffness is to measure shear wave velocity using bender element transducers (Fig. 4.21) implanted in the top cap and base pedestal to provide the maximum shear modulus G_0, i.e. the modulus at zero (or exceedingly small) strain. This is important because soil stiffness is now known to vary with strain level and operational

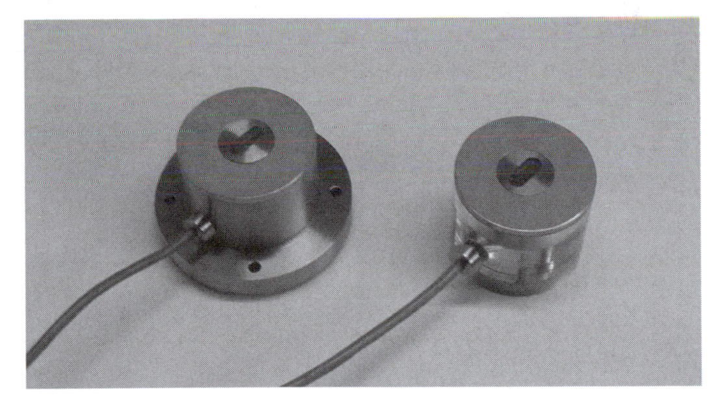

Fig. 4.21 Bender elements inserted in triaxial cell top cap and base pedestal (permission of GDS Instruments Ltd)

LIVERPOOL JOHN MOORES UNIVERSITY
Aldham Robarts L.R.C.
TEL 0151 231 3701/3634

strains around real civil engineering structures (such as tunnels, retaining walls and foundations) are often very small and so operational stiffness is just below the maximum value for stiff soils (around say 0·85 G_0), and probably about half the maximum value for soft soils (i.e. around 0·5 G_0).

Consolidation tests

Overview

As pointed out by Clayton *et al.* (1995b), consolidation tests are frequently required to assess the amount of volume change to be expected of a soil under load, for example beneath a foundation, and to allow prediction of the time that consolidation will take. The effects of predictions based on consolidation test results can be very serious, for example leading to the use of piling beneath structures to minimize predicted settlement, or the use of sand drains or stage construction for embankments to speed up the predicted rate of consolidation. It is therefore important to appreciate the limitations of the commonly available test techniques.

Three types of apparatus are in common use for consolidation testing. These are:

- the oedometer (Terzaghi, 1923; Casagrande, 1936)
- the triaxial apparatus (Bishop and Henkel, 1962)
- the hydraulic consolidation cell (Rowe and Barden, 1966).

Probably most soil mechanics laboratories in the world have some kind of Casagrande oedometer, and so we will concentrate our attention on this 'universal' apparatus. For how the results from this test are interpreted by Terzaghi's consolidation theory and other means, refer to the following Short Course Notes: Consolidation Theory.

Casagrande oedometer test

The apparatus (Fig. 4.22) consists of a cell which can be placed in a loading frame and loaded vertically. In the cell the soil sample is laterally restrained by a steel ring, which incorporates a cutting shoe used during specimen preparation. The top and bottom of the specimen are placed in contact with porous discs, so that drainage of the specimen takes place in the vertical direction when vertical stress is applied; consolidation is then one-dimensional.

The most common specimen size is 76 mm diameter by 19 mm high, since this allows the highly disturbed edges of a 102 mm diameter sample to be pared off during preparation of the test specimen. Where the specimen preparation process may be prevented by the presence of stones, the specimen diameter must be equal to that of the sampler.

BS 1377: Part 5: 1990, clause 3 (British Standards Institution, 1990), gives a standard procedure for the test. In this procedure the specimen is

Short Course Notes: Consolidation Theory

Terzaghi's theory of consolidation

The process of consolidation is illustrated by the piston and spring analogy shown below. At the instant the load is applied, because the system is not allowed to drain, the spring cannot deform and the loading is carried by the excess pore water pressure. If, now, slow drainage is allowed, water will leak out and the load will be transferred from the water to the spring until finally, after sufficient time has elapsed and the spring has deformed sufficiently to carry the applied loading, the deformation ceases and the excess pore water pressure is zero. In the corresponding soil element, the spring is replaced by the soil structure, and the rate of water expulsion is governed by the permeability of the soil and the length of the drainage path.

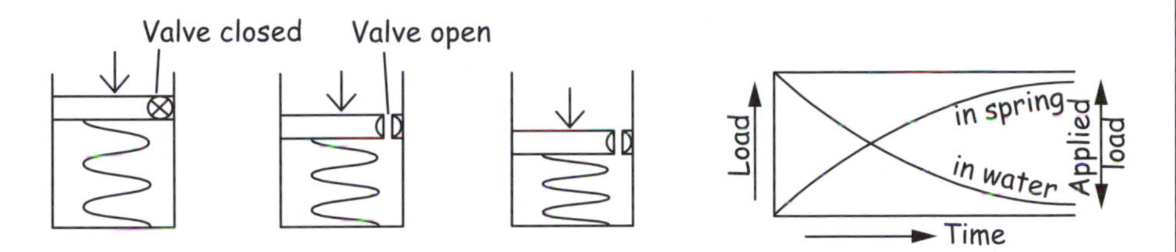

It is necessary, then, to calculate both the magnitude and the rate of the consolidation settlement and in practice, use is generally made of the Terzaghi theory of one-dimensional consolidation, which considers the situation shown here.

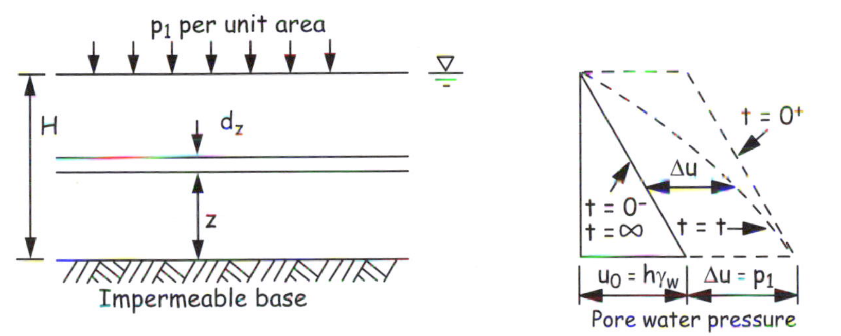

The main assumptions on which the theory is based are:

- the soil is saturated
- the water and the clay particles are incompressible
- Darcy's law is valid
- for a change in void ratio corresponding to a given increment in effective stress, the permeability, k, and the coefficient of volume change, m_v, remain constant

- the time taken for the clay to consolidate depends entirely on the permeability of the clay
- the clay is laterally confined
- the flow of water is one-dimensional
- effective and total stresses are uniformally distributed over any horizontal section.

These assumptions correspond to the oedometer test in the laboratory and to a clay layer in the field subjected to uniform global loading, that is, a uniformly distributed loading applied over an infinite area. Based on these assumptions, the governing differential equation relating excess pore water pressure, position and time can be derived as:

$$\frac{\partial u}{\partial t} = c_v \frac{\partial^2 u}{\partial z^2} \qquad (1)$$

where

$$c_v = \frac{k}{m_v \gamma_w} = \text{coefficient of vertical consolidation,}$$

and

$$m_v = \frac{\Delta V}{V} / \Delta p = \frac{\Delta H}{H} / \Delta p = \text{coefficient of volume compressibility,}$$

Equation (1) must be solved for the following boundary conditions:

$\Delta u = p_1$ for $t = 0$ and $0 < z < H$
$\Delta u = 0$ for $t > 0$ and $z = H$
$\Delta u = 0$ for $t = \infty$ and $0 < z < H$

The solution may be expressed in the form:

$$\Delta u = \frac{4p_1}{\pi} \sum_{N=0}^{N_{2300}} \frac{(-1)^N}{2N+1} e^{-(2N+1)^2 \pi^2 (T_v/4)} \cos\left[\frac{(2N+1)\pi z}{2H}\right]$$

where

$$T_v = \frac{c_v t}{H^2}$$

an independent dimensionless variable known as the Time Factor.

It follows that the degree of consolidation, U, at any time (equal to the settlement at that time, δ_t, expressed as a percentage of the total final settlement, δ_c, at the end of consolidation, is given by:

$$U = \frac{\delta_t}{\delta_c} = 1 - \frac{8}{\pi^2} \sum_{N=0}^{N=\infty} \frac{1}{(2N+1)^2} e^{-(2N+1)^2 \pi^2 (T_v/4)}$$

Solutions relating U and T_v for the various distributions of initial excess pore water pressure, and single and double drainage are given below. Using these solutions, it should be noted that the total thickness of the consolidating layer is always used in the computations. The solution relating U and T_v is then taken, corresponding to the drainage conditions of the particular problem.

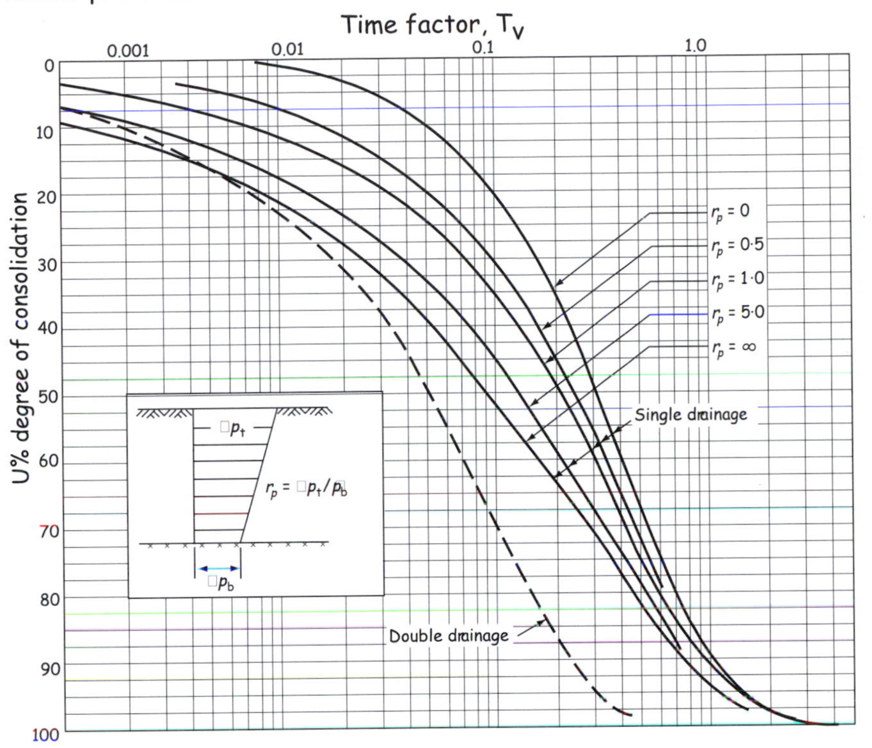

Settlement prediction from test results

A typical relationship between voids ratio and effective pressure is shown below (a) and between voids ratio and the logarithm of effective pressure (b).

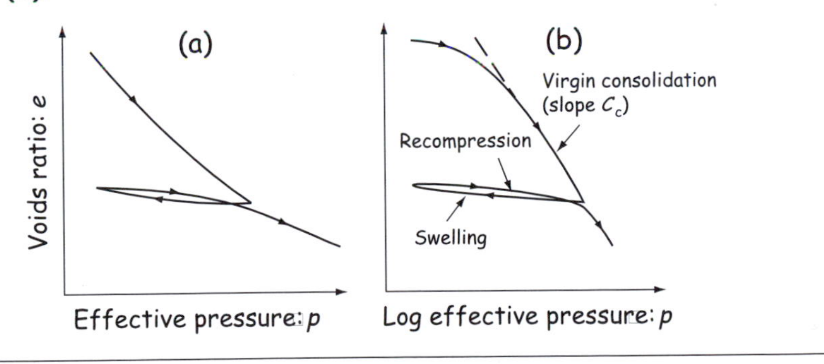

The final consolidation settlement can be calculated using any of the following expressions

$$\delta_c = \frac{\Delta e}{1+e_0} H \qquad \delta_c = m_v H \Delta p \qquad \delta_c = \frac{C_c}{1+e_0} H \log_{10} \frac{p'_0 + \Delta p}{p'_0}$$

where C_c = compression index and is the slope of the e - log p' line. It is convenient to use C_c when dealing with normally consolidated clays and m_v for overconsolidated clays.

Correction of measured compressibility

It is important, particularly for stiff clays having low compressibility, to correct the measured compressibility for the initial compression in the oedometer, otherwise values very much on the high side may result; that is, the deformation(d_0 - d_s) (see fig) must be excluded when determining m_v, d_0 is the corrected zero reading and d_s is the observed initial reading.

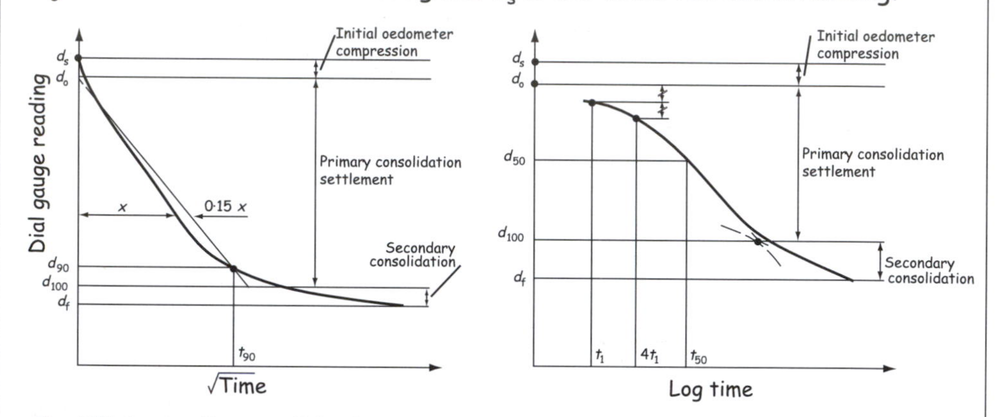

Coefficient of consolidation

The coefficient of consolidation for vertical flow, c_v, is calculated using either the square root of time plot or the logarithm of time plot; the procedures are illustrated above. The parameter c_v is used to scale up time from the small-scale laboratory test model to the full-scale field prototype. Note from

$$T_v = \frac{c_v t}{H^2}$$

that real time t depends on the square of the drainage path H. For example, if in the lab it takes 2 h to reach 90% consolidation for the 19mm test specimen (double drainage), then for a 1.9 m thick layer of clay in the field (also double drainage, say) the corresponding elapsed time would be over two years.

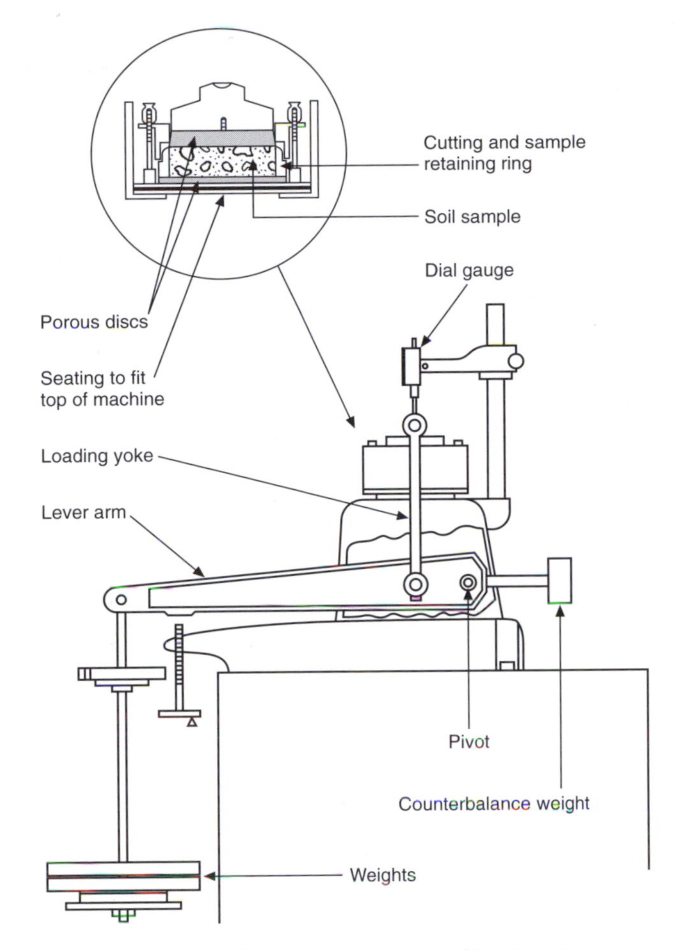

Fig. 4.22 Casagrande-type front-loading consolidation test apparatus

subjected to a series of pre-selected vertical stresses (e.g. 6, 12, 25, 50, 100, 200, 400, 800, 1600, 3200 kPa), each of which is held constant while dial gauge measurements of vertical deformation of the top of the specimen are made, and until movements cease (after normally 24 h). Dial gauge readings are taken at standard intervals of time after the start of the test (i.e. 0, 15 and 30 s, 1, 2, 4, 8, 15, 30 and 60 min, 2, 4, 8 and 24 h – chosen so that the square root of time is a convenient number). At the same time that the first load is applied, the oedometer cell is flooded with water, and if the specimen swells the load is immediately increased through the standard increments until swelling ceases.

Swelling pressures in stiff plastic over-consolidated clays are of considerable importance to the foundations of lightly loaded structures, and the technique suggested by BS 1377: Part 5: 1990, clause 4 allows an assessment of them to be made. The procedure involves balancing the swelling

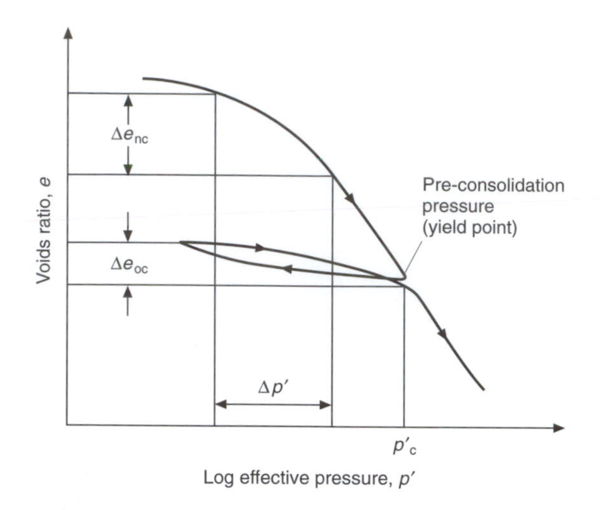

Fig. 4.23 e $-$ *log* p′ *curve showing how a stress change of* $\Delta p'$ *causes small deformations* Δe_{oc} *for the unloaded (over-consolidated) portion of the curve and much larger deformations* Δe_{nc} *for the first-time loaded (normally-consolidated) portion of the curve*

pressure once the water is added by keeping the dial gauge reading unchanged by the careful application of weights to the hanger.

Pre-consolidation pressure

It can be seen from Fig. 4.23 that the settlement for a given load increment is much greater for a normally-consolidated clay than for an over-consolidated clay. It is vital therefore to determine whether or not the clay is over-consolidated. This can be determined from the following.

- A knowledge of the geological history of the clay, that is, whether previous overburden has been removed or the groundwater table is now higher than in the past.

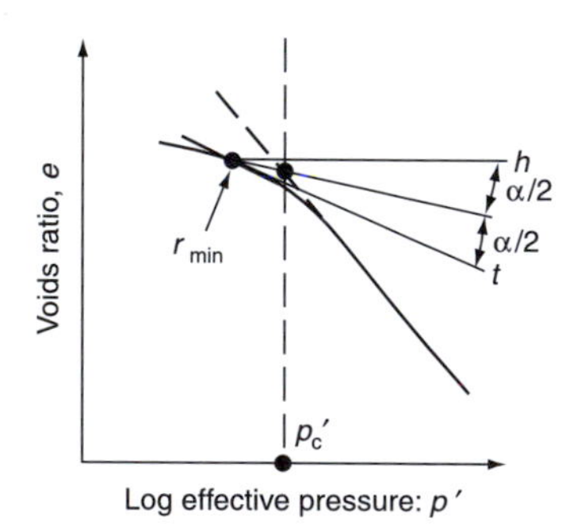

Fig. 4.24 Construction for determining the pre-consolidated pressure p'_c*, after Casagrande (1936)*

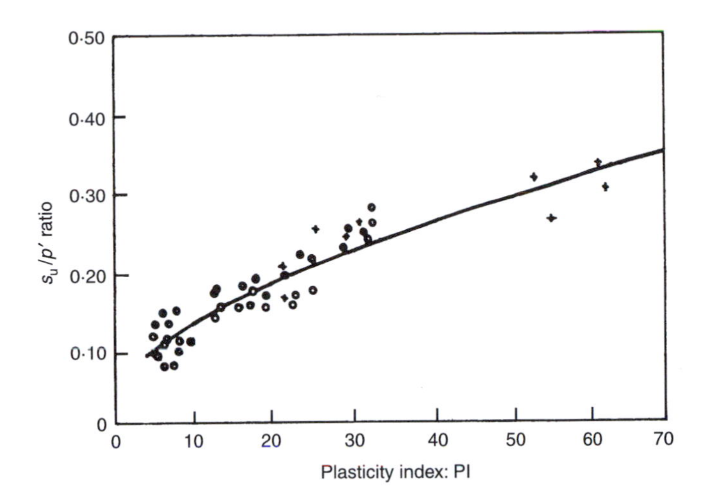

Fig. 4.25 Relationship between s_u/p' *and plasticity index, after Bjerrum and Simons (1960)*

- The Casagrande construction, illustrated in Fig. 4.24, which gives the pre-consolidation pressure p'_c (the greatest effective pressure the clay has carried in the past). If $p'_c > p'_0$, the clay is over-consolidated. If $p'_c = p'_0$, the clay is normally-consolidated.
- Comparing the measured undrained shear strength with that to be expected for a normally-consolidated clay having a similar plasticity index (see Fig. 4.25). If the measured strength is greater than that anticipated for a normally-consolidated clay the clay is probably over-consolidated.
- Comparing C_c corresponding to p'_0 with the value predicted for a normally-consolidated clay, $C_c = 0{\cdot}009$ (LL–10). If C_c at p'_0 is less than that expected for a normally-consolidated clay, the clay is probably over-consolidated.
- A determination of the liquidity index (LI), of the clay where

$$\text{LI} = \frac{w_c - \text{PL}}{\text{LL} - \text{PL}}$$

and LL is the liquid limit and PL is the plastic limit, determined according to BS 1377. Normally-consolidated clays have a liquidity index varying from about 0·6 to 1·0, and over-consolidated clays have a liquidity index varying from 0 to about 0·6. This gives a rough guide only.

Skempton and Bjerrum's correction

A most important contribution to settlement analysis was made by Skempton and Bjerrum (1957). They pointed out that an element of soil underneath a foundation undergoes lateral deformation as a result of applied loading and that the induced pore water pressure is, in general, less than the increment in vertical stress on the element, because it is dependent on the value of Skempton's pore pressure parameter, A. The consolidation of a clay results from the dissipation of pore water pressure. But a given set of stresses will set up different pore water pressures in different clays if the A values are different. Thus two identical foundations carrying identical loads, resting on two clays with identical compressibilities, will experience different consolidation settlement if the A values of the clay are unequal. This is true despite the fact that no difference would be seen in oedometer test results. For the special case of the oedometer test, however, where the sample is laterally confined, then, irrespective of the A value, the pore water pressure set up is equal to the increment in vertical stress (Simons and Menzies, 1974). Skempton and Bjerrum (1957) proposed that a correction factor should be applied to the settlement, calculated on the basis of oedometer tests and showed that the factor was a function of the geometry of the problem and the A value: the smaller the A value, the smaller the correction factor. The factor μ is shown in Fig. 4.26 and the field settlement is then equal to μ

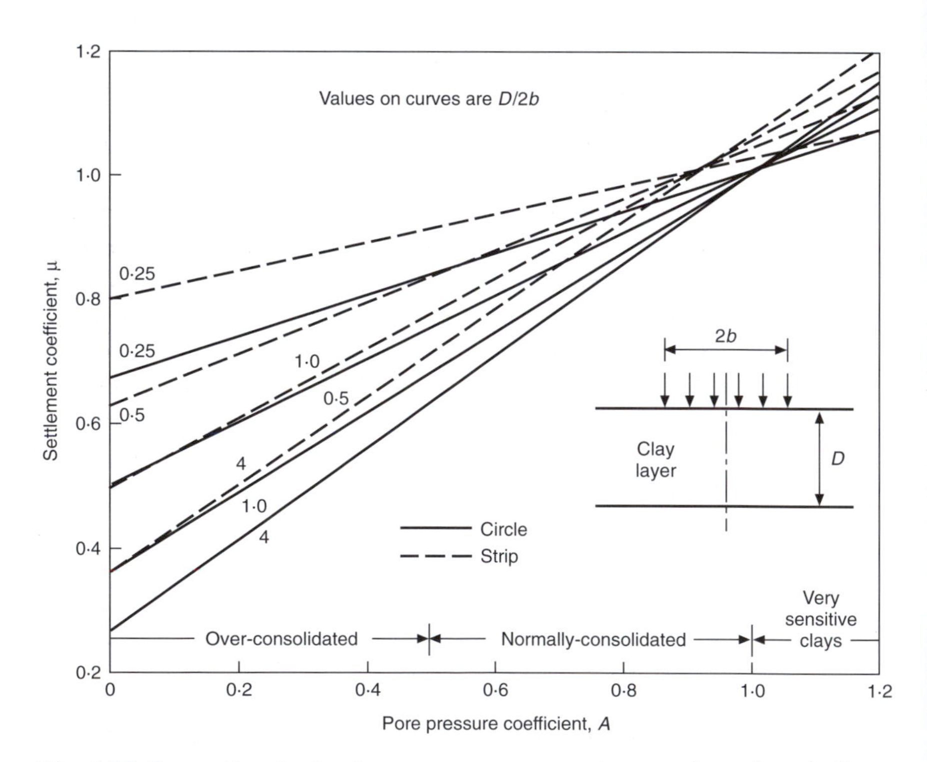

Fig. 4.26 Correction factor for pore pressures set up under a foundation, after Skempton and Bjerrum (1957)

times the settlement calculated on the basis of oedometer tests (see Simons and Menzies, 2000).

For heavily over-consolidated clays, A values less than one would be expected and the Skempton–Bjerrum correction factor is usually applied in such cases. It should be noted that in the working range of stress for normally-consolidated clays, particularly when there is no question of over-stressing occurring, A values of less than one can arise, and the correction factor should then be applied.

Continuous consolidation test equipment

In the Casagrande consolidation apparatus, the load increment tests are very slow because the hydraulic gradient causing pore water to flow out of the test specimen is itself reduced by the flow giving an exponential decay in flow rate. Accordingly, long periods (typically 24 h) must elapse before flow and hence consolidation for that stage is complete at which time effective stresses equal total stresses. From a test production point of view, therefore, continuous consolidation testing in back-pressured test apparatus is more attractive. Continuous consolidation testing may be classified as:

- controlled hydraulic gradient (Lowe *et al.*, 1969)
- continuous rate of loading (Janbu *et al.*, 1969; Aboshi *et al.*, 1970)
- constant rate of strain (Smith and Wahls, 1969; Wissa *et al.*, 1971).

The advantage of controlled hydraulic gradient continuous consolidation testing is illustrated in Fig. 4.27(a). It can be seen that pore pressure gradients and hence effective stress and thus strain gradients are nearly uniform. In addition, the excess pore pressure can be computer-controlled to give automatic drained testing rate. In contrast, as shown in Fig. 4.27(b) in the conventional oedometer the absence of back-pressure means that pore pressure is always atmospheric (zero gauge pressure) at the top and bottom of the test specimen and so effective stress and strain variations are large. One consequence of this is that a given pressure like the extremely important pre-consolidation pressure p'_c (see above) will be approached, equalled and then exceeded in different parts of the soil at different times (Fig. 4.27(b)).

Ring shear test

Residual strength

If a specimen of clay is placed in a shearing apparatus and subjected to displacements at a very slow rate (drained conditions) it will initially show increasing resistance with increasing displacement. However, under a given effective pressure, there is a limit to the resistance the clay can offer, and this is termed the 'peak strength', s_f. With further

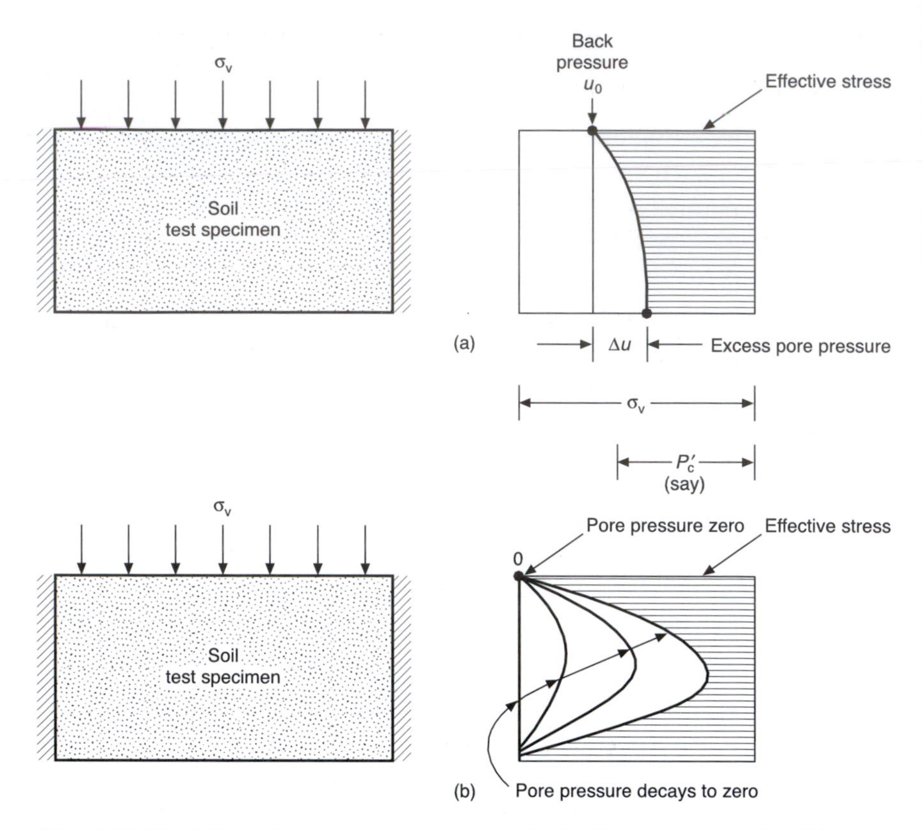

Fig. 4.27 Variations in pore pressure, vertical effective stress (and hence vertical strain) with time for a saturated clay test specimen for (a) controlled hydraulic gradient back-pressured oedometer, and (b) conventional load increment oedometer (Casagrande-type)

displacement the resistance or strength of clay decreases. This process, which Skempton (1964) refers to as 'strain softening', is not without limit because ultimately a constant resistance persists, regardless of the magnitude of displacement. This value of ultimate resistance is termed 'residual strength', s_r.

If several similar tests are conducted under different effective pressures, the peak and residual strengths when plotted against the effective normal pressure as shown in Fig. 4.28 will show a straight-line relationship, at least within a limited range of normal stress. Peak strength can therefore be expressed by

$$s_f = c' + \sigma' \tan \phi'$$

and the residual shear strength by

$$s_r = c'_r + \sigma' \tan \phi'_r$$

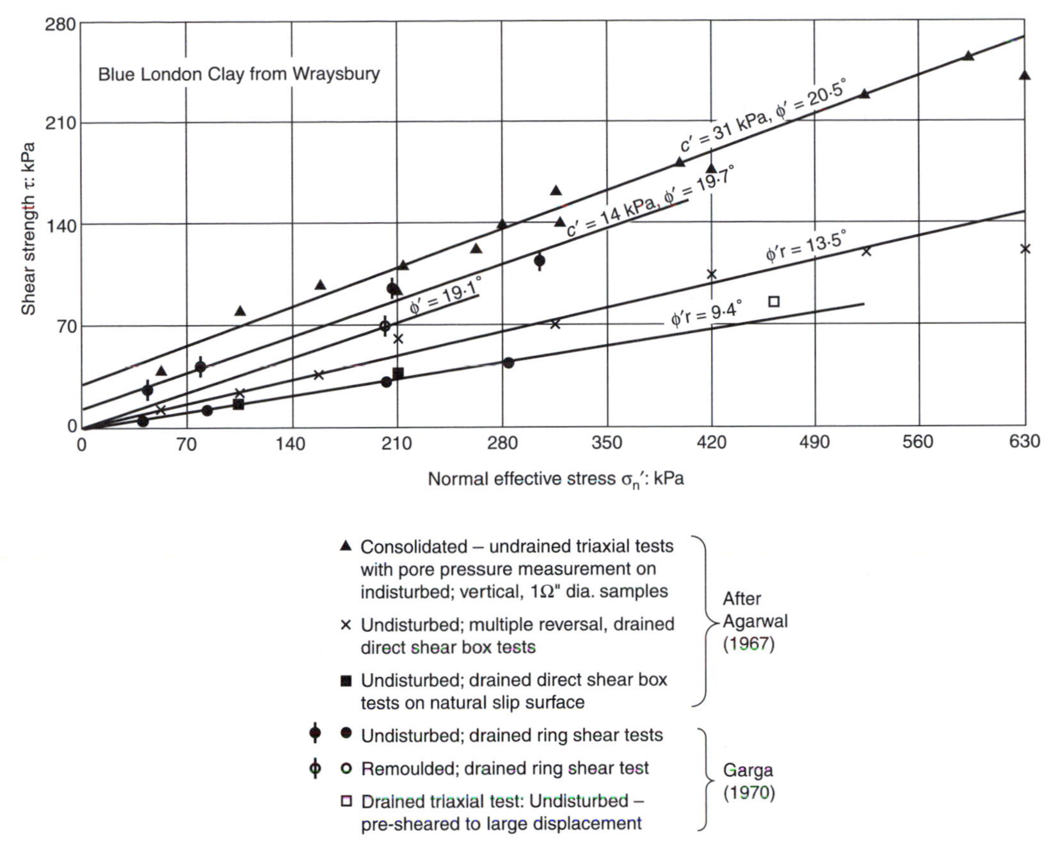

Fig. 4.28 Strength–effective stress relationships for blue London Clay from Wraysbury, after Bishop et al. *(1971)*

The value of c'_r is generally very small, but even so may exert a significant influence on the calculated factor of safety and depth of the corresponding slip surface. Thus, in moving from peak to residual, the cohesion intercept approaches zero. During the same process the angle of shearing resistance can also decrease. During the shearing process, over-consolidated clays tend to expand, particularly after passing the peak. Thus the loss of strength in passing from peak to residual is partly due to an increase in water content. A second factor that equally contributes in the post-peak reduction of the strength is the development of thin bands or domains in which the clay particles are orientated in the direction of shear, as noted by Skempton (1964).

In general, the difference between peak and residual strength depends on soil type and stress history and is most marked for heavily over-consolidated fissured clays. For normally-consolidated clays this difference is generally small. Thus, the concept of residual strength is of particular importance in the case of the long term stability of slopes of over-consolidated fissured clay.

Determination of residual strength

The residual shear strength is not only of practical importance in relation to the analysis of long term stability of slopes, natural or man-made, but it may also be considered to be a fundamental property of the particular soil. Therefore, it is important in the laboratory to measure accurately residual strength. It is commonly assessed by:

- reversing shear box tests
- triaxial tests
- ring shear tests
- back-analysis of a field failure on a pre-existing failure surface where post-slip movements occur and where post-slip piezometric levels are known.

Figure 4.28 compares multiple reversing direct shear box tests with test results from triaxial and ring shear apparatus. Clearly the ring shear apparatus is best suited to measure residual strength.

Ring shear test apparatus

The ring and rotational shear tests are the only tests in which very large and uniform deformations can be obtained in the laboratory and have been used in soil mechanics for many years to investigate the shear strength of clays at large displacements (Tiedemann, 1937; Haefeli, 1938). Several designs of the apparatus and results have also been

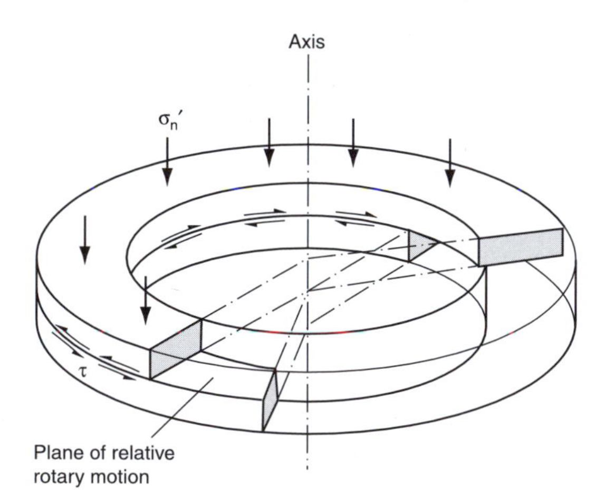

Fig. 4.29 Diagrammatic view through a test specimen of the ring shear apparatus, after Bishop et al. *(1971)*

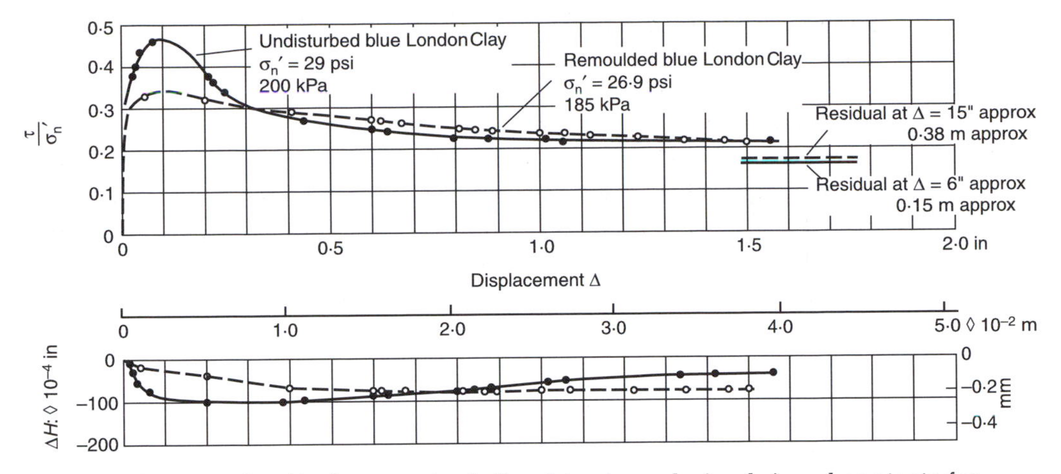

Fig. 4.30 Stress ratio–displacement relationships from drained ring shear tests for undisturbed and remoulded blue London Clay from Wraysbury, after Bishop et al. *(1971)*

reported by De Beer (1967), Sembelli and Ramirez (1969), La Gatta (1970), Bishop *et al.* (1971) and Bromhead (1979).

The ring shear apparatus described by Bishop *et al.* (1971) may be used to determine the full shear strength displacement of an annular soil specimen subjected to a constant normal stress, confined laterally and ultimately caused to rupture on a horizontal plane of relative motion. The apparatus may be considered as a conventional shear box extended round into a ring (Fig. 4.29). Consequently large displacements (e.g. 1 m) may be obtained in one direction so that the residual strength may be accurately determined (e.g. see Fig. 4.30).

The simplicity of design of the Bromhead ring shear apparatus (Bromhead, 1979; see Fig. 4.31(a) and Fig. 4.31(b)) together with its ease of use compared with the complex ring shear apparatus of Bishop *et al.* (1971) has ensured that this is the most widely used ring shear apparatus in both research and commercial testing laboratories.

Comment (Bromhead, 1979)

In his book, *The Stability of Slopes*, Bromhead (1979) remarks that peak strength and its measurement were far better understood than residual strength in geotechnical circles. This is due to a number of factors, including

- the apparatus for measuring peak strength is commonplace in industrial laboratories as well as research establishments, and is in daily use
- many peak strength tests are *undrained* and hence are quick and cheap, leading to their routine use

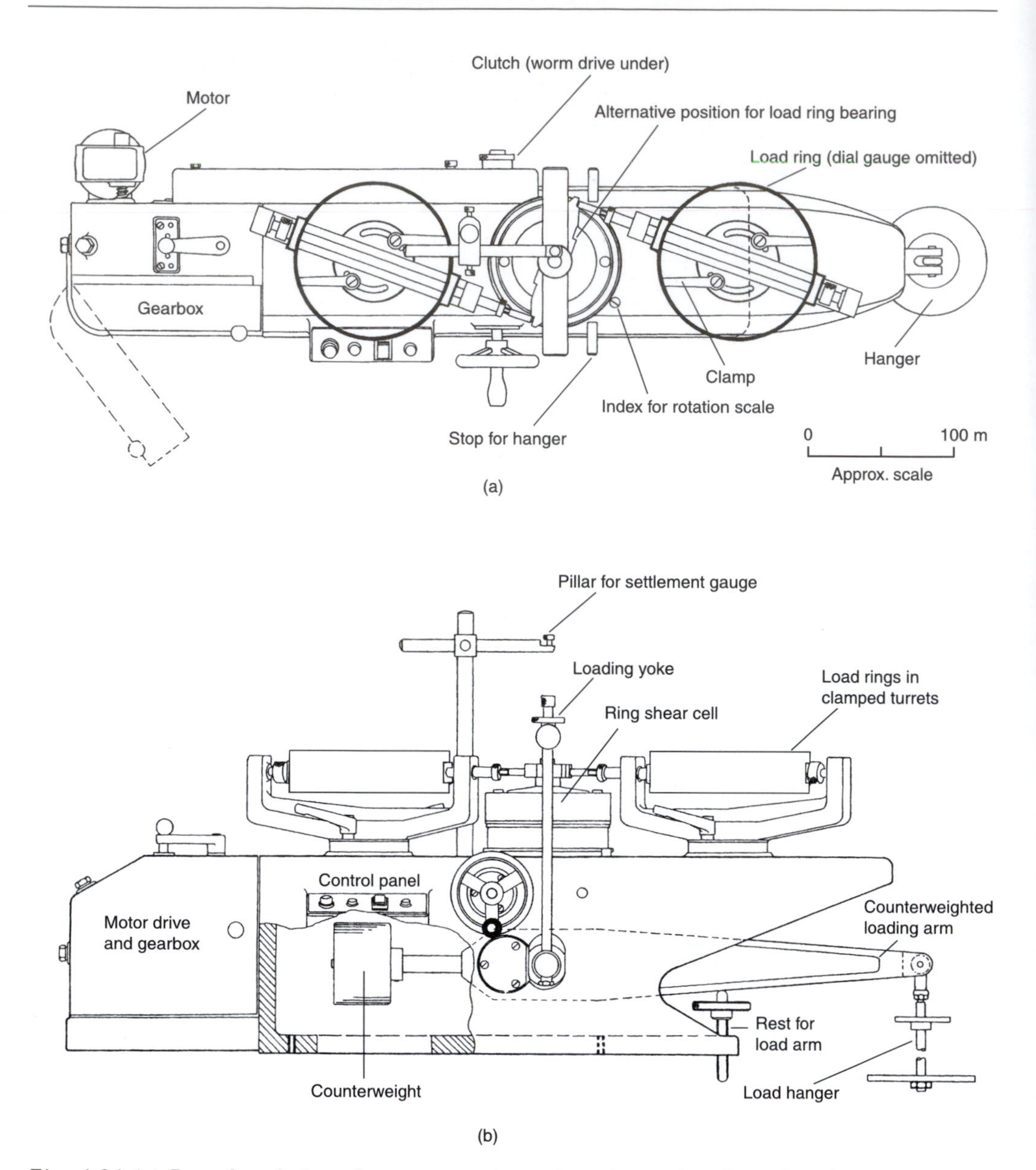

Fig. 4.31 (a) Bromhead ring shear apparatus: plan view, after Bromhead (1979). The torque is measured by the proving rings. These are mounted in turrets so that they can be swung out of the way for sample preparation (the test cell is removable). Two stops are provided on the torque arm to allow easy setting of the radius at which the proving rings act. (b) Bromhead ring shear apparatus: elevation and general layout. Ordinarily, this apparatus is mounted on a combined stand and small table, but is equally at home on a workbench, after Bromhead (1979)

- residual strength as a concept is relatively recent, and is viewed as being applicable to 'landslides' and not as an aspect of behaviour with far wider ramifications.

Furthermore, the view is prevalent that simple peak strength measurements in the laboratory genuinely represent field strengths, whereas residual strength measurement is in some way still 'experimental'. In fact the opposite is true!

The essence of peak strength measurement in the laboratory is to test a representative soil sample in such a way that it preserves its fabric. Peak strength testing, however, is beset with major problems, including:

- disturbance, which usually *decreases* the measured strength
- failure to follow the correct stress path in the test, which can either decrease or increase the measured strength relative to the field
- unrepresentative sampling, which *increases* the strength relative to that which is likely to be operative in the field
- poor testing technique, leading to partial or inadequate drainage in 'undrained' or 'drained' tests, respectively, so affecting the measured strengths
- failure to take into account progressive failure in brittle soils (e.g. by doing multi-stage tests on them) so that their behaviour is modified by systematic changes in the soil fabric as the test progresses
- errors in choosing strain rates, which cannot later be rectified.

In comparison, residual strength tests are measuring a soil property that is largely independent of the stress path followed, and in the ring shear apparatus it is possible to allow for, or rectify, many of the factors listed above because of the unlimited strain capacity of the machine.

Definitions

In summary, the following definitions should be noted.

- *The residual strength* is the drained strength after sufficient movement has occurred on a failure surface to orientate the clay particles into a parallel position.
- *The residual factor* is

$$R = \frac{s_f - \bar{s}}{s_f - s_r}$$

where $\bar{s}$ is the average or operational strength back-analysed from a field failure, s_f is the peak failure strength obtained from a drained peak strength test (e.g. a drained triaxial test) and s_r is the residual strength obtained from a drained ring shear test.

- *The brittleness index* is

$$I_B = \frac{s_f - s_r}{s_f}$$

where s_f is the peak failure strength obtained from a drained peak strength test (e.g. a drained triaxial test) and s_r is the residual strength obtained from a drained ring shear test.

Field tests

Shear vane

Overview

The field shear vane is a means of determining the in situ undrained shear strength. This consists of a cruciform vane on a shaft (Fig. 4.32). The vane is inserted into the clay soil and a measured increasing torque is applied to the shaft until the soil fails as indicated by a constant or dropping torque by shearing on a circumscribing cylindrical surface. The test is carried out rapidly. Now, if s_{uv} is the undrained shear strength in the vertical direction, and s_{uh} is the undrained shear strength in the horizontal direction, then the maximum torque is

$$T = \frac{\pi D^2}{2}\left(Hs_{uv} + \frac{D}{3}s_{uh}\right)$$

where H is the vane height and D is the vane diameter, and assuming peak strengths are mobilized simultaneously along all vane edges.

This equation in two unknowns, s_{uv} and s_{uh}, can only be solved if the torque is found for two vanes with different height to diameter ratios. It

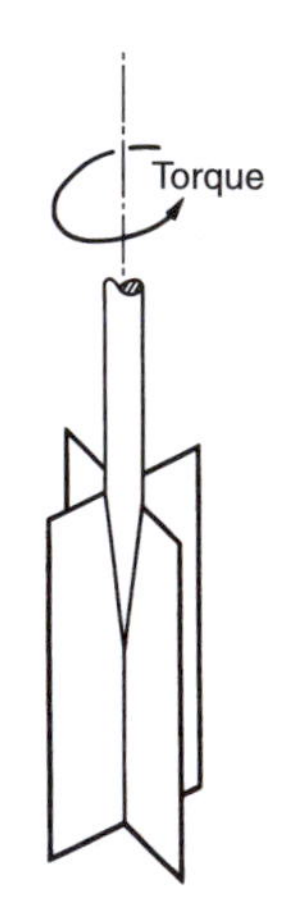

Fig. 4.32 Representation of the shear vane

is often incorrectly assumed that the soil is isotropic and

$$s_{uv} = s_{uh} = s_u$$

whence,

$$T = \frac{\pi D^2}{2}(H + D/3)s_u = ks_u$$

where k is a geometrical constant of the vane.

The in situ shear vane may be used in inspection pits and down boreholes for the extensive determination of in situ strength profiles as part of a site investigation programme. Some types of shear vane equipment have the extension rods in an outer casing with the vane fitting inside a driving shoe. This type of vane may be driven to the desired depth and the vane extended from the shoe and the test carried out. The vane may then be retracted and driving continued to a lower depth.

It must be emphasized that the in situ vane provides a direct measure of shear strength and because the torque application is hand operated, it is a relatively rapid strength measure, therefore giving the undrained shear strength.

Correlations of vane strength of soft clays with embankment stability

By comparing the stability of embankments which had failed with the predicted stabilities based on in situ vane measurements, Bjerrum (1972) derived an empirical correction factor based on the plasticity index to enable the strength measurements made by conventional rectangular vanes to be factored to give a realistic forecast of field stability, using a traditional limit analysis based on a rigid–plastic shear stress–displacement relationship which does not vary with orientation.

Bjerrum (1972) considered 14 case histories when formulating his correction factor μ_B. As shown in Table 4.4, we have added a further 15 case histories. Noting that

$$\mu_B = (s_u)_{field}/(s_u)_{vane} = 1/(F)_{vane}$$

enables μ_B to be plotted against plasticity index I_p% in Fig. 4.33 where a hyperbola is fitted.

Bjerrum's correction factor, correlated as it is with plasticity index, i.e. with data obtained from tests on remoulded soils, is linked to a parameter that cannot measure inherent anisotropy. This may account for some of the scatter in the data used by Bjerrum in obtaining the relationship between his correction factor and plasticity index. The best-fit curve that resulted may well have eliminated the unrelated effect of anisotropy, in which case the correction factor does not correct for anisotropy but may well accommodate the effects of testing rate and progressive failure. If this is so then Bjerrum's correction factor must be applied to correct for testing

Table 4.4 Correlation between plasticity index and factor of safety predicted from in situ shear vane measurements for embankments on soft clay which have failed

No	Literature reference and location	Factor of safety, F	Plasticity index, I_p
1 [a]	Parry and McLeod (1967) Launceston	1·65	108
2 [a]	Eide and Holmberg (1972) Bangkok, A	1·61	85
3 [a]	Eide and Holmberg (1972) Bangkok, B	1·46	85
4 [a]	Golder and Palmer (1955) Scrapsgate	1·52	82
5 [a]	Pilot (1972) Lanester	1·38	72
6	Eide (1968) Bangkok	1·5	60
7 [b]	Peterson *et al.* (1957) Seven Sisters	1·5	59
8 [a]	Pilot (1972) Saint André de Cubzac	1·4	47
9 [a]	Dascal *et al.* (1972) Matagami	1·53	47
10 [a]	Pilot (1972) Pornic	1·2	45
11	Serota (1966) Escravos Mole	1·1	40
12	Roy (1975) Somerset	1·2	36
13	Roy (1975) Brent Knoll	1·37	36
14 [a]	Wilkes (1972) King's Lynn	1·1	35
15 [a]	Lo and Stermac (1965) New Liskeard	1·05	33
16 [a]	Pilot (1972) Palavas	1·30	32
17 [b]	Stamatopoulos and Kotzias (1965) Thessalonika	≃1·0	30
18 [b]	La Rochelle *et al.* (1974) Saint-Alban	1·3	25
19	Flaate and Preber (1974) Jarlsburg	1·1	25
20	Flaate and Preber (1974) Aulielava	0·92	23
21	Flaate and Preber (1974) Nesset	0·88	22
22	Flaate and Preber (1974) Ås	0·80	20
23	Flaate and Preber (1974) Presterudbakken	0·82	17
24 [a]	Pilot (1972) Narbonne	0·96	16
25 [a]	Ladd (1972) Portsmouth	0·84	16
26 [a]	Haupt and Olson (1972) Fair Haven	0·99	16
27	Flaate and Preber (1974) Skjeggerud	0·73	11
28	Flaate and Preber (1974) Tjernsmyr	0·87	8
29	Flaate and Preber (1974) Falkenstein	0·89	8

[a] Quoted by Bjerrum (1972).
[b] Estimated average values.

rate and progressive failure and a separate further correction factor must be applied to correct specifically for anisotropy. For an illustration of the nature of progressive failure, see Short Course Notes: Progressive Failure.

In order to make some estimate of this correction factor, Menzies (1976a) considered the simplified bearing capacity model given in Fig. 4.34. The soil was taken to be weightless and to fail on a circular arc. To simplify the analysis further the centre of rotation was taken above the edge of the uniformly loaded strip. The analysis is similar to that given by Raymond (1967). It was assumed that the difference between the vertical undrained shear strength s_{uv} and the horizontal undrained shear strength

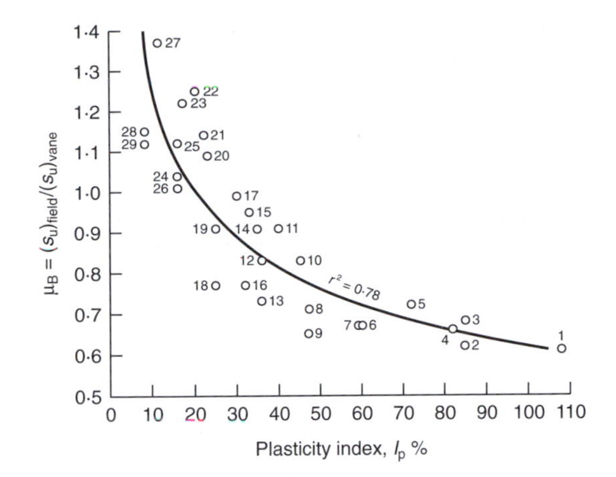

Fig. 4.33 Correlation between plasticity index and empirical shear vane correction factor μ_B for failed embankments on soft clay (Menzies and Simons, 1978)

s_{uh} may be distributed as the square of the direction cosine (Timoshenko, 1934; Casagrande and Carillo, 1944; Lo, 1965; Menzies, 1976) giving the undrained shear strength in any direction α as

$$s_{u\alpha} = s_{uh}[1 + (R - 1)\cos^2 \alpha]$$

where

$$R = s_{uv}/s_{uh}$$

is the degree of undrained strength anisotropy.

A factor was obtained which corrects for the influence of strength anisotropy on conventional shear vane measurements used to predict field bearing capacity as

$$\mu_A = \frac{[(R + 1)2\theta - (R - 1)\sin 2\theta]/\sin^2 \theta}{2{\cdot}37(2R + 1/3)}$$

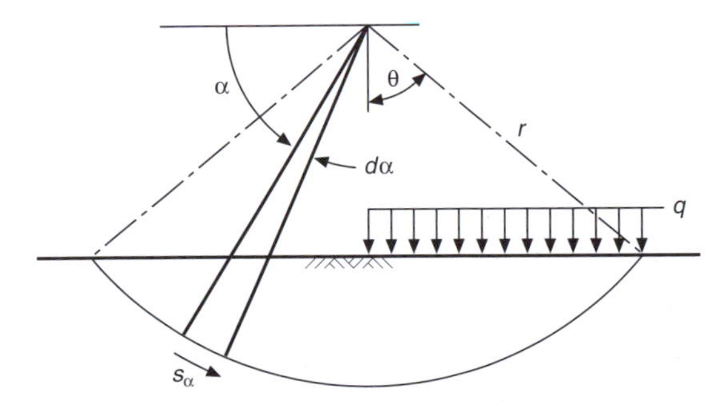

Fig. 4.34 Simplified bearing capacity configuration (Menzies, 1976a)

Short Course Notes: Progressive failure

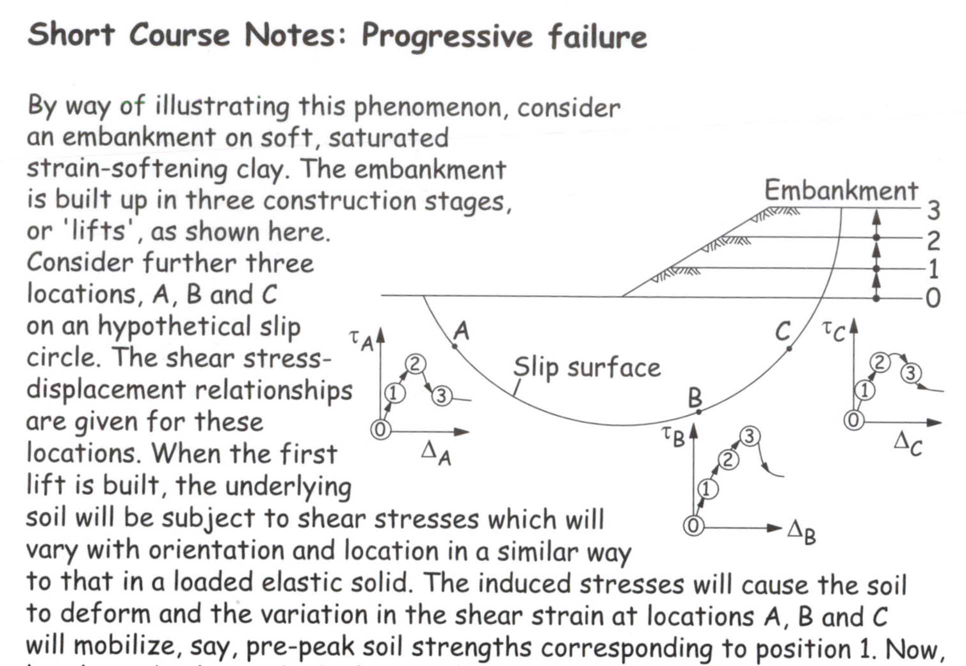

By way of illustrating this phenomenon, consider an embankment on soft, saturated strain-softening clay. The embankment is built up in three construction stages, or 'lifts', as shown here. Consider further three locations, A, B and C on an hypothetical slip circle. The shear stress-displacement relationships are given for these locations. When the first lift is built, the underlying soil will be subject to shear stresses which will vary with orientation and location in a similar way to that in a loaded elastic solid. The induced stresses will cause the soil to deform and the variation in the shear strain at locations A, B and C will mobilize, say, pre-peak soil strengths corresponding to position 1. Now, let the embankment be built up with a second lift of fill, again increasing shear stresses in the ground. Let peak strength be attained at location A which is now within a zone of failure or plastic yield at the toe of the slip which has not yet fully formed. Locations B and C are still within a pre-failure zone at pre-peak stresses represented by position 2.

Now, a third lift is constructed and further shear stresses applied to the soil with consequent shear strains developing. At location A this associated deformation exceeds the peak value and the load-carrying capacity of this location is diminished according to the strain-softening relationship of the soil. This shedding of load puts additional shear stress on the pre-failed regions with the consequence that a further failure zone develops at location C where the peak stress is exceeded. In turn, the load shedding from location C causes location B to fail, also attaining peak shear stress at position 3. In this way a strain-softening soil fails in a progressive way.

Conventional stability analyses do not model progressive failure. To take the peak strength as acting on all elements simultaneously around the slip surface overestimates the factor of safety. On the other hand, to take the residual shear strength as acting simultaneously around the slip surface underestimates the factor of safety and is clearly inappropriate in any event as the residual strength holds on an established shear surface, i.e. after failure has occurred.

The rigid-plastic shear stress-displacement relationship implied by a traditional analysis clearly does not accommodate this strain softening behaviour (see figure). Using the residual factor (Skempton, 1964) goes some of the way towards compensating for this effect but the value of the residual factor needs to be estimated from field failures and the analytical model implies, inappropriately, a rigid-plastic idealization of the shear stress-displacement relationship.

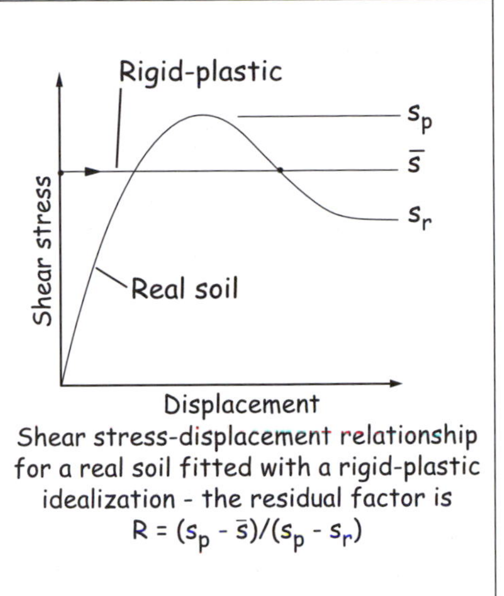

Shear stress-displacement relationship for a real soil fitted with a rigid-plastic idealization - the residual factor is $R = (s_p - \bar{s})/(s_p - s_r)$

where θ is given by

$$\tan\theta = 2\theta k$$

where

$$k = \frac{(R+1) - [(R-1)\sin 2\theta]/2\theta}{(R+1) - (R-1)\cos 2\theta}$$

The relationship between μ_A and R is plotted in Fig. 4.35.

In order that the conventional shear vane strength $(s_u)_{vane}$ may be used in a traditional limit analysis, i.e. assuming the soil has a rigid–plastic shear strength–displacement relationship which does not vary with

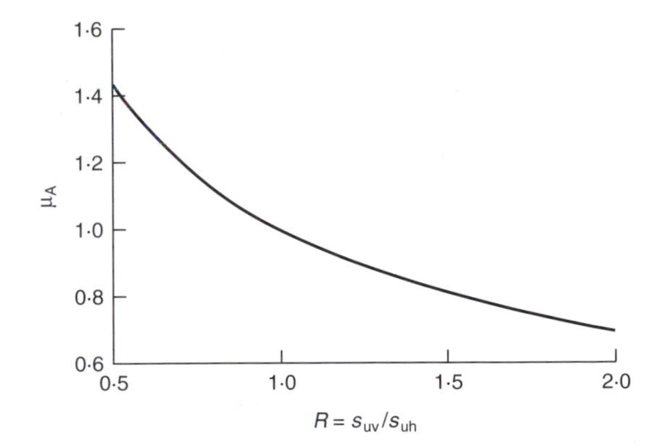

Fig. 4.35 Variation of anisotropy correction factor μ_A with degree of strength anisotropy R *(Menzies, 1976)*

direction, it is suggested that $(s_u)_{vane}$ should be corrected to give the field strength $(s_u)_{field}$ as follows:

$$(s_u)_{field} = \mu_A \mu_B (s_u)_{vane}$$

where μ_A is the correction factor for strength anisotropy given in Fig. 4.35 and μ_B is Bjerrum's correction factor for the effects of testing rate and progressive failure given in Fig. 4.33. This equation is similar to that proposed by Bjerrum (1972).

Case study: the Brent Knoll trial embankment, UK

The Brent Knoll trial embankment was built as part of the testing programme for the geotechnical design of the M5 motorway. This motorway from Birmingham to Exeter forms part of the principal road network linking major development areas in Britain. The section of motorway from Clevedon 10 km south of Bristol, to Huntworth 2 km south of Bridgwater, is built on the soft alluvial clays of the Somerset Levels where the depth of alluvium is typically about 26 m. These sediments are so soft that the stability of all embankments was low and large settlements were expected. The site of the trial embankment is 9 km SSE of Weston-Super-Mare and 1·6 km from East Brent. The National Grid reference co-ordinates of the site are ST 3653.

Construction of the trial bank started at the end of October 1967 and continued for three months. It was 189 m long and 70 m wide, with side slopes of about 1 : 3, and was built of compacted quarry waste with not more than 15% passing a 75 mm BS sieve, except for the bottom 0.46 m where the fines were limited to less than 10% in order to form a drainage layer. After compaction by at least eight passes of a vibrating roller, the fill had a bulk unit weight of about 23 kN/m^3. The intention was to construct the bank to a height of 9·1 m with a final crest width of 15 m. This was the expected profile of the motorway interchange embankments. A plan of the embankment is shown in Fig. 4.36.

In January 1968, with 7·9 m of fill in place a 61 m length of the bank slipped with a 1·2 m slump at the crest, the failure extending to 20 m beyond the toe of the bank. Numerous tension cracks opened up in the unslipped fill generally parallel to the slump at the crest. The opposite side of the bank appeared to be on the verge of failure, the side slope becoming slightly S-shaped and a construction peg near the toe of the bank moving outwards approximately 0·5 m. A section through the slip is shown in Fig. 4.37.

To prevent further movement of the slip and additional damage to the instrumentation, a 1·2 m thick, 15·2 m wide berm was immediately placed along the toe of the bank in the area of the slip. Construction then ceased.

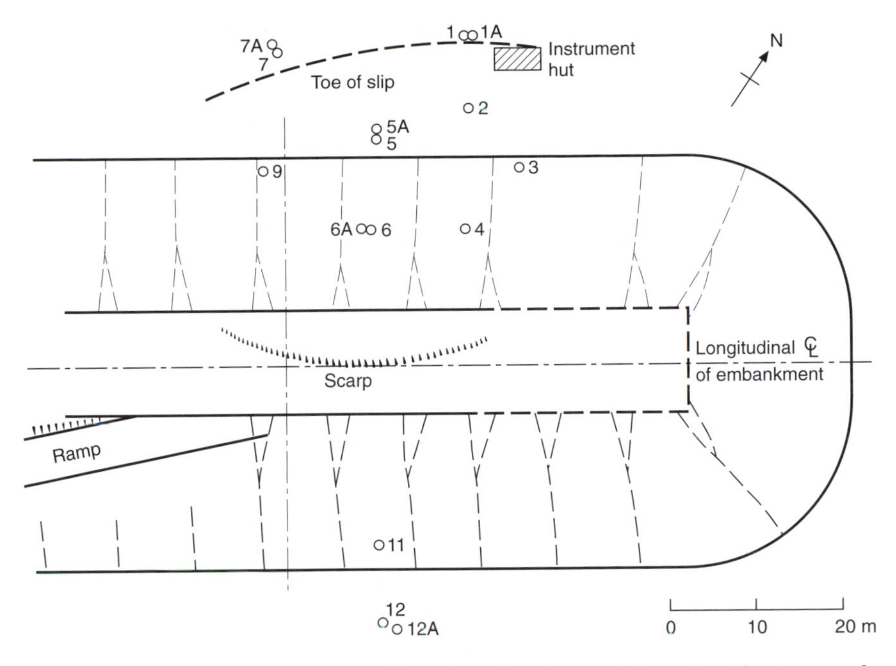

Fig. 4.36 Plan of the Brent Knoll trial embankment showing the toe and scarp of the slip failure (Menzies and Simons, 1978)

Four borings (2, 3, 4 and 9) were made in the slipped area and continuous samples were taken. These were examined on site for signs of failure zones. Two 150 mm thick bands of disturbed clay were generally found in these holes and, although it was very difficult on site to differentiate between sample disturbance and failed material, a general

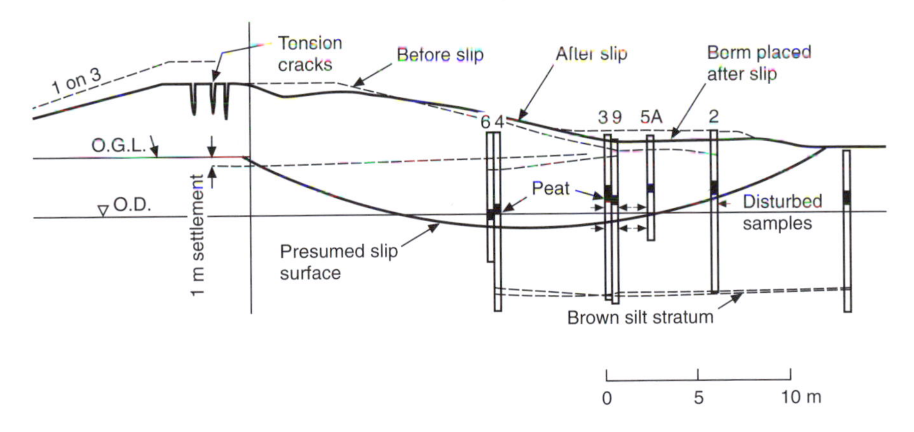

Fig. 4.37 Cross-section through the Brent Knoll trial embankment after failure showing the presumed slip surface and borehole elevations (Menzies and Simons, 1978)

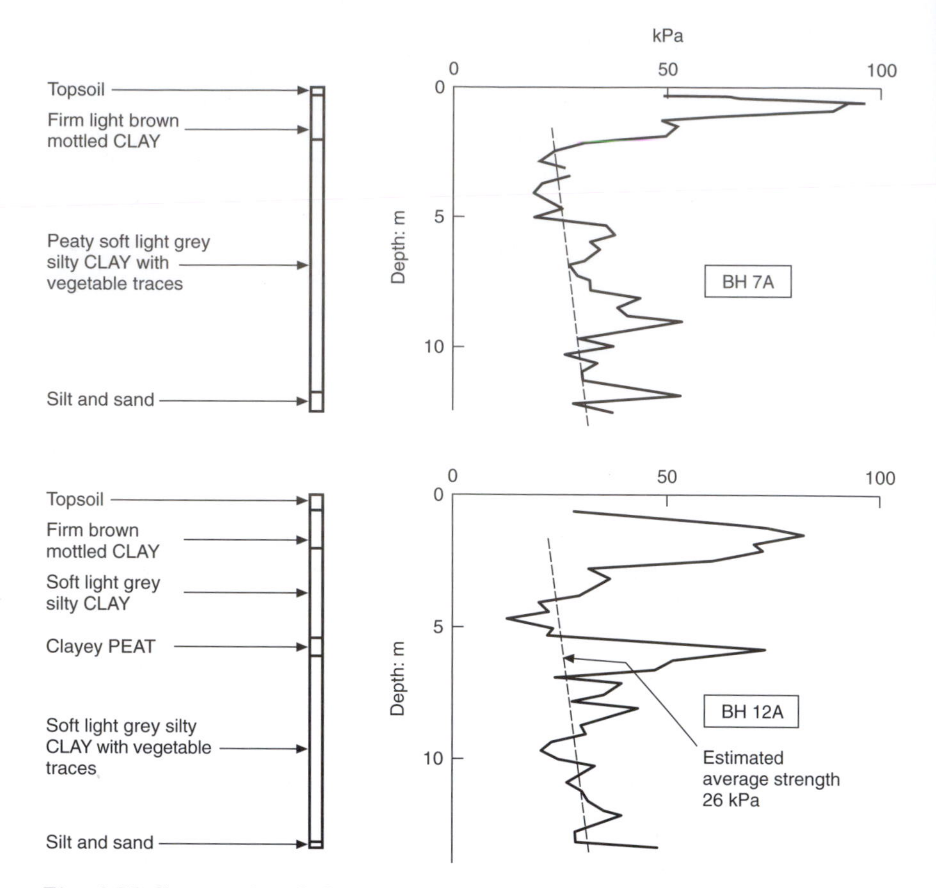

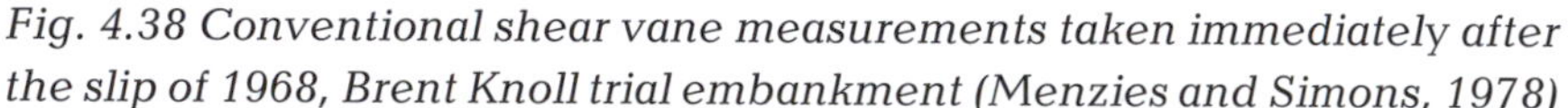

Fig. 4.38 Conventional shear vane measurements taken immediately after the slip of 1968, Brent Knoll trial embankment (Menzies and Simons, 1978)

pattern developed when these positions were plotted on a cross-section (Fig. 4.37). The evidence from borehole 9 is particularly conclusive as overlapping piston samples were taken in this hole.

Standard vane tests were carried out in three boreholes outside the bank (boreholes 1A, 7A and 12A). The results of two of these are shown in Fig. 4.38. It was found that the ground became disturbed very easily and that the vane tests had to be executed well below the bottom of the hole.

During July and August of 1972, the failed trial embankment was levelled to a height of approximately 2·5 m above original ground level. In November 1975 further field testing was carried out. A test pit was excavated by digger to a depth of 2·5 m alongside the south-west corner of where the instrument hut had been sited. Two block samples were removed for subsequent laboratory testing. The block samples were cut with thin wire from the excavated clay retained in the digger bucket,

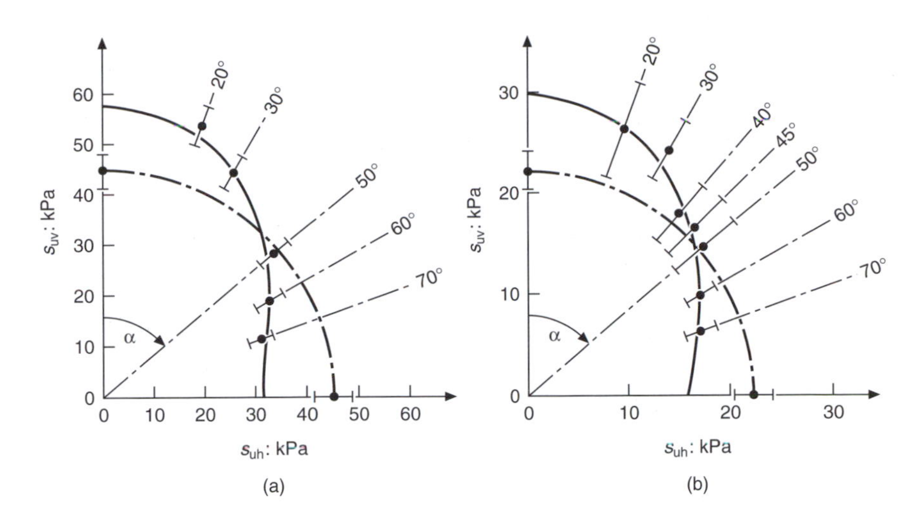

Fig. 4.39 Polar diagram showing the distribution of undrained shear strength measured by diamond-shaped vanes for (a) cores from the bed of the North Sea – points show range of values, (b) in situ tests on the east bank of the New Blind Yeo, Clevedon, Somerset – points show standard deviation. The solid line distributes the difference between the vertical and horizontal strengths in proportion to the square of the direction cosine. The broken line is the circular arc representing the conventional vane measurement (Menzies and Mailey, 1976)

trimmed on site to approximate cubes of side 400 mm, immediately wrapped in aluminium foil and sealed with melted wax.

From the floor of the test pit in situ shear vane tests were carried out using a conventional rectangular vane and a set of diamond vanes. The concept of the diamond-shaped shear vane was introduced by Aas (1967) who carried out shear tests using a diamond vane making an angle of 45° to the axis of rotation. Menzies and Mailey (1976) extended this concept by using diamond-shaped shear vanes of different angles to the axis of rotation, enabling the variation of shear strength with direction to be measured as shown in Fig. 4.39. The vanes used are shown in Fig. 4.40.

The vanes used at Brent Knoll are shown in Fig. 4.41. In common with the conventional vane, which was 140 mm long and 70 mm in diameter, the diamond vanes were symmetrical and cruciform in cross-section. The vanes were used in conjunction with a MHH B 800 torque indicator and were designed to give the same torque assuming strength isotropy. It was assumed that a diamond vane making an angle α with the vertical vane axis of rotation shears the soil on planes making an angle α with the vertical direction thus giving a measure of the undrained shear strength in the orientation α with the vertical.

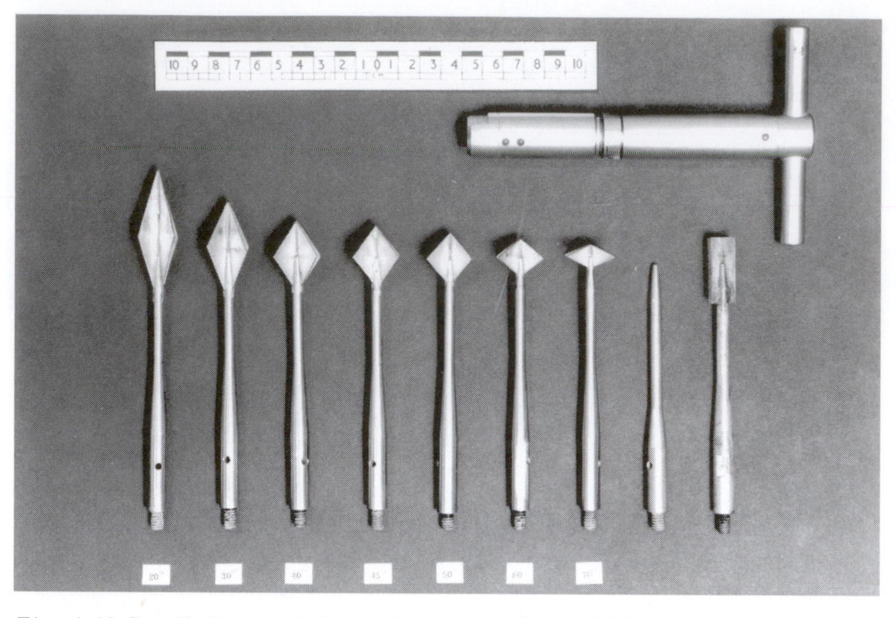

Fig. 4.40 Small diamond-shaped, rectangular and bladeless vanes and Geonor torque head. The bladeless 'vane' provided measurements of shaft friction. (Menzies and Mailey, 1976)

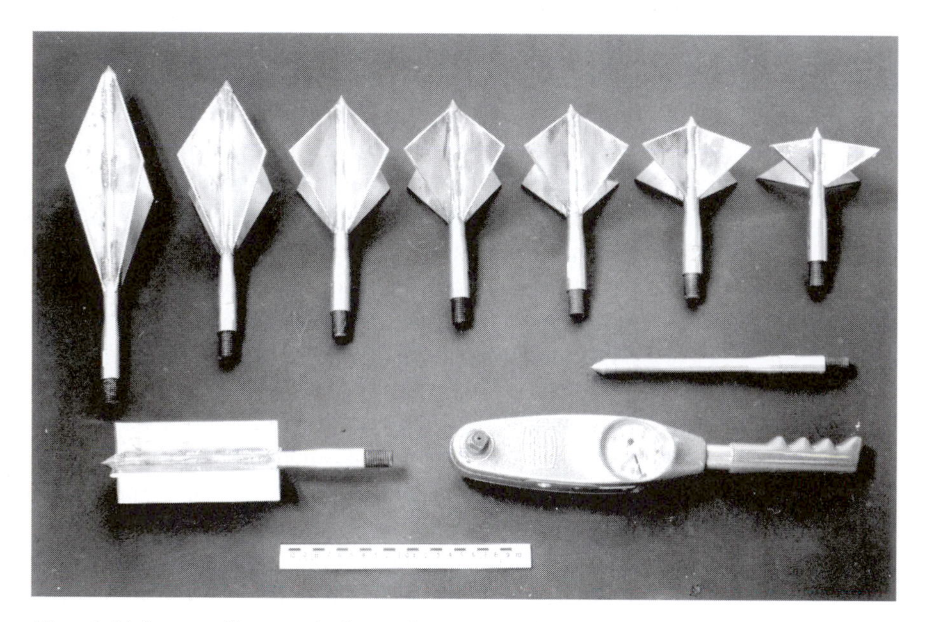

Fig. 4.41 Large diamond-shaped, rectangular and bladeless vanes and MHH torque head. Diamond vanes, left to right, are 20°, 30°, 40°, 45°, 50°, 60° and 70° with respect to the vertical axis of rotation. (Menzies and Simons, 1978)

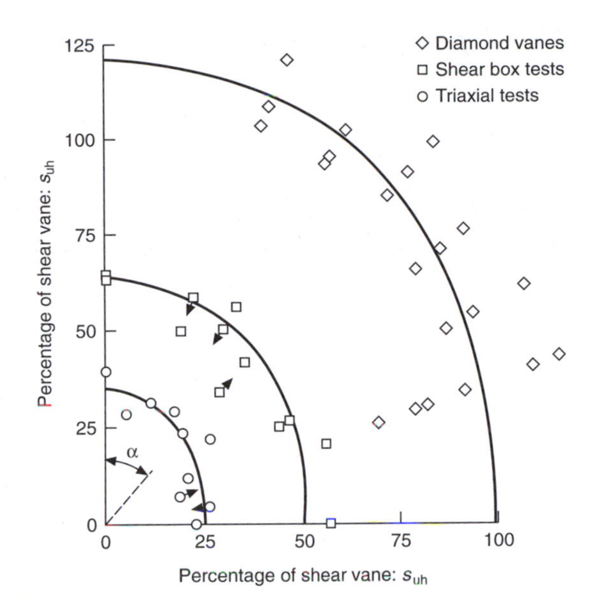

Fig. 4.42 Polar diagram of undrained shear strength measured by diamond vane, triaxial and direct shear tests as a percentage of the strength in the horizontal plane as measured by the diamond vane (inferred by curve fitting). The arrows show the direction of corrections for the effect of water content variations. (Menzies and Simons, 1978)

The conventional and diamond shear vanes were progressively inserted in the base of the test pit and readings of strength with depth were carried out at different depths up to 2·0 m below the bottom of the pit. The excavation of the pit and the vane testing was carried out over a period of 8 h. A polar diagram showing the variation of strength with direction measured by the diamond vanes is given in Fig. 4.42. The curve fitted to the points has the difference between the vertical and horizontal strength distributed as the square of the direction cosine.

In the laboratory, triaxial test specimens 38 mm in diameter by 76 mm long and prismoidal direct shear test specimens 60 mm by 60 mm by 20 mm were trimmed from the blocks and tested in undrained shear. Polar diagrams showing the variation with direction of undrained shear strength for the direct shear test and the triaxial test are also shown in Fig. 4.42. The polar diagram for the undrained triaxial test is not directly comparable to those for the shear vane and the direct shear tests. In the triaxial test results the direction given is of the test specimen axis and not of the failure plane. Distinct failure planes were not detected in the triaxial test specimens and so it was not possible to give the results with respect to the orientation of a sheared surface.

It may be noted that there are significant disparities in average strength magnitudes between different types of test. Similar disparities were observed by Madhloom (1973) who carried out undrained tests on a soft clay of plasticity index 35% from King's Lynn, Norfolk. England. The direct shear test gave an undrained shear strength some 60% greater than that given by the triaxial compression test.

For the soft clay of Brent Knoll some of this variation of strength with type of test may be attributed to the effects of type of soil distortion during the test, testing rate, time between sampling and testing, size of test specimen or zone, and degree of sample disturbance. The degree of anisotropy measured by the different types of test varies from about 1·1 to 1·4, the average being 1·26.

Using a conventional total stress stability limit analysis assuming strength isotropy, the field strength was deduced from the slip circle shown on Fig. 4.37 and found to be 19 kPa. The conventional shear vane measurements taken immediately after the slip in 1968 are given in Fig. 4.38. The conventional rectangular shear vane indicated an average undrained shear strength in the vicinity of the slip of about 26 kPa. Correcting this value of shear strength using the data given in Table 4.5 gives a factor of safety of 1·02.

Given the various assumptions and engineering judgements made, the prediction of the factor of safety $F = 1{\cdot}02$ must be considered as fortuitously exact compared with $F = 1{\cdot}00$ for the failed embankment. The correction factor for anisotropy, μ_A, is itself only an approximation and the empirical vane correction factor, μ_B, is based on stability data which were undoubtedly influenced by strength anisotropy to various unknown extents. Accordingly, it may be expected that applying both correction factors will over-correct for anisotropy and thus lead to conservatively low values of field strength. Nevertheless, making the correction

Table 4.5 Stability data relating to the Brent Knoll trial embankment

Parameter	Value
Natural water content, w (%)	60
Liquid limit, w_I (%)	64
Plastic limit, w_P (%)	28
Plasticity index I_P (%)	36
Empirical correction factor, μ_B	0·84
Anisotropy correction factor, μ_A	0·89
No vane correction, F	1·37
With vane correction	
$\mu_B F$	1·15
$\mu_A \mu_B F$	1·02

for anisotropy together with the empirical vane correction leads to a good estimate of the field strength in this particular case.

> The Brent Knoll Trial Embankment case study illustrates how measured strength from a test model (in this case the shear vane) when empirically correlated with field performance (via Bjerrum's correction factor) and combined with realistic analytical interpretation (anisotropy correction factor) can help to achieve a good estimate of field stability.

Plate loading test

Overview

Traditionally, plate loading tests were carried out to determine strength and compressibility of soils that were difficult to sample and test in a representative way. Plate loading tests are now carried out to determine stiffness parameters in materials that are difficult to sample such as granular soils and fractured rock. Plate loading tests also provide 'ground truth' for benchmarking stiffness found by seismic methods (these are discussed later in this chapter). In fissured clays the stiffness of the fissures is sufficiently close to the stiffness of the intact clay not to reduce the mass stiffness significantly. For fissured clays, therefore, small-strain laboratory tests may be carried out to obtain stiffness parameters. For rocks, however, the presence of discontinuities significantly reduces the stiffness of the rock mass. Intact blocks of rock may have a high stiffness but the stiffness of the rock mass may only be 10% of this stiffness depending on the frequency and orientation of the discontinuities. Accordingly, plate loading tests are now used in the assessment of the properties of rocks, particularly weak rocks (Ward *et al.*, 1968; Hobbs, 1975; Lake and Simons, 1975; Powell *et al.*, 1990; CIRIA, 1994) along with seismic methods. Plate loading tests represent the best method for providing mass compressibility parameters in weak rock and certain types of soil (e.g. sands, gravels, boulder clay and fill material). Although expensive, such tests afford the greatest degree of similitude between the test and the field prototype. The interpretation of the plate loading test is, however, complicated by the fact that the rigidity and geometry of the plate results in different elements of ground beneath the plate following different stress paths. This results in problems extrapolating from the test scale to that of a full-scale foundation.

Method

The equipment used is shown in Fig. 4.43(a) and Fig. 4.43(b). Plate loading tests may be carried out in shallow pits or in a borehole. The test involves loading a circular plate (rectangular or square plates are

(a)

Fig. 4.43 Plate test. (a) Photograph of the rig used by Matthews (1996).

also used) and monitoring the settlement. Techniques for carrying out these tests are described by CP 2001: 1957, Burland and Lord (1970); ASTM D1194-72, Tomlinson (1980); BS 1377: 1990 (Part 9), Clayton *et al.* (1995); and BS 5930: 1999, British Standards Institution (1999). In the test the soil or rock surface to be tested is carefully cleaned by removing all loose and softened material and the plate is bedded into this surface using sand–cement mortar or Plaster of Paris. The cleaning operation is of particular importance as any loose compressible material left on the test surface will lead to errors in the determination of modulus E. Adequate cleaning may be difficult or impossible when tests are carried out down a borehole.

Load is applied to the plate via a load cell and a hydraulic jack. The jack may bear against beams supporting kentledge or reaction may be provided by tension piles or ground anchors installed on either side of the load position. In weak rocks such as chalk tension piles are necessary

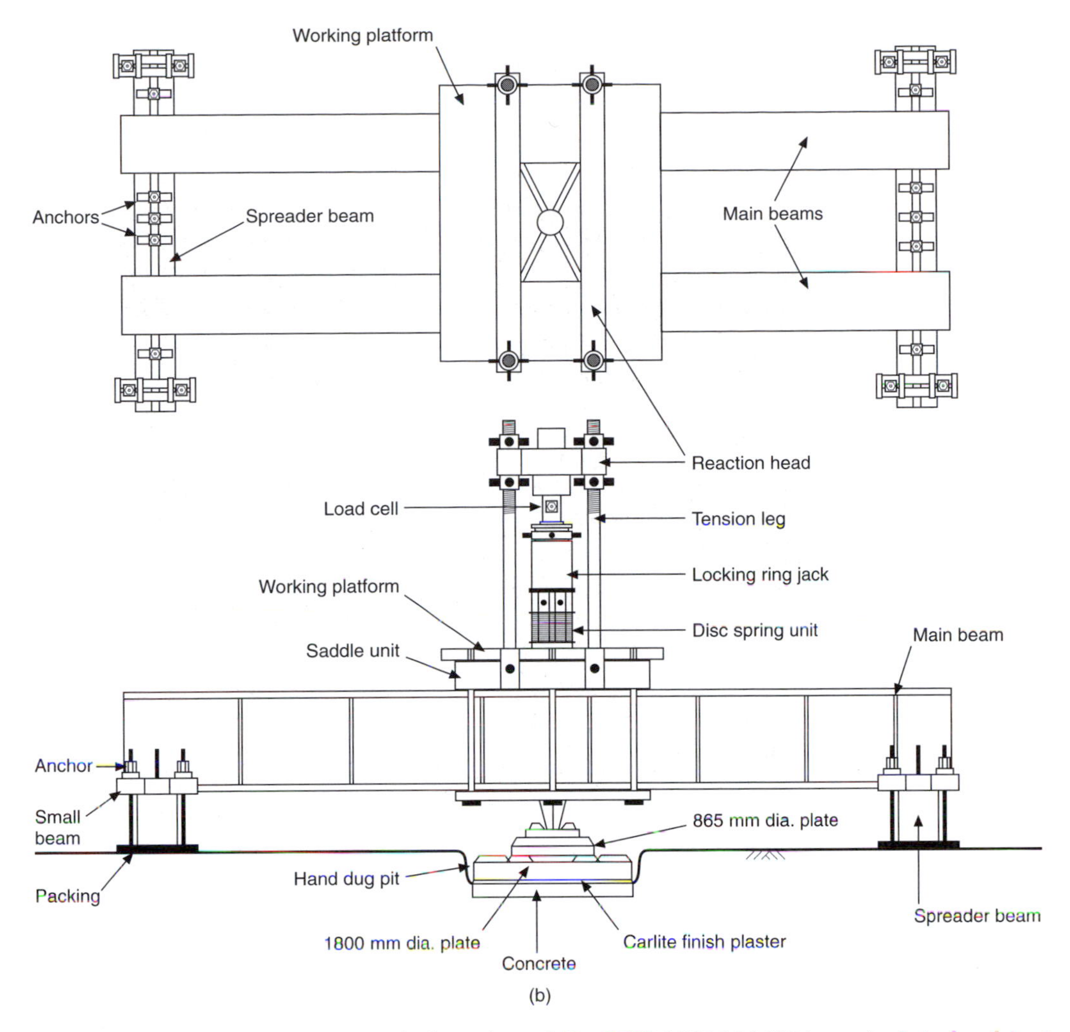

Fig. 4.43 Plate test. (b) Plan and elevation of the BRE 5000 kN (50 tonne) plate load test equipment, surface mounted

if it is intended to study the post-yield behaviour with an 865 mm or larger diameter plate. When using tension piles it is important that they are positioned outside the zone of influence of the plate and hence do not influence the results. BS 1377: 1990 (British Standards Institution, 1990) recommends that tension piles should not be positioned less than three times the plate diameter from the centre of the plate. When kentledge is used, the maximum plate size practicable may be considered to be about 1 m diameter, since such a plate loaded to two and a half times a design pressure of 200 kPa will require about 40 tonnes of kentledge. The manner and rate at which the load is applied to the plate gives rise to two categories of test.

- *Constant rate of penetration test (CRP)*. The load is applied in a controlled manner such that the selected rate of penetration is uniform and continuous. These tests are normally carried out at such a rate that only the undrained or immediate settlement behaviour is obtained. The rate of penetration in CRP tests carried out on the chalk at Luton was 2.5 mm per minute (Powell *et al.* 1990).
- *Maintained incremental load test (ML)*. Load is applied to the plate in successive increments of about one fifth of the design loading, with a certain time interval between the application of each increment. Burland and Lord (1970) suggest that in tests on chalk the time interval between increments may be based on the rate of settlement. They adopted a settlement rate of 0·005 mm per half hour (0·0002 mm/min). The time interval between loading increments was never less than 15 min. Clayton *et al.* (1995) recommend a rate of settlement of 0·004 mm per minute measured for a period of at least 60 min. Since most geomaterials exhibit time dependent settlement characteristics the values of E will be influenced by the time interval between loading increments together with the magnitude of the increment. Load increments are applied either until shear failure of the soil occurs, or more commonly the plate pressure reaches two or three times the design bearing pressure proposed for the full-scale foundation.

Settlement of the plate is measured using dial gauges or linear displacement transducers reading to 0·05 or 0·01 mm. In order to measure any tilt Clayton *et al.* (1995) suggest using four gauges on the perimeter of the largest plate. These instruments are normally supported on rigid uprights driven (or grouted) firmly into the ground at a distance of at least twice the plate diameter from the plate centre. An example of such an arrangement is shown in Fig. 4.44. A reference beam for mounting gauges should be adequately protected from temperature changes.

When the test is conducted down a borehole the vertical settlement of the plate is transferred to the ground surface by an Invar tape and a reference beam system described by Ward *et al.* (1968). The reference beam needs to be supported on stable concrete plinths positioned outside the zone of influence of the plate. This means that the reference beam needs to be at least some 5 m long and hence is prone to movements associated with temperature changes. These movements can be minimized by protecting the beam from direct sunlight. Nevertheless, regular measurements of the beam movement should be made using precise levelling during each loading stage. It is important to level both sides of the beam since such long beams have a tendency to twist.

When the test is performed in a shallow pit the settlement of the plate may be monitored using precise levelling using deep benchmarks

Fig. 4.44 Photograph showing layout of dial gauges and beams for precise levelling for plate settlement measurement, and also showing stiffening plates used on the loading plate (Matthews, 1993)

situated outside the zone of influence of the plate. A combination of dial gauges and precise levelling should be used for surface plate tests. The dial gauges will not give a measure of absolute plate movement but will provide a close back-up at minimal cost. Clearly the time taken to perform a round of levels means this technique is unsuitable for CRP tests.

The stiffness measured using a plate loading test is sensitive to plate diameter. Clayton *et al.* (1995) suggest that as a general rule of thumb the plate diameter should never be less than either six times the maximum soil particle size or six times the maximum intact block size. BS 1377: 1990 recommends for plate tests on fissured clays that the plate should be more than five times the average block size. Lake and Simons (1975) suggested the use of a 600 mm diameter plate on chalk which had an intact block size in the range 50–100 mm, based on predicted and observed settlements of a building at Basingstoke,

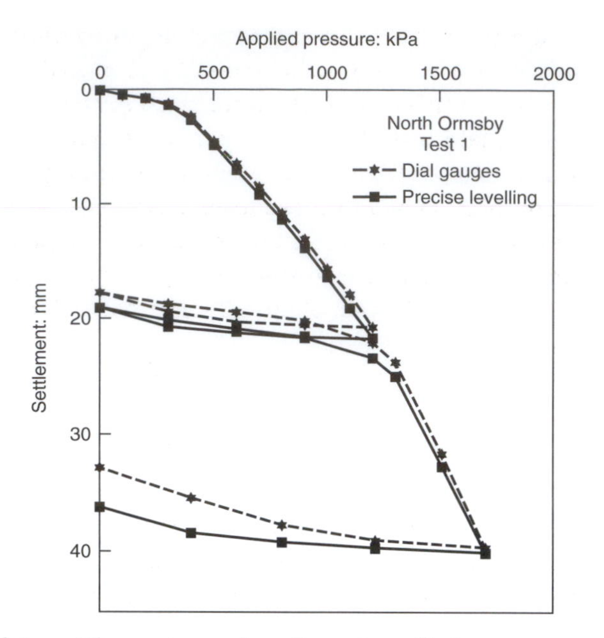

Fig. 4.45 Plate settlement as a function of applied pressure, comparing measurements made by dial gauges and precise levelling (Matthews, 1993)

England. Although the plates used on the chalk in the UK range in diameter between 140 mm and 1710 mm, the preferred size is 865 mm.

Details of the equipment, setting it up, and the testing procedure are given by Burland and Lord (1970) and Marsland (1971). Where large plates are used they should be made as rigid as possible by stacking successively smaller spreader plates above them and below the load cell (Fig. 4.44). Thus a 1 m plate will typically have 0·75 m, 0·50 m and 0·30 m diameter plates above it. The minimum plate size should be 0·30 m.

Typical results of a plate loading test on chalk are shown in Fig. 4.45. For plate tests intended to give elastic moduli values for soils or rocks, BS 5930: 1999 recommends the use of the equation for a uniformly loaded rigid plate on a semi-infinite elastic isotropic solid, with no stiffness increase with depth (Poulos and Davis, 1974) for the determination of Young's modulus E:

$$E = \frac{\pi q D}{4} \frac{(1 - \nu^2)}{\rho}$$

where q is the applied pressure between the plate and the ground, D is the plate diameter, ν is the Poisson's ratio for the soil/rock mass and ρ is the settlement of the plate under the applied pressure q. For granular soils and soft rocks, Poisson's ratio will normally be between 0·1 and 0·3, and so the term $(1 - \nu^2)$ has a relatively small effect. Where plate tests are carried out in the stressed zone of a proposed foundation, the value of q can be taken as the vertical foundation stress to be applied at the level of the plate test. Alternatively, a safety margin can be incorporated by taking q to be 50% (for example) higher than the estimated applied stress.

The plates used in these tests, however, are not perfectly rigid and in general the tests are often carried out in pits that are comparatively larger than the plate, at shallow depth or at depth in shafts. This necessitates the introduction of an influence factor I in the above equation, i.e.

$$E = \frac{\pi q D}{4} \frac{(1 - \nu^2)}{\rho} I$$

The influence factor is governed by:

- the flexibility of the plate
- the embedment ratio Z/B of the plate, where Z is depth below ground level and B is the width/diameter of the pit or shaft
- partial loading of the base of the pit or shaft given by D/B.

For a flexible circular plate used in tests at the ground surface $(Z/D = 0)$ the modulus E is given by (after Giroud, 1972).

$$E = 0{\cdot}85 \frac{qD(1 - \nu^2)}{\rho_{\text{ave}}}$$

where ρ_{ave} is the average settlement, equal to the actual settlement at a radius ratio of about 0·75 from the plate centre. In terms of the maximum settlement, i.e. at the centre, the equation is:

$$E = 1{\cdot}00 \frac{qD(1 - \nu^2)}{\rho_{\max}}$$

If the plate is located in a shaft or pit below the ground surface then the equations for E must be multiplied by a depth reduction factor μ_0. Values of the depth reduction factor have been determined for weak rock using finite element analysis by Burland (1970), Pells and Turner (1979) and Donald *et al.* (1980). These various results are summarized in Fig. 4.46 assuming the entire area of the shaft or pit is loaded (i.e. $D/B = 1$). If the plate only occupies part of the base area of the pit the reduction is less. Pells (1983) notes that a full parametric solution is not available, but quotes a particular solution for a flexible circular footing for $Z/B = 2{\cdot}5$ and $\nu = 0{\cdot}25$. This solution can be approximated to a linear relationship

$$\mu_1 = 1 - (1 - \mu_0)D/B$$

where μ_1 is the depth reduction factor corrected for partial loading of the base of the shaft or pit of minimum dimension B with a plate of diameter D.

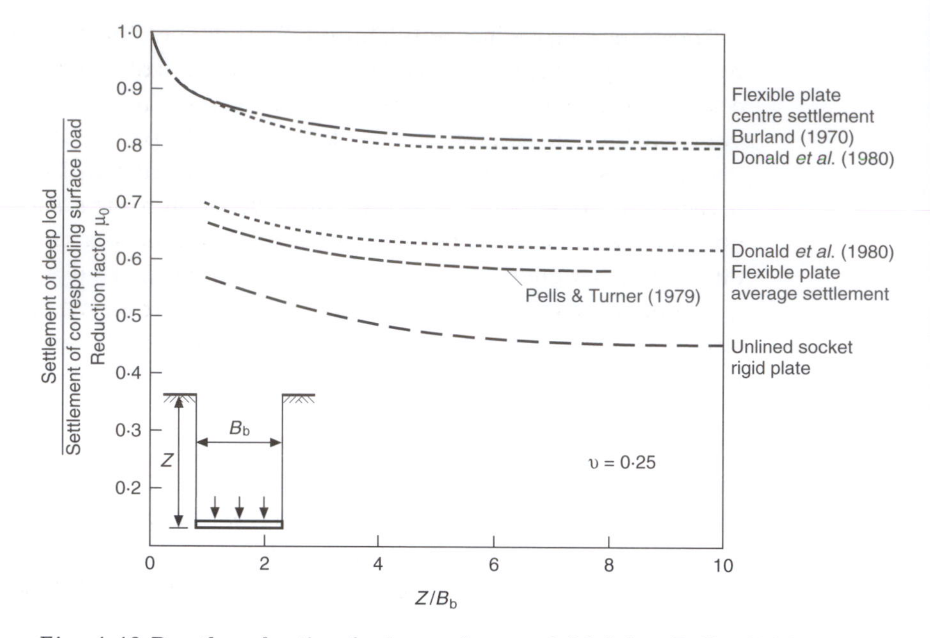

Fig. 4.46 Depth reduction factor μ_0 for $\nu = 0{\cdot}25$ (after Pells, 1983)

Because Young's modulus E is obtained from load–displacement measurements, an assumption regarding the constitutive model is necessary. In most cases the elastic model described above is employed, i.e. homogeneity is assumed. Most geomaterials, however, often display a steady increase in modulus with depth. This imposes a serious limitation on the interpretation of plate loading tests if homogeneity is assumed in conjunction with surface settlement measurements. The effect of non-homogeneity depends on the diameter of the plate or foundation and on the ratio E_0/k where E_0 is the modulus at the surface and k is the rate of increase of E with depth. It is not possible to determine E_0 and k from measurements of plate settlement alone (Hillier, 1992). Since plates of increasing diameter will stress the ground to greater depths the effect of non-homogeneity causes the average E determined from plate settlements to be sensitive to the plate diameter even when it exceeds the average block size by some five or six times. This sensitivity of observed load–settlement behaviour to plate size has been observed in the results of plate loading tests reported by Lake and Simons (1975) and Hodges (1976). E_0 and k may be determined from the results of a suite of plate loading tests in which plates of various diameters are used. This methodology is not cost effective. An alternative to using plates of differing diameters is to use a single plate at a number of different depths in order to determine a profile of modulus with depth.

These techniques for determining E_0 and k all involve performing a number of plate loading tests, which will prove expensive. An attractive alternative is to make measurements of settlement at a number of discrete points below the centre of the plate. An underplate settlement measurement system has been developed by Marsland and Eason (1973) to investigate the effects of bedding disturbance and expansion of the soil on load–settlement behaviour of large-diameter plates. The system allows the settlement at four points beneath the plate (set at 150 mm in a 38 mm diameter axial borehole) to be recorded via a transducer system housed in the plate (see Fig. 4.47). In this way the settlement distribution beneath the plate may be measured and the corresponding vertical strains calculated.

In order to determine modulus values from under-plate strain data, it is necessary to assume a distribution of total stress beneath the centreline of the plate. An axially symmetric stress system may be considered for the ground between each pair of measuring points. From a suitable starting point (e.g. the vertical overburden pressure), the (secant) Young's modulus for each element of ground can be calculated from:

$$E = \frac{\Delta\sigma_{\mathrm{v}} - 2\nu\Delta\sigma_{\mathrm{h}}}{\Delta\varepsilon_{\mathrm{v}}}$$

where $\Delta\sigma_{\mathrm{v}}$ is the assumed change in vertical stress, $\Delta\sigma_{\mathrm{h}}$ is the assumed change in horizontal stress, $\Delta\varepsilon_{\mathrm{v}}$ in the change in vertical strain and ν is Poisson's ratio.

At the simplest level the stress changes can be estimated on the basis of solutions available in the literature for perfectly elastic materials and idealized geometries. Finite element analysis of plate loading tests carried out by Hillier (1992) indicates that serious errors in modulus arise if under-plate strain measurements are made within half a plate diameter of the underside of the plate, when such a simplistic approach is taken for the determination of the stress changes. The device described by Marsland and Eason (1973) was designed for use in stiff clays and relies on the anchor points digging into the borehole wall. When used in weak rock such effective anchoring cannot be achieved and slippage often occurs which significantly reduces the accuracy and resolution of this device.

Accurate settlement predictions of full-scale structures based on plate loading test data requires knowledge of the modulus depth profile. The average E derived from surface measurements of plate settlement alone can give rise to significant errors in settlement predictions when the ground is non-homogeneous. If the plate is the same size as the foundation or the ground has a uniform stiffness such errors do not arise. Thus extrapolation from the test scale to full-scale has always been problematic.

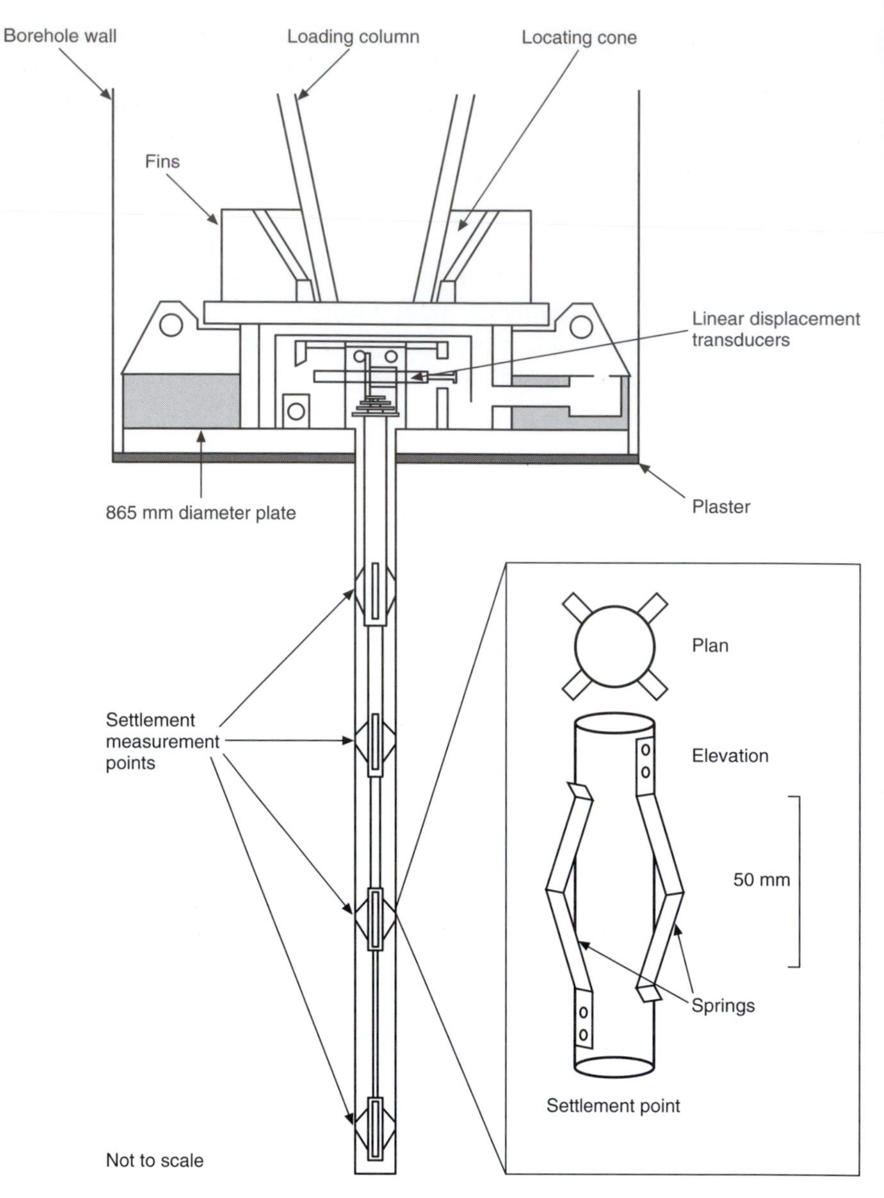

Fig. 4.47 Schematic diagram showing 865 mm diameter loading plate with four-point sub-plate ground-deformation measuring system (after Marsland and Eason, 1973)

Terzaghi and Peck (1948) proposed:

$$\rho_B = \rho_1[2B/(B+1)]^2$$

where ρ_B is the settlement of a footing of width B and ρ_1 is the settlement of a 1 ft plate. This leads to a maximum settlement ratio of four, however big the footing. Work carried out by Bjerrum and Eggestad (1963), D'Appolonia *et al.* (1970), Sutherland (1975) and Levy and Morton (1975) indicates that extrapolation of settlement from small plates to large loaded areas on granular soils is rather unreliable, and therefore the plate loading test on granular material should be regarded as giving a modulus of compressibility value for the soil immediately beneath the test location. Elastic stress distributions indicate that the soil will only be significantly stressed to a depth below the plate of about 1·0–1·5 times the width of a square or circular loaded area.

In making settlement predictions for foundations on chalk it is sometimes assumed that:

$$\frac{S_f}{S_p} = \left(\frac{B_f}{B_p}\right)^{\alpha}$$

where S_f is the settlement of the foundation, S_p is the settlement of the plate, B_f is the width/diameter of the foundation, B_p is the width/diameter of the plate and α is an empirical factor. The empirical factor α may be determined from tests using different size plates (Lake and Simons, 1975). However Hobbs (1975) shows that except at values of α approaching one for relatively small foundations the above equation does not adequately model the condition of steadily increasing modulus with depth, which is the only condition apart from homogeneity in which this procedure can be realistically applied.

The results of these measurements are normally plotted in two forms – a time–settlement curve and a load–settlement curve. Owing to the natural variability of soil a single test will rarely be sufficient but, due to the relatively high cost of the test, performing many tests will not be viable. The number of tests that should be carried out depends on both the soil variability and the consequences of poor data on geotechnical design. Tests should not normally be carried out in groups of less than three. In order to allow assessments of variability any plate testing should be carried out at the end of a site investigation, or as part of a supplementary investigation.

Other plate loading tests

Smaller plate loading tests are routinely carried out down-hole in some countries (for example, South Africa) as part of investigations which rely upon visual description. This combination has proved particularly valuable above the water table in hard, gravelly or unsaturated and saprolitic soils, all of which can be very difficult to sample and test. The plate test is carried out across a large-diameter hole which is formed using an auger-piling rig. An engineer or geologist is lowered down the hole to describe the ground and produce a borehole record. Test depths are then selected, and diagonally opposed faces are hand trimmed to provide flat areas upon which the small-diameter (100, 200 or 300 mm) plate test will bear. Details can be found in Wrench (1984).

Another adaptation of the plate test is the 'skip test'. Here a heavy-duty waste-disposal skip is used to simulate the relatively low levels of loading produced, for example, by low-rise housing. This type of test is now the subject of a standard (BS 1377: Part 9: 1990, clause 4.2: Determination of the settlement characteristics of soil for lightly loaded foundations by the shallow pad maintained load test (British Standards Institution, 1990)). Settlements are measured using levelling.

Standard penetration test (SPT)

Historical perspective

In their book *Site Investigation*, Clayton *et al.* (1995b) describe the historical background to the SPT. The standard penetration test, commonly known as the SPT, is carried out in a borehole, by driving a standard 'split spoon' sampler (Fig. 4.48) using repeated blows of a 63·5 kg (140 lb) hammer falling through 762 mm (30 inch). The hammer is operated at the top of the borehole, and is connected to the split spoon by rods. The split spoon is lowered to the bottom of the hole, and is then driven a distance of 450 mm (18 inch), and the blows are counted, normally for each 76 mm (3 inch) of penetration. At the end of driving, the split spoon is pulled from the base of the hole, and the sample is preserved in an airtight container. The penetration resistance (N) is the number of blows required to drive the split spoon for the last 300 mm (12 inch) of penetration. The penetration resistance during the first 150 mm (6 inch) of penetration is ignored because the soil is considered to have been disturbed by the action of boring the hole.

The term standard penetration test was first used by Terzaghi at the 1947 Texas Soil Mechanics Conference. In the USA, site investigation holes were traditionally made by wash boring. In the 19th century, soil type was identified from the cuttings which were flushed to the top of the hole. In 1902 Colonel Charles R. Gow introduced a 1 inch diameter open-drive sampler, which was driven into the ground by repeated blows of a 110 lb hammer (Fletcher, 1965). In subsequent years the American site investigation industry developed variations on this small-diameter tube sampler. By 1947, therefore, a number of different diameter tube samplers were in use (for example, see Hvorslev (1949)). Terzaghi recognized that by counting the blows necessary to drive a tube sampler, additional information on the consistency or density of the soil could be obtained, and at very little extra cost. What he was advocating was a procedure very similar to that currently used in the UK when taking a U100 undisturbed sample, namely the routine recording of penetration resistance.

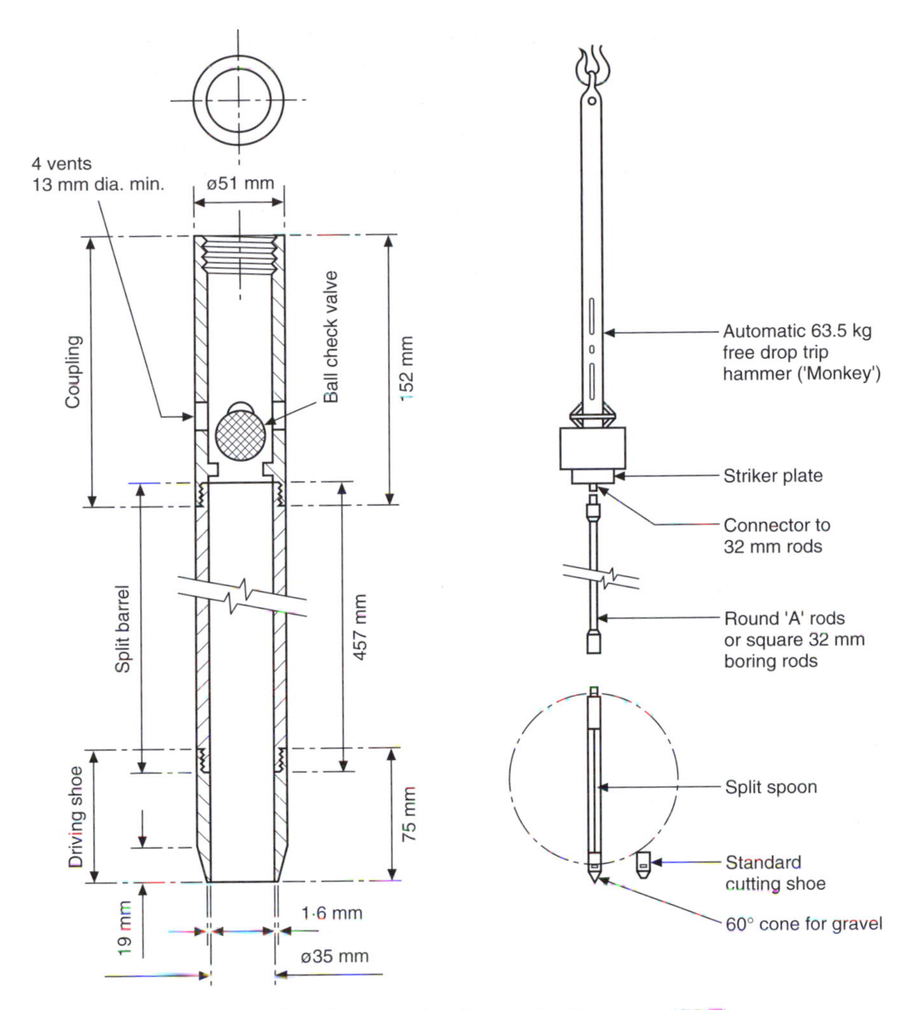

Fig. 4.48 Equipment for the standard penetration test (SPT)

There have been major efforts to unify SPT equipment and practice. In the early 1980s de Mello conceived of the idea of a series of International Reference Test Procedures (IRTPs) which would be distinct from international standards in that they would provide an acceptable way in which international practices could be brought closer, rather than mandatory procedures (which some countries might be unable to adopt). The International Reference Test Procedure for the SPT was published by the ISSMFE in 1988 (see Decourt 1990). National standards are available in many countries, the most commonly followed being the British Standard (BS 1377: Part 9: 1990), the American Standard (ASTM D1586 1984), and the Japanese Standard (JIS-A219 1976). CIRIA Report CP/7 (Clayton 1993) gives the procedures and standards adopted around the world, as well as describing in detail the test, its strengths and weaknesses, and its uses for geotechnical design.

Clayton *et al.* (1995b) point out that correlations between SPT *N* values and soil or weak rock properties are wholly empirical, and depend upon an international database of information. Because the SPT is not completely standardized, these correlations cannot be considered particularly accurate in some cases, and it is therefore important that users of the SPT and the data it produces have a good appreciation of those factors controlling the test, including:

- variations in the test apparatus
- disturbance created by boring the hole
- soil type into which the SPT is driven.

Is it time to retire the SPT?

At the15th International Conference on Soil Mechanics and Geotechnical Engineering, held in Istanbul in August 2001, Prof. Paul Mayne of Georgia Tech gave a presentation called 'Soil property characterization by in situ tests' in which he asked the question 'Is it time to retire the SPT?' Mayne (2001) contrasted the enormous developments during the past century of the telephone, aeroplane and automobile with the non-development in the same period of the SPT, a point he emphasized by showing Fig. 4.49. He also showed the image given in Fig. 4.50 to demonstrate the impossibility of using just one number (the SPT *N* value) to give satisfactory correlations with the large range of parameters required in present-day design. Not only was he making a case for abandoning the

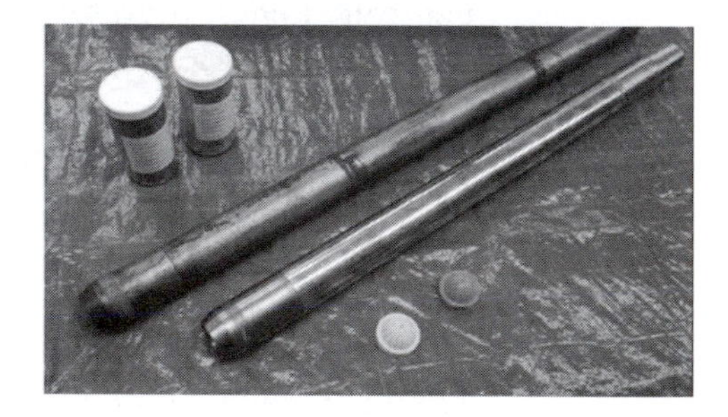

Geotech Test
1902

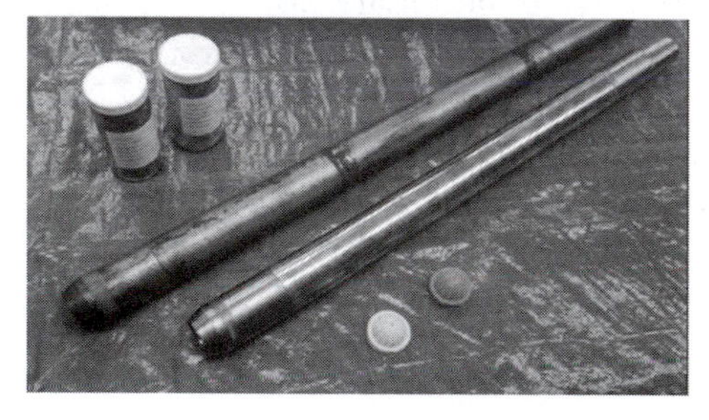

1902 – Colonel Charles Gow
of Raymond Pile Company

Geotech Test
2002?

Fig. 4.49 The standard penetration test: 1902 and 2002? (Mayne, 2001)

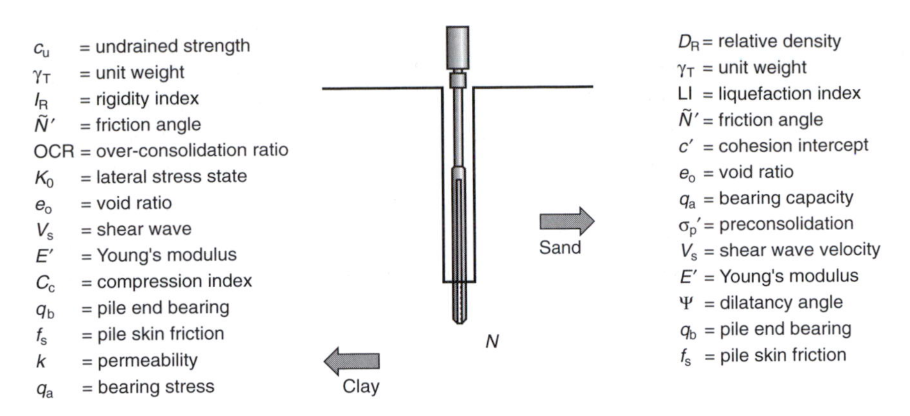

Fig. 4.50 The SPT N *value: is one number enough? (Mayne, 2001)*

undeveloped SPT, but he was also making a case for enhanced in situ tests that make use of developments in technology to enable them to measure a wider range of ground parameters (e.g. see Fig. 4.54). These enhanced in situ tests include the following:

- cone pressuremeter
- seismic piezocone
- dilatocone
- seismic dilatometer
- resistivity cone

The seismic piezocone (Mayne, 2001) is considered later in this chapter.

The cone penetration test (CPT)

Historical background

As pointed out by Clayton *et al.* (1995b), the cone penetration test (CPT) is carried out in its simplest form by hydraulically pushing a 60° cone with a face area of $10\,cm^2$ (35·7 mm diameter) into the ground at a constant speed ($2 \pm 0{\cdot}5\,cm/s$) whilst measuring the force necessary to do so. Most commonly, however, a friction cone is used. The shear force on a $150\,cm^2$ 'friction sleeve' with the same outer diameter as the cone and located immediately above the cone, is then also measured. Both electrical and mechanical means of measuring cone resistance and side friction are currently used, with the shape of the cone differing considerably according to the method in use. The cone is driven from ground surface, without making a borehole, using a special mobile hydraulic penetrometer rig.

The CPT was developed in Holland in 1934 and was originally used as a means of locating and evaluating the density of sand layers within the soft deltaic clays of that country for driven pile design. The mechanical Delft

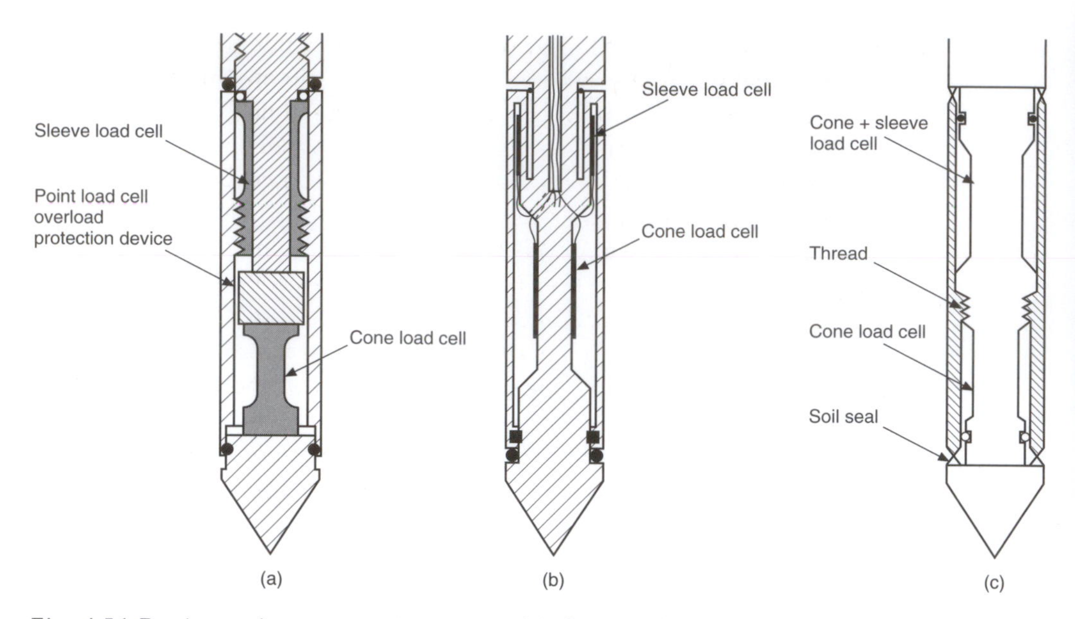

Fig. 4.51 Designs of cone penetrometers. (a) Cone resistance and sleeve friction load cells in compression. (b) Cone resistance load cell in compression and sleeve friction load cell in tension. (c) Subtraction type cone penetrometer. After Lunne et al. *(1997b)*

cone described by Vermeiden (1948) was developed by the Delft Laboratory for Soil Mechanics and is in widespread use in Holland and many other parts of the world. Its development overcame the major problem of the original cone, where soil particles could become lodged between the cone and the bottom face of the rods. The value of the Delft cone was increased very significantly by Begemann, who introduced the mechanical friction cone above the Delft mantle (see Begemann (1965)). The electric cone (Fig. 4.51), where measurements are made using strain gauges or transducers located immediately above the cone, was first developed in 1948 but only came into widespread use in the late 1960s. Measurement of the pore pressures developed at the cone end during penetration first took place in the late 1960s and early 1970s. Other developments and enhancements of the cone have also taken place and continue to this day.

Because the Delft and Begemann friction cones are mechanical, they are rugged and simple to use and maintain. They can give reliable results provided the equipment is properly maintained, and the testing carried out with care. Against this, however, they have a system of measurement which can lead to serious errors, some of which have been described by Begemann (1969) and de Ruiter (1971). Because friction develops between the inner rods and the inside wall of the outer rods, the cone resistance should always be measured while the inner rods are moving relative to the outer rods in order to keep this friction to a minimum. Pushing the inner and outer rods at the same time as measuring cone resistance will result in large irregular variations in rod friction, and noticeable decreases in the measured cone resistance after the penetration is stopped to allow the addition of rods.

At high cone resistances, loads as high as 10 tonnes may need to be applied to the cone. At 30 m depth the compression of the inner rods may be of the same order as the 8 cm stroke used in a Delft cone and, although the top of the inner rods is pushed downwards by the correct amount, the cone will not then advance ahead of the outer rods. This effect will obviously be more serious when a Begemann cone is in use, because the available stroke is only 4 cm. In addition, in deep soft soils, corrections should be made to mechanical cone data to compensate for the mass of the rods.

Electric cone testing

Clayton *et al.* (1995b) point out that electric cones are more expensive, both in terms of cone manufacture and data logging and recording. They have the advantages, however, of being simpler to use, of measuring forces close to their point of application (and therefore without the frictional and rod-shortening effects described above), and of providing almost continuous data with respect to soil depth. Figure 4.51 shows schematic diagrams of the more common types of electric cone. Cone resistance is measured as a standard procedure and side friction measurement is also extremely common. In addition, the following measurements may be available, depending upon the cone type used:

- cone inclination, to check that the cone is not drifting out of vertical
- pore pressure (in the 'piezocone')
- soil resistivity (used, for example, in pollution studies)
- ground vibration, using three-component geophones (in the 'seismic cone')
- gamma-ray backscatter (for density determination)
- pressuremeter values (see later)
- sound (the 'acoustic' penetrometer).

The speed and convenience with which the electric cone may be used has led to its widespread adoption in many countries, although mechanical cones are still common.

In their book *Cone Penetration Testing in Geotechnical Practice*, Lunne *et al.* (1997b) point out that cone resistance q_c and sleeve friction f_s are derived from measurements on electrical strain gauge 'load cells' or force transducers. Different arrangements are used by different

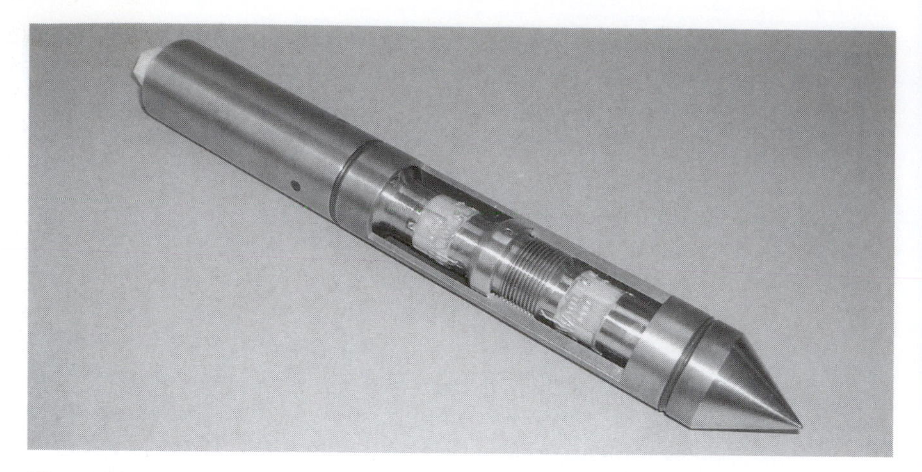

Fig. 4.52 Photograph of a cut-away subtraction cone (permission of GeoMil Equipment BV)

manufacturers. Figure 4.51 illustrates the main design types. In Fig. 4.51(a) cone resistance and sleeve friction are measured by two independent load cells both in compression. In Fig. 4.51(b) the sleeve friction load cell is in tension. In Fig. 4.51(c) the sleeve friction load cell is in compression and records the summation of the loads from both the cone resistance and sleeve friction. The sleeve friction is obtained from the difference in load between the friction and cone resistance load cells. This cone is often referred to as the 'subtraction cone'. A modern subtraction cone is shown in section in Fig. 4.52.

Electric cone data can be processed as penetration is carried out, to produce not only plots of cone resistance and sleeve friction, but also to provide estimates of soil type and soil parameters. This gives the engineer the opportunity to make decisions regarding both the design of a ground investigation and the design of the civil engineering works even while testing is proceeding.

The piezocone

The measurement of pore water pressure during cone testing is particularly suited to testing in soft, primarily cohesive, deposits. A porous element is included in the apparatus, with an electronic pore pressure transducer mounted in a cavity behind it.

Piezocones are usually referred to in 'short hand' by the acronym 'CPTu' (*u* is the symbol for pore water pressure). A modern CPTu cone is shown in section in Fig. 4.53. Note the porous insert in the cone to allow the measurement of pore water pressure. Note also the 'on-board' electronics. The major applications of the piezocone are as follows.

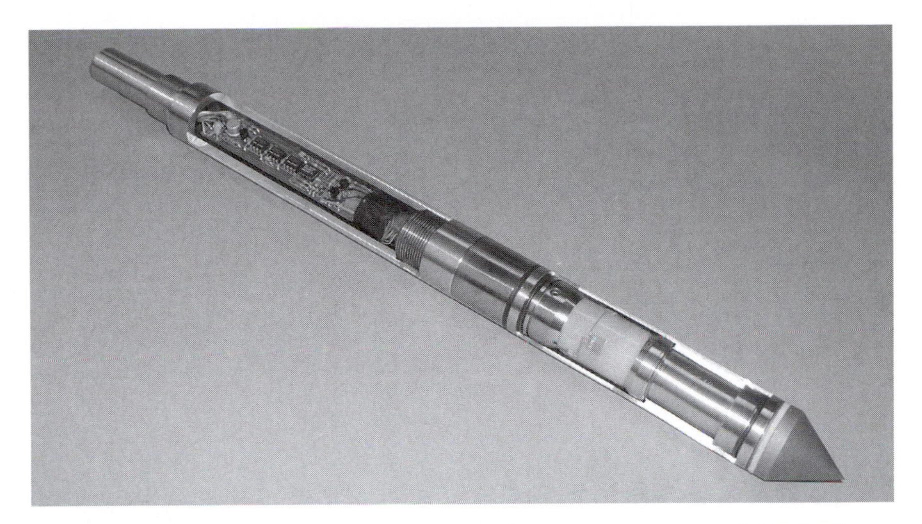

Fig. 4.53 Photograph of a cut-away piezocone (permission of GeoMil Equipment BV)

- *Profiling.* The inclusion of a thin pore-pressure-measuring element allows the presence of thin granular layers to be detected within soft cohesive deposits. Such layers are of great importance to the rate of consolidation of a soft clay deposit.
- *Identification of soil type.* The ratio between excess pore pressure and net cone resistance (see below) provides a useful (although soil-type-specific) guide to soil type.
- *Determining static pore pressure.* Measurements of the static pore pressure can be made in granular soils (where dissipation is rapid), and estimates can be made in clay, either when the cone is stopped to add rods, or by deliberately waiting for full dissipation of the excess pore pressures set up by penetration.
- *Determination of in situ consolidation characteristics.* In clays, the horizontal coefficient of consolidation, c_v, can be determined by stopping the cone and measuring pore pressure dissipation as a function of time (Torstensson, 1977; Acar *et al.*, 1982; Tavenas *et al.*, 1982).

The seismic cone

Seismic cones contain either one or two three-component geophone arrays, mounted internally, some distance behind the friction sleeve. Where two arrays are used, the vertical distance between the arrays will be of the order of 1 m or more. In recent years it has proved a valuable tool for determining the benchmark value of very small strain stiffness G_0 by means of either parallel cross-hole testing or, more normally (because it is considerably more economical), down-hole testing.

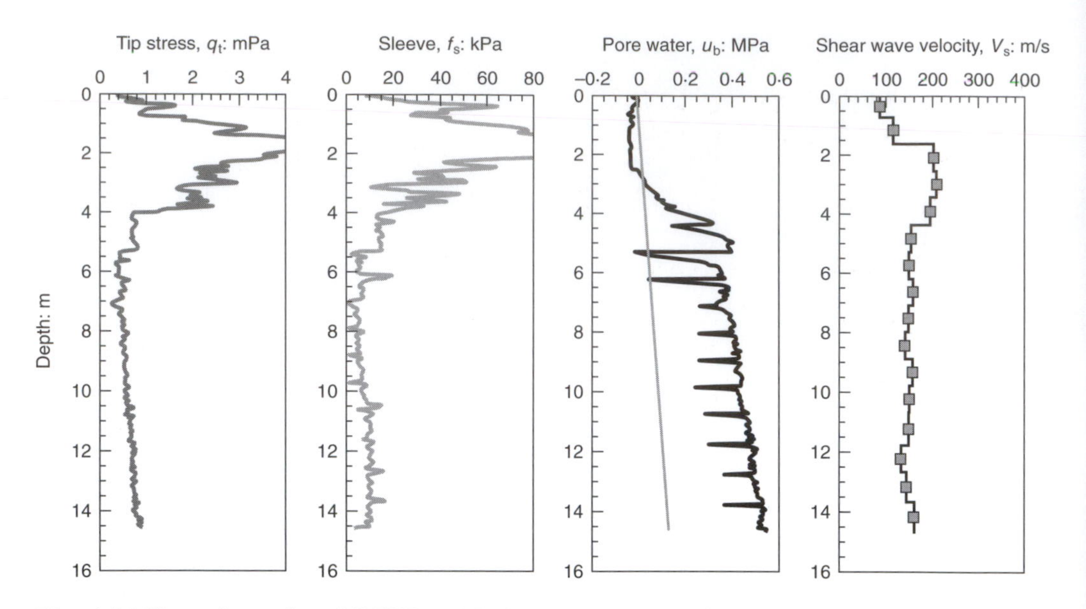

Fig. 4.54 Plotted results of SCPTu with dissipation at Amherst test site (Mayne, 2001)

The shear wave velocity V_s is measured by the seismic cone. Mayne (2001) pointed out that shear wave velocity:

- is a fundamental measurement in all solids (steel, concrete, wood, soils, rocks)
- provides small-strain stiffness represented by shear modulus: $G_0 = \rho V_s^2$ (alias $G_{dyn} = G_{max} = G_0$)
- applies to all static and dynamic problems at small strains ($\gamma_s < 10^{-6}$)
- applies to undrained and drained cases

and that G_0 needs a reduction factor for relevant (operational) strain levels. Mayne (2001) advocated the use of the seismic piezocone (or 'SCPTu' where 'S' is for 'seismic') and gave results of SCPTu testing on the Amherst test site (Fig. 4.54) and Opelika test site (Fig. 4.55). For the Opelika test site (Fig. 4.55) note not only the large amount of data from the one SCPTu test, but also the good correlation between the different seismic methods. This demonstrates the reliability of seismic methods to measure consistently the soil property of shear wave velocity.

Pressuremeter test

Overview

Pressuremeters are devices for carrying out in situ testing of soils and rocks for strength and stiffness parameters. They are generally cylindrical, long with respect to their diameter, with part of this length being covered by a flexible membrane. Pressuremeters enter the ground by pushing, by

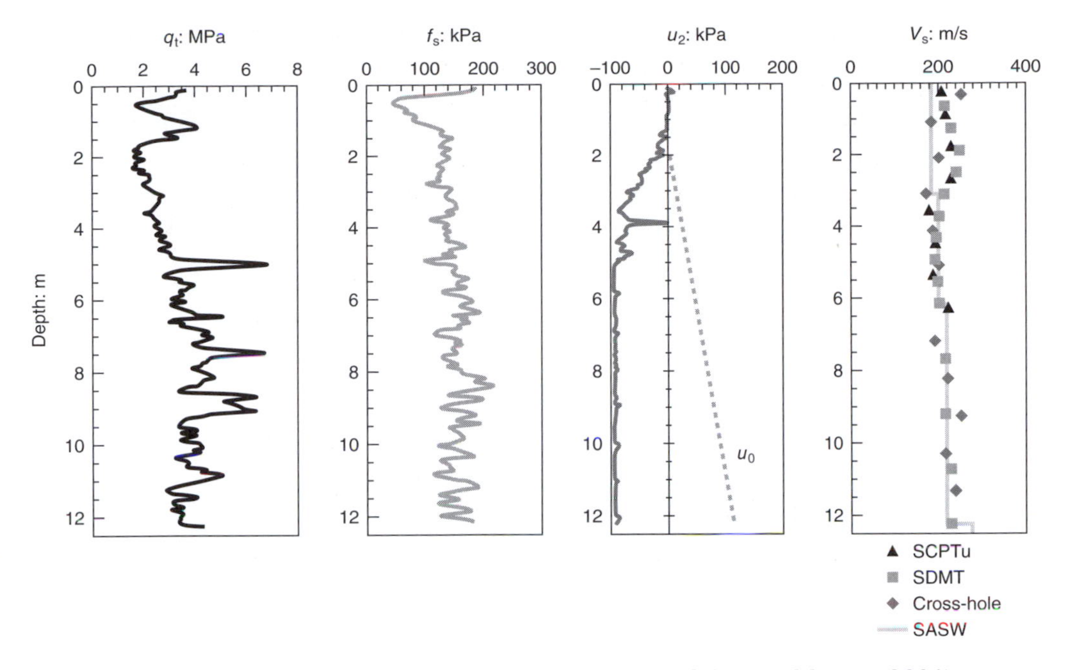

Fig. 4.55 Plotted results of SCPTu at Opelika test site, Alabama (Mayne, 2001)

pre-boring a hole into which the probe is placed, or by self-boring (Fig. 4.56) where the instrument makes its own hole. Once in the ground, increments of pressure are applied to the inside of the membrane forcing it to press against the material and thus forming a cylindrical cavity. A test consists of a series of readings of pressure and the consequent displacement of the cavity wall (Fig. 4.57), and the loading curve so obtained may be analysed using rigorous solutions for cylindrical cavity expansion and contraction. It is the avoidance of empiricism that makes the pressuremeter test potentially so attractive. The test is usually carried out in a vertical hole so the derived parameters are those appropriate to the horizontal plane.

Insertion

Interpretation of the pressuremeter test results must take account of the disturbance caused by the method used to place the probe in the ground. The least disruptive of the methods is self-boring where disturbance is often small enough to lie within the elastic range of the material and is therefore recoverable. This is the only technique with the potential to determine directly the in situ lateral stress σ_{h0}, the major source of uncertainty when calculating K_0, the coefficient of earth pressure at rest. All methods, however, allow the confining stress to be inferred.

The disturbance caused by pre-boring and pushing is never recoverable. However, for any pressuremeter test it is possible to erase the

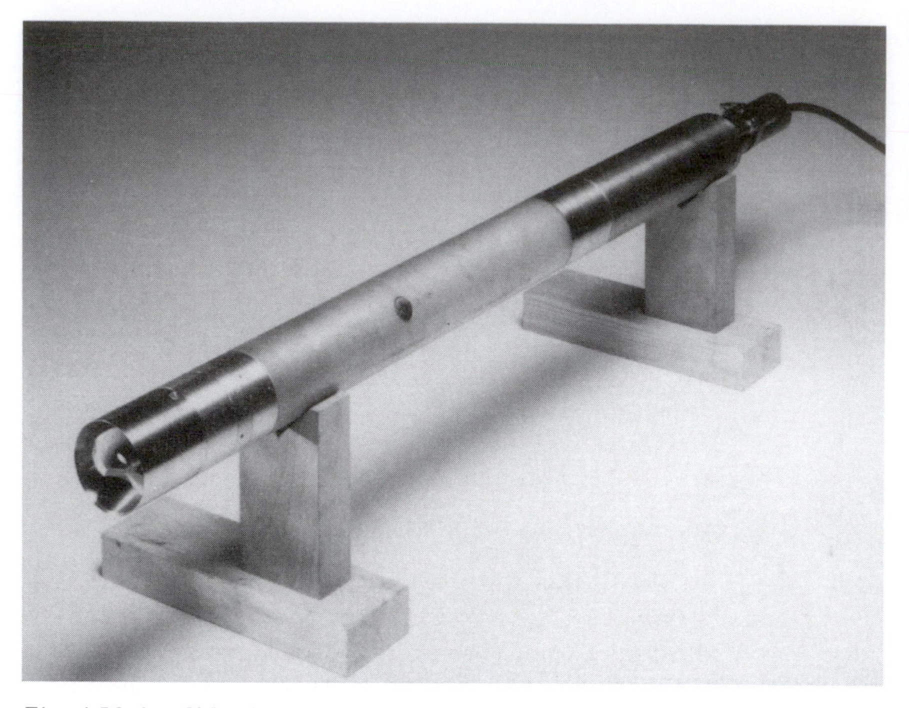

Fig. 4.56 A self-boring pressuremeter, approximately 1·25 m × 0·08 m (permission of Cambridge Insitu Ltd)

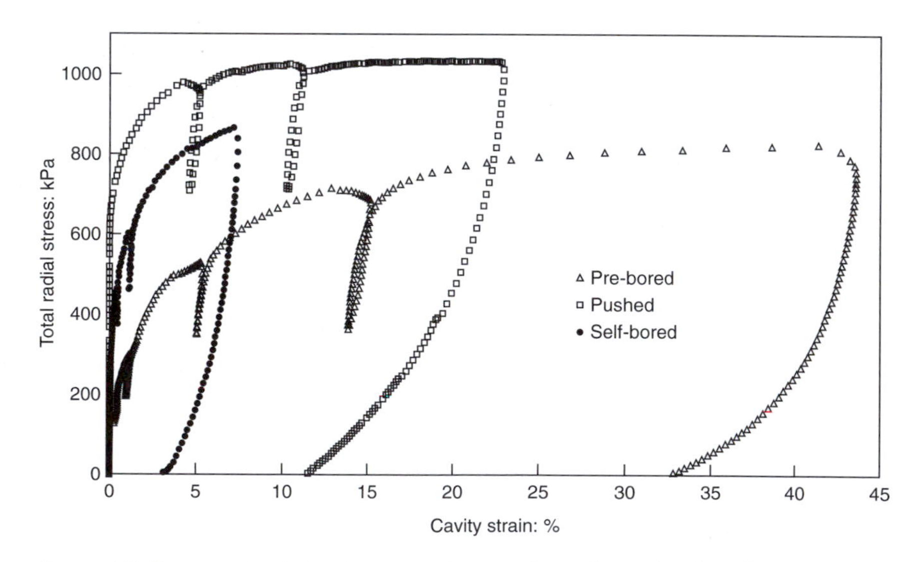

Fig. 4.57 Stress–strain test curves for pre-bored, pushed and self-bored pressuremeters in Gault clay at about 5 m depth (permission of Cambridge Insitu Ltd)

previous stress history by taking the material to a much higher stress than it has previously seen, and then to reverse the direction of loading. The point of reversal is a new origin and the stress–strain response will be that due to the undisturbed properties of the material. Figure 4.57 shows data from the three types of test. The tests were carried out at the same location (a heavily over-consolidated Gault clay site) at similar depths and give similar results for strength and stiffness. Although the loading paths appear very different there are similarities in the unloading paths and whenever a small rebound cycle is taken. These cycles are of particular importance. No matter how disturbed the material prior to insertion, all types of pressuremeter test have the potential to make repeatable measurements of shear stiffness as well as measurements of the reduction of stiffness with increasing strain. The methods of pre-boring, pushing and self-boring are now considered in detail.

Pre-boring

A hole is formed in the ground by conventional drilling tools and the instrument is subsequently placed in the pre-formed hole. The major drawback of this method is the complete unloading of the cavity wall that takes place in the interval between removing the boring tool and pressurizing the probe. The material must be capable of standing open and so the method is best suited to rock. As Fig. 4.57 indicates it is possible to make a test in stiff clay. Comparing the pre-bored curve with the self-bored curve, however, shows how much further the cavity has to be expanded before the influence of insertion disturbance can be erased. The method can be used in dense sand if drilling mud is used to support the open borehole but it is unlikely to be suitable for loose sands. The Ménard pressuremeter, widely used in France, is an example of a pre-bored device. In the UK the high-pressure dilatometer (the terms 'dilatometer' and 'pressuremeter' are interchangeable in this context) is available but usage tends to be restricted to rocks and difficult materials such as boulder clay with gravel layers. An example of a high-pressure dilatometer is shown in Fig. 4.58. A pre-bored operation will require the assistance of a drilling rig. Unlike the other insertion methods, however, if the hole is cored then it is possible to make laboratory tests on material that is directly comparable to that tested by the pressuremeter. Pre-bored pressuremeter testing in a vertical hole has been carried out to depths greater than 500 m and depths of 200 m are routine.

Pushing

As the name suggests, pushed-in pressuremeters are forced into the ground so raising the state of stress in the surrounding soil. A special case of this approach is the cone pressuremeter (CPM) where a 15 cm^2

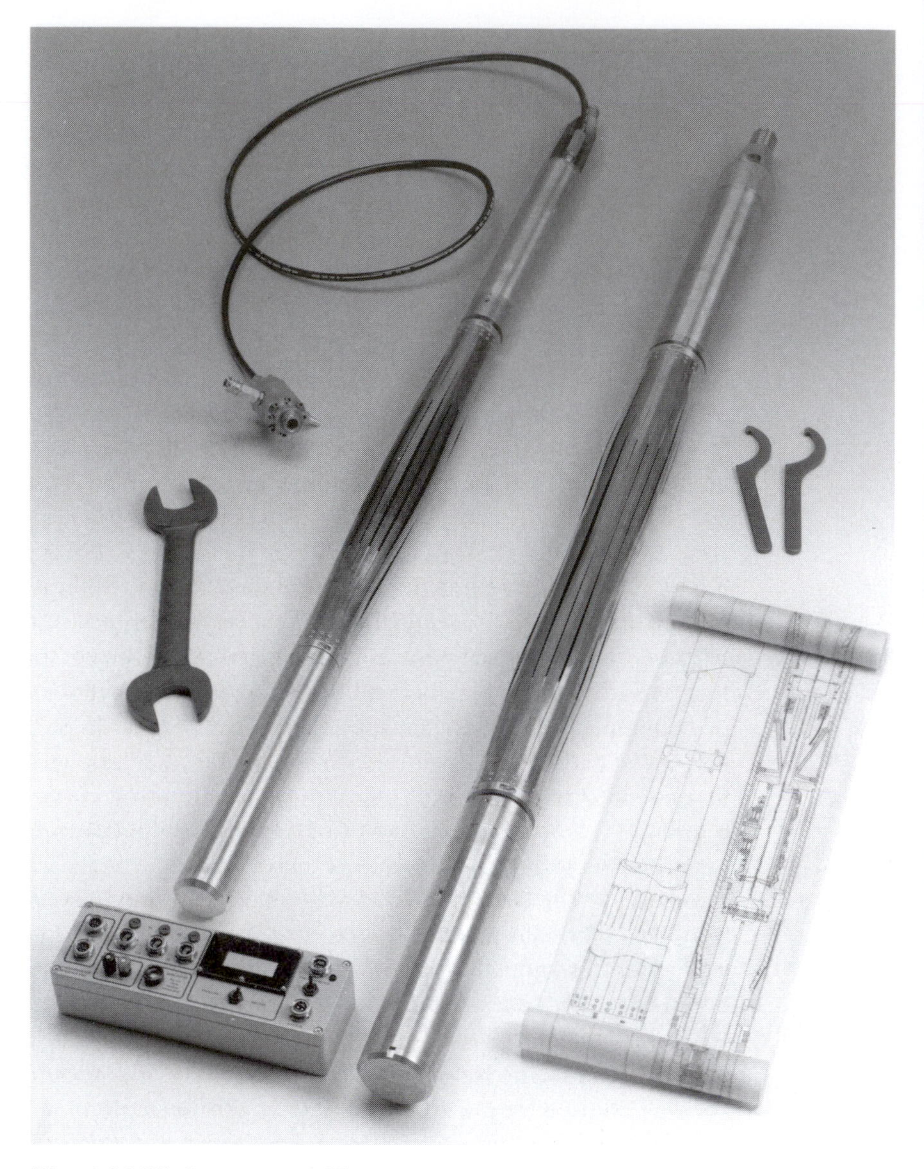

Fig. 4.58 High-pressure dilatometers, 73 mm and 95 mm in diameter (permission of Cambridge Insitu Ltd)

cone is connected to a pressuremeter unit of the same diameter. The disturbance caused to the material is significant and the only parameter that can be obtained from the loading path is the limit pressure of the soil. The 'pushed' curve in Fig. 4.57 is an example of a CPM test and shows a clear plateau after the cavity has been expanded by about 15%. Strength parameters are derived from the contraction curve and stiffness parameters from the response of small rebound cycles. The method is fast and can make a test in any material into which a cone can be inserted. The coupling of the profiling capability of the cone with the ability to make direct measurements of strength and stiffness is especially attractive. However, as Fig. 4.57 indicates, the stresses required to make a satisfactory test are much higher than for the other methods, and at these levels of stress it is probable that crushing of the soil particles is taking place. This may be a significant factor especially for tests in sand. Furthermore, a jacking force of 10 tonnes or more is not unusual in pushing the probe and this may present difficulties.

Self-boring

Figure 4.59 shows a schematic diagram of the Cambridge self-boring pressuremeter (SBP). The instrument is a miniature tunnelling machine that makes a pocket in the ground into which the device very exactly fits. The foot of the device is fitted with a sharp-edged internally tapered cutting shoe. When boring, the instrument is jacked into the ground, and the material being cut by the shoe is sliced into small pieces by a rotating cutting device. The distance between the leading edge of the shoe and the start of the cutter is important and can be optimized for a particular material. If too close to the cutting edge the ground suffers stress relief before being sheared. If the cutter is too far behind the shoe edge then the instrument begins to resemble a close ended pile. In stiff materials the usual setting is flush with the cutting shoe edge. The cutting device takes many forms. In soft clays it is generally a small drag bit whereas in more brittle material a rock roller is often used.

The instrument is connected to the jacking system by a drill string. This is in two parts – an outer fixed casing to transmit the jacking force and an inner rotating rod to drive the cutter device. The drill string is extended in one metre lengths as necessary to allow continuous boring to take place. All the cut material is flushed back to the surface through the instrument annulus without erosion of the cavity wall. Normally water is used but air and drilling mud have been applied successfully.

Self-boring is effective in materials from loose sands and soft clays to very stiff clays and weak rock. It will not operate in gravel and materials hard enough to damage the sharp cutting edge. In principle the probe can be made to enter the ground with no disturbance at all. In practice the probe causes a small, generally recoverable, degree of disturbance that must be assessed before deciding a value for the in situ lateral stress.

The SBP requires a modest amount of reaction. On some soft clay sites it is possible for the self-boring kit to operate without support from other drilling tools. The minimum interval between tests is one metre. Where tests are more widely spaced or in materials with occasional bands of hostile layers the SBP can be used in conjunction with a cable percussion

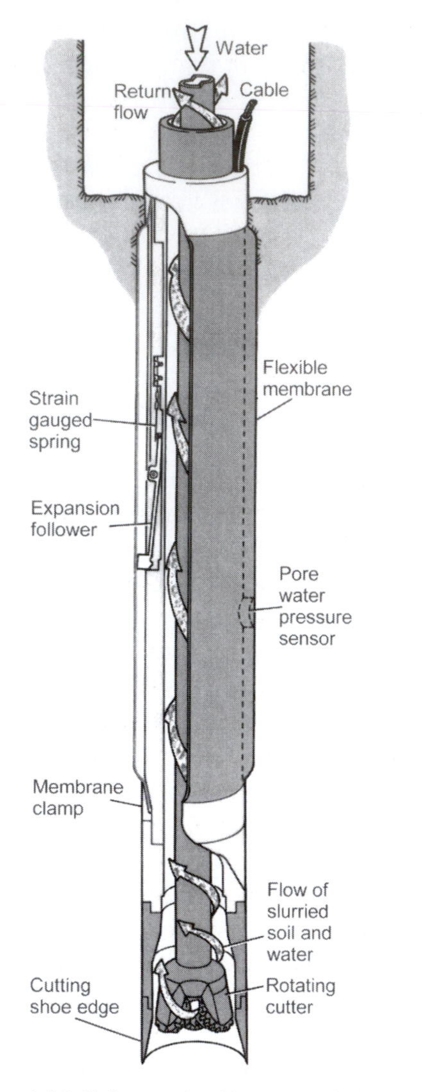

Fig. 4.59 Schematic diagram of the self-boring pressuremeter (permission of Cambridge Insitu Ltd)

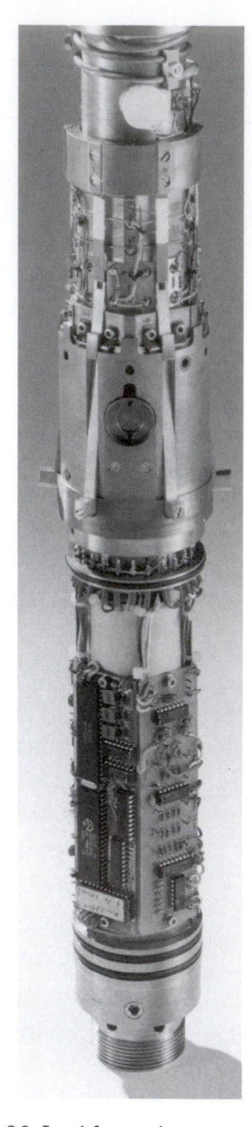

Fig. 4.60 Inside a six-arm self-boring pressuremeter (permission of Cambridge Insitu Ltd)

system or be driven by a rotary rig using special adaptors. Self-boring in a vertical hole is routinely carried out to depths of 60 m or more. The self-boring method is also used as a low-disturbance insertion system for other devices such as load cells and permeameters.

Construction and calibration

There are many designs of pressuremeter in current use, some of which are of complex construction. Figure 4.60 is a view of the inside of a six-arm

Cambridge self-boring pressuremeter. There are transducers for measuring the radial displacement of the membrane at six places and the total and effective pressure being applied to the cavity wall. The electronics for the signal conditioning including the conversion from analogue to digital is contained in the probe itself. Apart from supplying power, the output of the probe may be connected directly to the serial port of a small computer. This approach is necessary in order to obtain a high resolution free of noise. Pressuremeters with local instrumentation are able to resolve without difficulty displacements of 0·5 μm and pressure changes of 0·1 kPa.

Pressuremeters can be expanded using air or a non-conducting fluid such as light transformer oil. Automated systems for pressurizing the equipment are available. Automation allows the expansion of the cavity to occur at a constant rate of strain. It is conventional to log the output of the pressuremeter on a computer and to plot the loading curve in real time.

Meticulous calibration of the equipment is vital. The transducers must be calibrated regularly both for sensitivity and drift. Almost all pressuremeters suffer the drawback that the output of the transducers is governed by the movements and pressure on the inside of the membrane. What are required, of course, are the displacements and stresses acting on the cavity wall. The properties of the pressuremeter membrane can therefore be a significant source of uncertainty. It requires an amount of work to make it move, and an additional component to keep it moving. This is relevant to tests in soft soils. The membrane contribution may be estimated by carrying out membrane expansion tests in free air.

The other major influence on the measurements is system compliance, i.e. the contribution of the probe itself to the measured stiffness. This can be a significant source of error if the probe is used in very stiff soils or weak rock. This contribution may be estimated by inflating the instrument to full working load inside a metal sleeve of known elastic properties.

The importance of the various calibrations depends on the type of pressuremeter involved and where it is being used. For example, the contribution of the hose supplying pressure to the probe is highly relevant if volume changes are being measured at the surface, but is of no importance at all for a probe with internal instrumentation.

Advantages of pressuremeter testing

- A large number of fundamental soil properties are obtained from a single test.
- To derive these properties, no empirical correcting factors whatsoever are needed.

- Measurements are made in situ at the appropriate confining stress.
- A large volume of material is tested – a typical test loads a column of material 0·5 m high and extending to more than ten times the expanded cavity radius. This is the equivalent of at least 1000 triaxial tests on 38 mm samples.
- Representative loads are applied – in the example shown in Fig. 4.57 about 12 tonnes is being applied to the cavity wall.
- Results can be obtained quickly as all the data logging and most of the data processing is carried out by automated systems.
- Commercial operation has shown that the instruments, though more complex than conventional site investigation equipment, are reliable.
- There are many materials whose properties can only be realistically determined by in situ measurement.
- The pressuremeter test is particularly appropriate for predicting the performance of laterally loaded piles.

Disadvantages of pressuremeter testing

- The instrument will not penetrate gravels, claystones or the like, so generally pressuremeter testing requires support from conventional drilling techniques.
- Failure planes and deformation modes are not usually appropriate to those occurring in the final design. An estimate of the anisotropy of the material will be required in order to derive vertical parameters from lateral values.
- Many familiar design rules and empirical factors are based on parameters obtained from traditional techniques. It is not always possible to use them with pressuremeter derived values, even if the in situ parameters more accurately represent the true state of the ground (see the section on sampling disturbance at the start of this chapter).
- Only two stress paths can in practice be followed – undrained and fully drained.
- The instruments and their associated equipment are complex by conventional site investigation standards and can only be operated by trained personnel.
- Use of an inappropriate analysis to interpret a pressuremeter test can result in seriously misleading parameters.

Case study: Kilburn, London

Figure 4.61 is an example of an undrained self bored pressuremeter (SBP) test in London Clay, at 10·3 m below ground level. The curve of pressure versus displacement of the cavity wall is essentially the integrated shear stress–shear strain response. It is usual to report a test in terms of fundamental soil parameters for stiffness, strength and in situ lateral

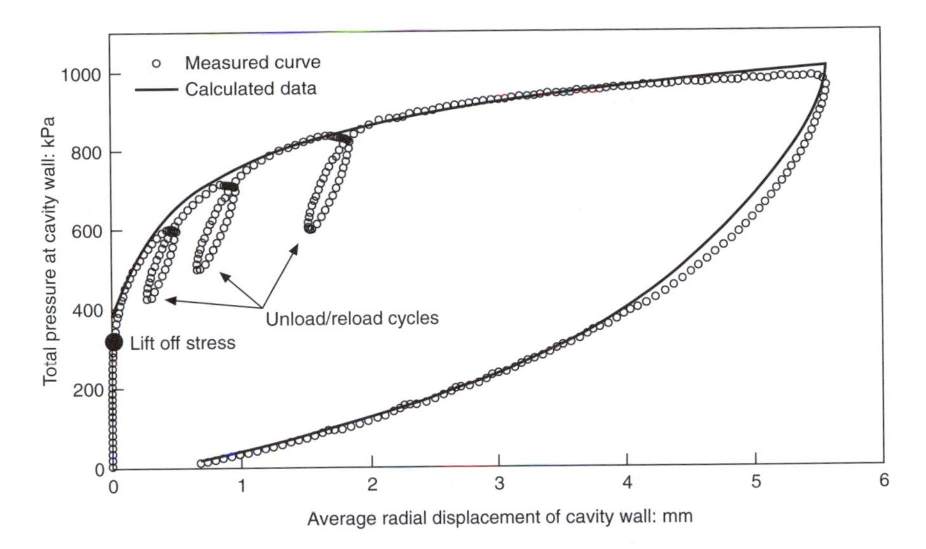

Fig. 4.61 Undrained SBP test showing measured and calculated stress–strain curves (permission of Cambridge Insitu Ltd)

stress as detailed below. Having obtained a set of parameters it is current practice to see if they recover the measured curve, as has been done in Fig. 4.61.

- *In situ lateral stress.* A preliminary value is obtained by inspection of the first part of the loading. Because the probe has been self-bored into position, the initial part of the test shows the pressure climbing without the cavity expanding, until a stress (marked 'lift off stress' in Fig. 4.61) is exceeded. If the probe has been inserted without significant disturbance of the surrounding soil then this stress should be close to the in situ lateral stress σ_{h0}. The assumption of minimal disturbance needs to be checked, and the final choice for σ_{h0} will be obtained by optimising the fit of the calculated curve to the measured data.
- *Undrained strength.* Palmer (1972) showed that for any part of the undrained loading curve the current mobilized shear stress τ is given by the following partial differential equation:

$$\tau = \gamma(\delta P/\delta\gamma)$$

where P is the total pressure at the cavity wall and γ is current cavity shear strain (Bolton and Whittle, 1999). Applying this to the test shown in Fig. 4.61 by taking tangents to the loading and contraction curve at regular intervals gives the response plotted in Fig. 4.62.

In Fig. 4.62 the strain axis is now current cavity shear strain and the shear stress–shear strain curves have been plotted under the field curve (ignoring the unload/reload cycles). The unloading

LIVERPOOL JOHN MOORES UNIVERSITY
LEARNING SERVICES

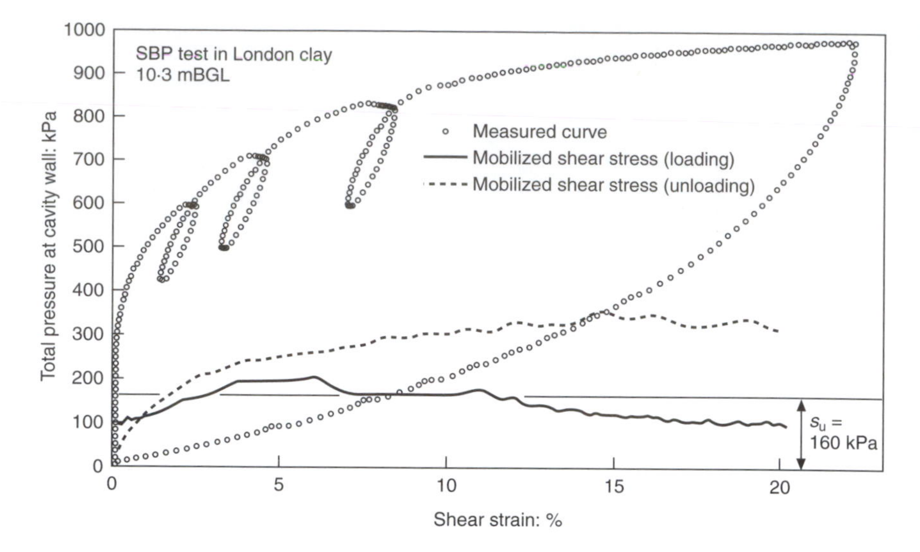

Fig. 4.62 Undrained SBP test showing measured stress–strain curve and mobilized shear stress, loading and unloading (permission of Cambridge Insitu Ltd)

strength is twice the value for the loading because the contraction starts with the material already at maximum strength in one direction. The loading response indicates a peak followed by a gentle reduction with increasing strain, a common feature of tests in London Clay.

The Palmer solution makes no assumptions concerning the shape of the shear stress–shear strain curve. For the purposes of curve matching it is helpful to use closed-form solutions where the shape of the shear stress–shear strain response is pre-determined. If perfect plasticity is assumed then a single value of shear strength must be derived. From inspection of both cavity expansion and contraction shear stress–shear strain curves in Fig. 4.62 a reasonable average value for this test appears to be 160 kPa.

Figure 4.61 shows a 'best fit' theoretical solution for this test supposing the deformation can be described by a non-linear elastic/ perfectly plastic solution (Bolton and Whittle, 1999; Whittle, 1999). The fit is compared with the measured field curve and for the most part is convincing. It is worth noting that the calculated curve suggests the best choice for σ_{h0} is remarkably close to that suggested by the 'lift off stress' marked in Fig. 4.61.

- *Secant modulus.* It is unsafe to depend on the initial part of the test for elastic parameters because of the uncertainty over insertion effects. It is better to take the material to a well developed plastic condition, thus

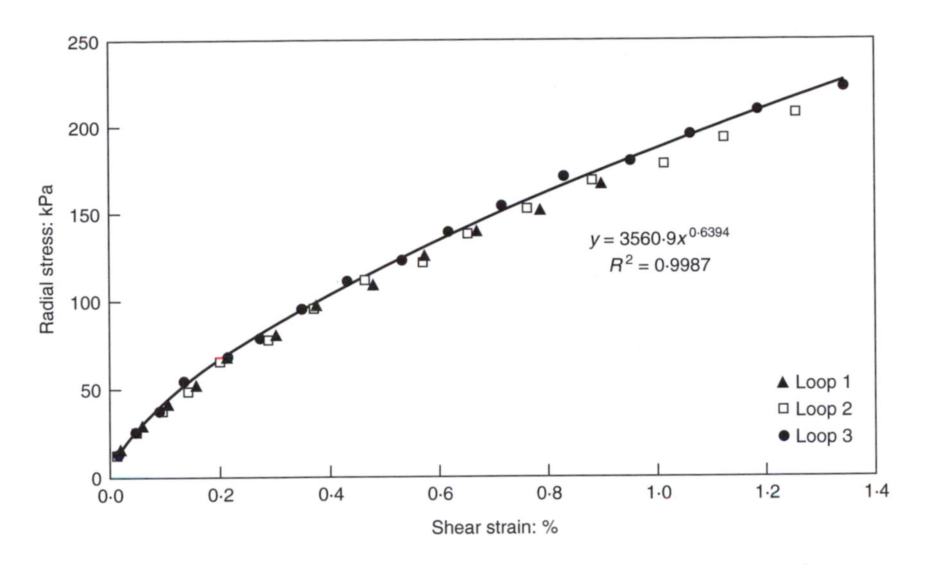

Fig. 4.63 Describing the stiffness–strain response using a power law (permission of Cambridge Insitu Ltd)

erasing any previous stress history, and then to unload a little of the applied stress before reloading once more from this new state. There are three such cycles in this example showing similar behaviour. They are also similar in form to the initial loading response following lift off, and also to the first part of the final unloading. A marked non-linear response is apparent.

Figure 4.63 shows the *reloading* data from the loops in Fig. 4.61 plotted using the loop turn-around point as an origin. The data from the loops tend to draw the same trend and a power law fits the trend with a high degree of correlation.

Using the relationship $G_s = \alpha\gamma^{\beta-1}$ the degradation of stiffness with strain response can be described for shear strains down to the limit of resolution of the equipment, about 10^{-4}. Secant shear modulus values for three magnitudes of shear strain are given in Table 4.6. Shear strains of 10^{-4} are not quite small enough to allow the maximum stiffness G_0 to be determined.

Table 4.6 Secant shear modulus using power law results given in Fig. 4.63

	Shear strain		
	$\gamma = 10^{-4}$	$\gamma = 10^{-3}$	$\gamma = 10^{-2}$
G: MPa	63·1	27·5	12·0

Seismic methods

Overview

During the 1980s and 1990s, careful back-analyses of the behaviour of the ground around constructions such as tunnels and excavations have repeatedly shown that the in situ stiffness of soils and rocks is much higher than was previously thought, and that the stress–strain behaviour of these materials is non-linear in most cases. Numerical analyses, using finite element and finite difference computations and field observations, have demonstrated that when margins of safety are adequate the strain levels in the ground around retaining walls, foundations and tunnels are small, and typically are of the order of 0·01% to 0·1%. Jardine *et al.* (1986) carried out finite element analyses to examine the levels of strain around a range of construction types in a material with a stiffness similar to that of a stiff low plasticity clay. They found (Fig. 4.64) that typical shear strain levels are less than about 0·1%. In cases where loading is light or when excavations are well restrained or in stiff ground, typical strain levels can be even less and perhaps as low as 0·01%.

Improved measurements in the laboratory have confirmed the non-linear stress–strain behaviour of soil and shown that stiffness is much higher when measured locally and at small strain levels than when determined using conventional laboratory techniques.

Seismic tests apply very small strains (10^{-6}–10^{-4}%) to the materials in which they are used (Auld, 1977). Because of this it has generally been

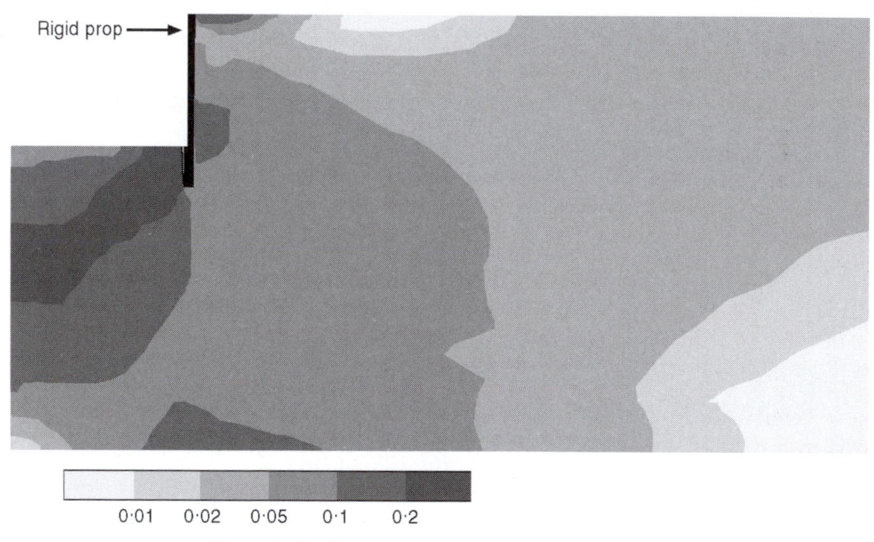

Fig. 4.64 Shear strains associated with a proposed cantilever retaining wall (after Jardine et al.*, 1986)*

Table 4.7 Strain limit of elastic behaviour for a number of geomaterials measured in a triaxial cell, from Heymann (1998)

Material	Strain at limit of elastic behaviour: %	Description and comments
Dogs Bay sand (Jovičić and Coop, 1997)	$<1 \times 10^{-3}$ [a]	Uniform, angular biogenetic carbonate sand
Leighton Buzzard sand (Park, 1993)	2×10^{-3}	Uniform, sub-rounded, quartz sand
Kaolinite (Mukabi *et al.*, 1991)	2×10^{-3}	Reconstituted clay
Berthierville clay (Smith, 1992)	$<2 \times 10^{-3}$ [a]	Soft silty clay
Bothkennar clay (Smith, 1992)	$<2 \times 10^{-3}$ [a]	Soft marine clay
Bothkennar clay (Heymann, 1998)	3×10^{-3}	Soft marine clay
Quensborough clay (Smith, 1992)	$<2 \times 10^{-3}$	Soft silty clay
Osaka Bay clay (Mukabi *et al.*, 1994)	1×10^{-3}	Stiff, over-consolidated clay
London Clay (Heymann, 1998)	2×10^{-3}	Stiff, over-consolidated, fissured clay
Vallericca clay (Georgiannou *et al.*, 1991)	$<10^{-2}$ [b]	Weakly cemented, over-consolidated clay
Calcarenite (Cuccovillo and Coop, 1997)	1×10^{-2}	Weak rock, carbonate sand cemented with calcite
Sandstone (Cuccovillo and Coop, 1997)	2×10^{-2}	Weak rock, quartz grains weakly bonded by iron oxide
High-density chalk Low-density chalk (Matthews, 1993)	5×10^{-3} 4×10^{-2}	Dry density = 1·94 Mg/m^3 Dry density = 1·35 Mg/m^3
Low-density chalk (Heymann, 1998)	2×10^{-3}	Dry density = 1·35 Mg/m^3
Cement-treated sandy soil (Tatsuoka and Shibuya, 1992)	1×10^{-2}	Hard soil/weak rock
Sagamihara mudstone (Kim *et al.*, 1990)	2×10^{-2}	Weak rock

[a] Linear stress–strain range not observed due to insufficient resolution of local strain measurement instruments.
[b] Limit determined from resonant column tests, but triaxial tests show this limit to be inconclusive.

Table 4.8 Examples of stiffness degradation for different geomaterials (after Heymann, 1998)

Material	$E_{0\cdot01}/E_0$	$E_{0\cdot1}/E_0$	$E_{1\cdot0}/E_0$
Intact chalk	0·87–0·93	0·42	failed
London clay	0·83–0·97	0·35–0·58	0·11–0·20
Bothkennar clay	0·75–0·81	0·36–0·55	0·11–0·21

considered that they give results relevant only to the linear–elastic phase of soil deformation (Fig. 4.1). Under dynamic loading the resonant column test typically shows that this linear range extends only to between 0·001% and 0·01% shear strain. The few measurements that have been made using the triaxial apparatus (Table 4.7) suggest that stiffness reduction starts at about 0·002% axial strain for soils and approximately 0·02% for weak rocks. Clayton and Heymann (2001) have used local strain instruments with greater accuracy and resolution and demonstrated that the linear range for a soft clay, a stiff clay and an intact weak rock, extended to a similar axial strain level in all cases (Table 4.7). Although the reduction of stiffness which accompanied increasing strain levels was found to be a function of stress path, the reduction of stiffness was fairly small at the strain levels generally found around structures (0·01–0·1%), as indicated by the values of stiffness normalized with E_0 shown in Table 4.8. This suggests that geophysical measurements of stiffness should have an important role in determining stiffness parameters for engineering calculations.

The realization that strain levels around constructions are small, and that field stiffnesses are much higher than previously measured in the laboratory has led to the reappraisal of small-strain stiffnesses derived from field seismic geophysical methods. Traditionally, geophysics has been used as an indirect means of targeting and dimensioning subsurface features. The new understanding of the stiffness behaviour of geomaterials has resulted in seismic methods now being used to provide design parameters. In particular there is a growing appreciation of the value of measuring maximum shear modulus G_{max} using seismic methods (Ballard and McLean, 1975; Abbiss, 1981; Tatsuoka and Shibuya, 1992) as part of a site investigation. Such methods allow stiffnesses to be determined on representative volumes of the ground and at the in situ stress state.

The measurement of stiffness using seismic methods

Seismic methods utilize the propagation of elastic waves through the ground (see Short Course Notes: Elastic Waves). There are two categories of seismic wave: body waves, comprising compressional (P) and shear (S)

Short Course Notes: Elastic Waves - definitions and terminology

The nature of elastic waves and the rules they observe (e.g. dispersive nature) make them particularly suited to substructure stiffness profiling. Seismic waves propagated by elastic solids can be classified into two types: body waves and surface waves.

Body waves

These can be primary or P waves. This is a longitudinal wave in which the directions of motion of the particles are in the direction of propagation. This motion is irrotational and the wave is one of dilation propagated with speed V_p. Compression and dilational waves are called in seismology the primary or P (or 'push') waves.

Body waves can also be secondary or shear or S waves. This is a transverse wave in which the direction of motion of the particles is perpendicular to the direction of propagation. The motion is rotational and propagated with speed V_s. Since $V_p > V_s$ the first waves to arrive from any disturbance will be P waves. In seismology, they are known as secondary or shear or S waves. A fluid such as pore water does not transmit shear waves. Accordingly, in soil, shear waves are conducted through the soil skeleton only.

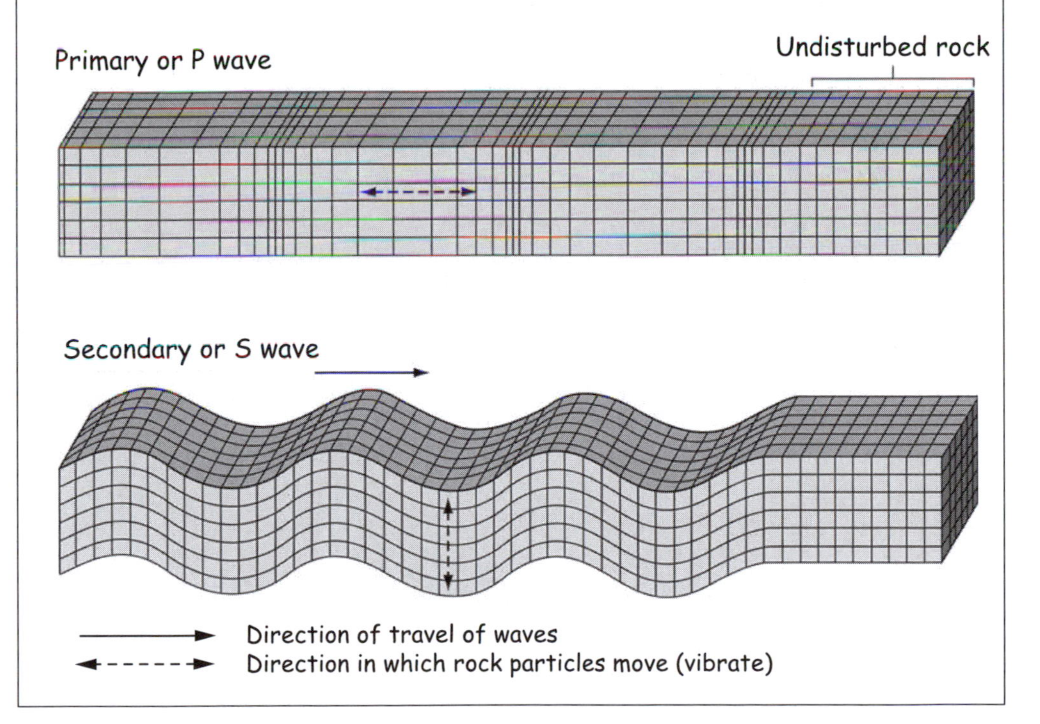

Surface waves

In a uniform, infinite medium only P and S waves appear. If the medium is bounded or non-uniform (as surface soils are) other simple types of waves appear. The most important are the surface waves that are propagated near the surface of a solid. Surface waves have depths of penetration depending on their wavelengths. In non-uniform media, surface waves travel at a velocity dependent on their frequency. In seismology this phenomenon is described as 'dispersive'. The analogy is with optics. The ground separates the radiation according to the wavelength like a prism disperses or separates white light according to wavelength.

Surface waves can be Rayleigh waves or Love waves. Rayleigh waves are waves beneath a free surface in the plane $z = 0$ whose amplitude diminishes exponentially in the z-direction being propagated along the x-axis. They are waves in which the particles of the medium move in vertical planes. The particle motion describes vertical ellipses in which the vertical axes are about 1.5 times the horizontal. At the highest points of the ellipses the particle motion is opposite to the direction of wave advanced.

If a solid is stratified so that the regions are of different materials, a type of Rayleigh wave appears. These waves are known as 'Love' waves and are horizontally polarized.

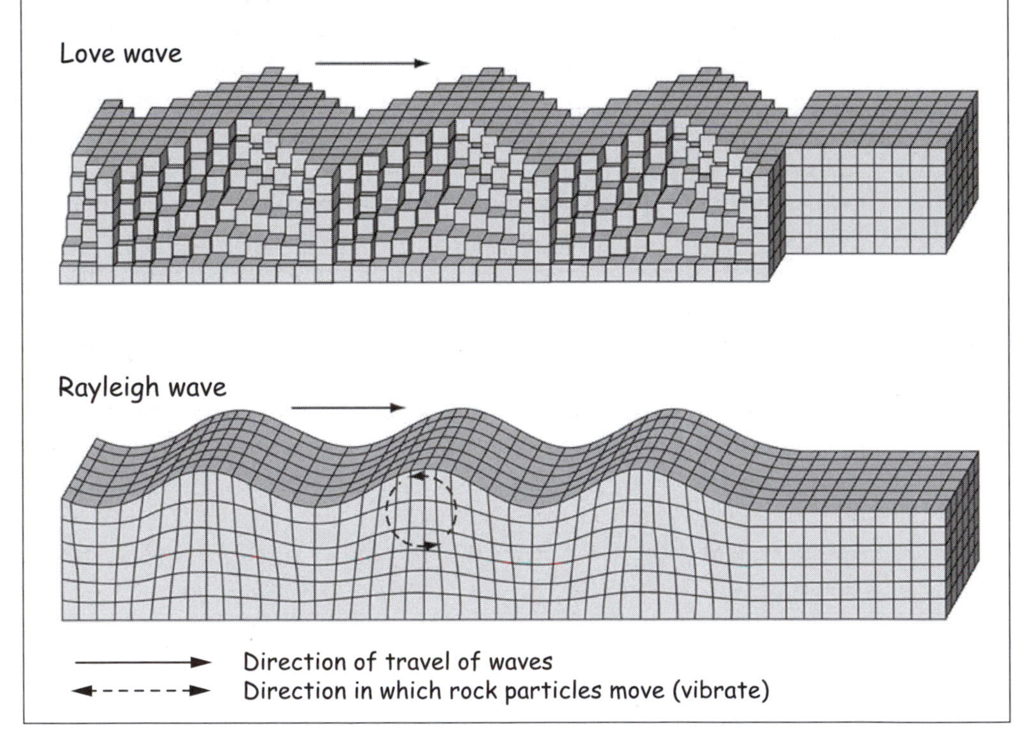

waves, and surface waves, which include Rayleigh (R) waves. The modes of propagation of these wave types are well known and are described in most texts on seismic methods (e.g. Telford *et al.*, 1990). The waves propagate at velocities which are a function of the density and elastic properties of the ground.

In an isotropic elastic medium, the velocity of a compressional wave V_p is given by:

$$V_p = \left(\frac{K + \frac{4}{3}G}{\rho}\right)^{1/2}$$

and the velocity of a shear wave V_s is:

$$V_s = \left(\frac{G}{\rho}\right)^{1/2}$$

where K is the bulk modulus, G the shear modulus and ρ the density. According to the theory of elasticity, Young's modulus E is related to G and K thus:

$$K = \frac{E}{3(1 - 2\nu)}$$

and

$$G = \frac{E}{2(1 + \nu)}$$

where ν is Poisson's ratio. Thus G can be obtained from measurements of V_s alone, but V_s and V_p are needed to determine E, K and ν.

It should be noted that in saturated uncemented soils the propagation of P-waves will represent a short term undrained loading. In such a case the compressibility of the pore water will tend to dominate the compressibility behaviour of the soil. The result is that the measured P-wave velocity is likely to be close to that of water (i.e. 1500 m/s) since most of the energy will travel through the pore water and will not reflect the true undrained stiffness E_u of the soil. As the degree of cementation increases the rigidity of the mineral skeleton increases such that first arrival P-waves become more representative of the material. In saturated rock the elastic modulus E measured from the P-wave velocity will be representative of the stiffness of the mineral skeleton. Thus for stiffness measurements in soils only shear wave velocities should be used since these are not affected by the compressibility of the pore fluid.

Surface waves may also be used to determine shear stiffness in soils and rocks. Approximately two thirds of the energy from an impact source propagates away in the form of surface waves of the type first described by Rayleigh in 1885. Exploration geophysicists have traditionally regarded Rayleigh waves or 'ground roll' as a nuisance. Crucially, however, Rayleigh waves travel at speeds governed by the stiffness–depth profile of the near-surface material. Geotechnical engineers have long recognized that Rayleigh waves offer a useful non-invasive method of investigating the ground in situ (e.g. Hertwig, 1931; Jones, 1958; Heukolom and Foster, 1962; Abbiss, 1981). It can be shown from the theory of elasticity that the relationship between the characteristic velocity of shear waves V_s and Rayleigh waves V_r in an elastic medium is given by:

$$V_r = CV_s$$

The range of C is from 0·911 to 0·955 for the range of Poisson's ratio associated with most soils and rocks if anisotropy is ignored. The maximum error in G arising from an erroneous value of C is less than 10%.

The absolute magnitude of the strains associated with the propagation of seismic waves has not been measured, but it is thought that this is less than 0·001%. The maximum values of stiffness will therefore be measured using seismic methods (Fig. 4.1). For this reason the stiffness parameters measured in this manner are referred to as G_{max} and E_{max} (or G_0 and E_0, i.e. the values of G and E at 'zero' or infinitesimally small strain). These parameters provide a valuable benchmark against which other stiffness values measured in the laboratory or in situ can be compared.

Geotechnical engineers generally require stiffness measurements to be made at different depths in order to determine a stiffness–depth profile. Using direct methods of investigation this can be achieved by taking samples at different depths and subjecting them to appropriate laboratory tests or by conducting in situ loading tests (e.g. pressuremeter or plate loading tests). Stiffness is frequently determined indirectly using the SPT. This appears to be attractive since the SPT is carried out routinely as part of most site investigations. However, stiffness parameters are determined using empirical relationships (e.g. Wakeling, 1970; Kee and Clapham, 1971; Stroud, 1988) many of which have very limited accuracy (Clayton *et al.*, 1995b) – see also Fig. 4.73. Field seismic techniques allow stiffness to be determined on representative volumes of ground, and at the in situ stress state, and for this reason may provide valuable data that are unaffected by borehole sampling disturbance, or by penetration effects. In many cases the cost per stiffness measurement may be less using geophysical methods than for the direct methods outlined above.

The seismic methods employed to determine stiffness–depth profiles are described in Table 4.9. The seismic cone carries one or two three-component geophones which permit down-hole tests to be carried out at specified depths without the need for a borehole (Clayton *et al.*, 1995b). Typically, shear waves are generated at the ground surface using a hammer source, and shear wave velocities are determined as in the conventional down-hole method. The seismic cone has the advantage of providing both strength and stiffness data. The depth of penetration is limited, however, by the strength of the ground and any obstructions such as boulders, claystones or rock layers. Surface wave methods overcome these problems, however, because they are non-invasive and no penetration or boring is required.

Surface wave methods exploit the dispersive nature of Rayleigh waves: the speed of propagation of a Rayleigh wave travelling at the surface of inhomogeneous ground depends on its wavelength (or frequency) as well as the material properties of the ground. Measurements of phase velocity of Rayleigh waves of different frequencies (or wavelengths) can be used to determine a velocity–depth profile. Two distinct surface wave methods are available:

- *Spectral analysis of surface waves (SASW) method.* This method makes use of a hammer as an energy source. The field technique is described by Ballard and McLean (1975), Nazarian and Stokoe (1984) and Addo and Robertson (1992).
- *Continuous surface wave (CSW) method.* This method makes use of a vibrator as an energy source. The field technique is described by Ballard and McLean (1975), Abbiss (1981), Tokimatsu *et al.* (1991) and Matthews (1993).

The velocities determined over a range of frequencies will form a characteristic dispersion curve (phase velocity to wave length relationship) for the ground under investigation. This can be inverted (or interpreted) using a variety of different methods to give a velocity–depth profile from which the stiffness–depth profile can be determined.

Figure 4.65 shows results from an investigation carried out to provide stiffness parameters for the structural design of cut and cover tunnels and retaining walls associated with a major junction improvement scheme. Stiffness data were also required to assess the effects of construction on nearby buildings. The site is underlain by over 30 m of London Clay. The undrained strength of the top 15 m is generally below 75 kPa (i.e. firm) as a result of weathering and periglacial action (Gordon *et al.*, 1995). It will be seen from Fig. 4.65 that data obtained from seismic cone (SCPT), cross-hole and CSW surveys show good agreement and form an upper bound for the stiffness measurements. Figure 4.65 also shows stiffness values determined from laboratory tests at strains of 0·1% and 0·01%. In general the shear modulus at very small strain (i.e. from the seismic tests) is about twice that measured at 0·01% strain. This indicates that the drop in stiffness from very small strain levels to the lower bound of field operational strain levels is perhaps not as severe as indicated by the idealized model of stiffness–strain behaviour shown in Fig. 4.1. This reinforces the value of taking stiffness measurements using seismic methods.

Table 4.9 Seismic methods used for the determination of stiffness–depth profiles

Method	Diagram	Advantages	Disadvantages
Up-hole	Source Receiver	Only single borehole required. Tests can be carried out in all soil and rock types. Average velocity is measured in layered materials.	Need to install plastic casing to provide stable borehole.
Down-hole		Only single borehole required. Tests can be carried out in all soil and rock types. Average velocity is measured in layered materials. Higher energy sources (e.g. explosives) can be used without damaging the borehole.	Need to install plastic casing to provide stable borehole.
Seismic cone	Cone	No borehole required; probe is pushed into the ground. Provides other geotechnical parameters in addition to stiffness. Average velocity is measured in layered materials.	Penetration limited by strength of ground. Not suitable for rock.
Cross-hole		Can detect low-velocity (i.e. low-stiffness) layers, provided they are thick compared with borehole spacing. Tests can be carried out in all soil and rock types.	Quality of data diminishes at shallow depths. Maximum velocity is emphasized in thinly layered soils due to head waves.
Cross-hole tomography		Gives two-dimensional distribution of stiffness. Tests can be carried out in all soil and rock types.	Expensive. Artefacts make interpretation difficult. Specialist processing facilities required.
Refraction	Source Receiver	No borehole required.	Cannot detect low-velocity (low-stiffness) layers below higher velocity layers. Cannot detect thin layers. Problems with interpretation of continuous velocity increase with depth. Cannot use in situations of continuous velocity decrease with depth, although such cases are rare.

Reflection		No borehole required.	Expensive high-resolution seismic reflection is required for engineering surveys. Method only effective in layered ground.
SASW method	Hammer source	No borehole required. Field method is quick and relatively simple.	No selective control over the frequencies generated; therefore measurements are limited to those frequencies which can be generated in the medium by a given impulsive seismic source. It may be necessary to use a number of different impulsive energy sources.
CSW method	Vibrator	No borehole required. Selective frequency control of vibratory seismic source. Field method is relatively quick and simple. Preliminary stiffness–depth profile may be viewed on site.	Depth of investigation is currently limited to about 10 m in soft soils and 30 m in stiff soils unless large lorry-mounted vibrators are employed.

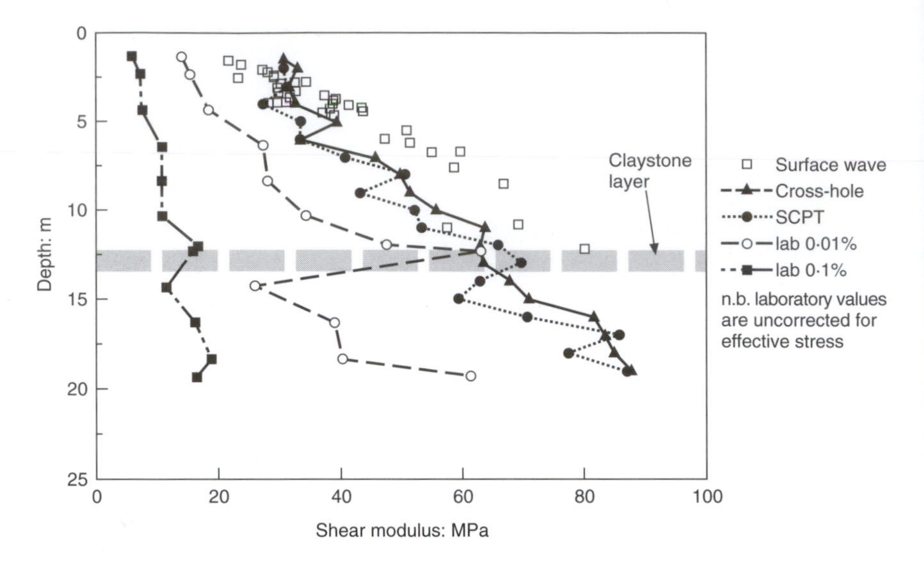

Fig. 4.65 Stiffness–depth and undrained strength profiles for London Clay (Matthews et al.*, 1996)*

Weak rocks such as chalk display more linear stress–strain behaviour than soils. Such materials are generally fractured, however, and the compressibility of the discontinuities will dominate the compressibility of the mass. In this case the stiffness of the intact rock is of little value in predicting ground deformation. In chalk it has been shown using large-scale loading tests and observations of full-scale foundations that the load–settlement behaviour is more or less linear elastic up to the yield stress which varies between 200 kPa and 400 kPa (Ward *et al.*, 1968; Burland and Lord, 1970; Matthews, 1993). It seems reasonable, therefore, to use stiffness–depth profiles determined using seismic methods directly in predictions of ground deformation such as foundation settlements. Matthews (1993) carried out nine large-diameter (1·8 m) plate loading tests on weathered chalks with similar discontinuity patterns but different intact stiffnesses. In each case the stiffness–depth profile beneath the plate locations was determined using surface wave geophysics, the SPT and visual assessment based on the work of Ward *et al.* (1968). Figure 4.66 shows a comparison between the observed and predicted settlement for these tests and indicates that the SPT either grossly overpredicts or underpredicts the settlement, whereas predictions based on geophysics yield reasonable agreement.

Field geophysical stiffness measurement methods

Well established field geophysical techniques suitable for geotechnical stiffness measurement include refraction (Grainger *et al.*, 1973; Abbiss,

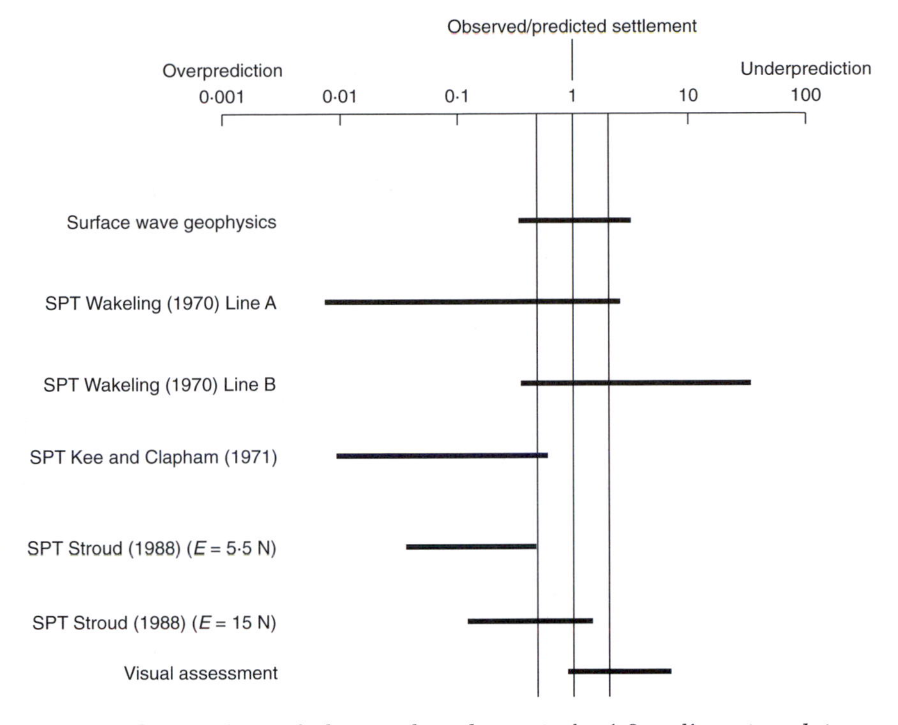

Fig. 4.66 Comparison of observed settlement of a 1·8 m diameter plate on weathered chalk loaded to 200 kPa average bearing pressure with predictions based on stiffness–depth profiles determined using a number of in situ methods (Matthews et al., *1997)*

1979), down-hole (Auld, 1977; Jacobs and Butcher, 1996; Ricketts *et al.*, 1996), cross-hole (Barton, 1929; Stokoe and Woods, 1972; McCann *et al.*, 1975; Ballard, 1976; Auld, 1977; Baria *et al.*, 1986; McCann *et al.*, 1996) and surface wave methods (Jones, 1958; Heukolom and Foster, 1962; Abbiss, 1981; Nazarian and Stokoe, 1984; Matthews *et al.*, 1996). These are illustrated in Table 4.9. All the techniques require a measurement of the travel time of either a shear or Rayleigh wave between at least two receiver (geophone) positions, and knowledge of the distance between those positions. Travel times are usually obtained from geophone traces. Each geophysical method has advantages and disadvantages which relate to:

- the cost and complexity of the equipment
- the plan area required to conduct a survey
- ability to penetrate to the required depth
- simplicity of interpretation of field data
- ability to detect layers of different thickness
- sensitivity to background noise.

The choice of an appropriate survey method to provide geotechnical stiffnesses for design is dependent upon these factors, and is discussed below.

Equipment

Geotechnical surveys generally aim to provide profiles of stiffness at a variety of locations across a site. Such profiles are derived from measurements of the variation in shear wave velocity with depth. In order to make these measurements it is necessary to input pulses of energy to the ground which are rich in shear waves relative to their P-wave energy. Some common types of source are illustrated in Fig. 4.67. Refraction and down-hole measurements routinely involve generating a horizontally

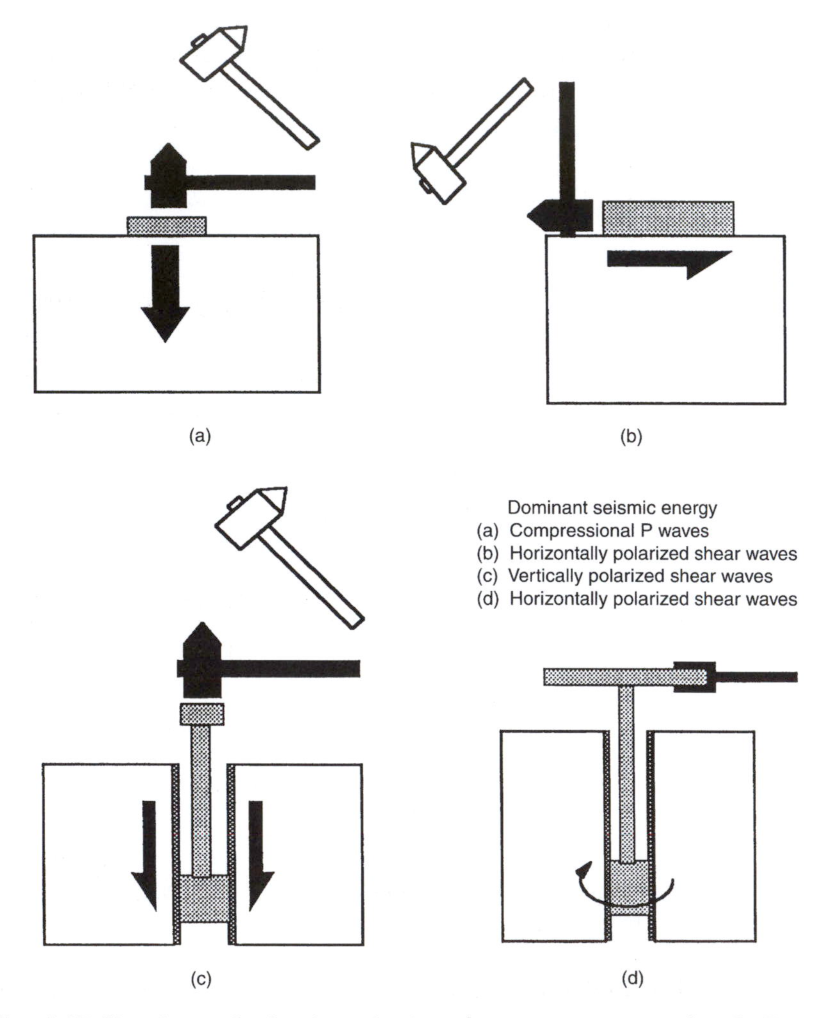

Fig. 4.67 Simple methods of producing shear wave energy for shallow seismic surveys (after Clayton et al., *1995b)*

polarized shear wave at the surface, and detecting the arrival of this at different locations on the ground surface or at different depths in one or more boreholes. The energy source commonly used is a sledge hammer delivering a horizontal blow to the end or side of a wooden or metal block on which a vehicle is parked. For deep investigations a larger energy source may be required. Auld (1977) describes a system that uses a 125 kg pendulum which strikes a metal plate mounted at the end of a shallow pit. Although rich in shear waves these energy sources also produce a certain amount of compressional waves, thus making identification of the first arrival of shear waves on the seismic records difficult. To overcome this problem a second record is made at each depth in which the sledge hammer blow is delivered from the opposite direction. The resulting shear wave motion recorded on the seismic record is reversed whereas the first-arriving compressional wave does not show such a reversal. Figure 4.68(a) shows two seismic records produced in this way.

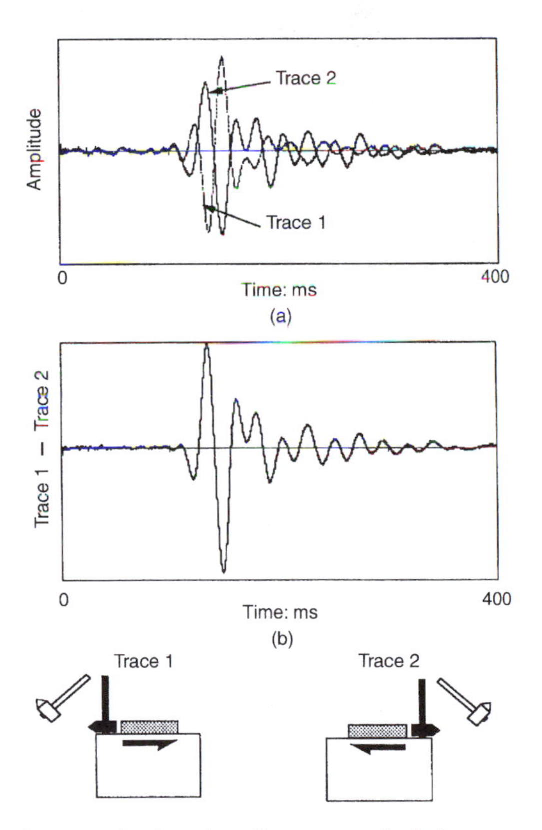

Fig. 4.68 Seismic records showing the reversal of shear wave events in response to reversing the direction of motion of the seismic energy source (Matthews et al., *2000a)*

The reversal of the shear waves is easily identified, but the first shear wave arrival is not clearly defined. This may be enhanced by subtracting one trace from the other as shown in Fig. 4.68(b).

Sliding hammer sources (e.g. Auld, 1977) are favoured for most cross-hole surveys. These are clamped to the borehole wall. The hammer unit is usually hydraulically clamped and the seismic energy is generated by dropping or raising a shuttle causing it to impact against a static anvil. The impact of the shuttle against the anvil causes a small vertical shear displacement of the borehole casing resulting in the generation of seismic energy rich in vertically polarised shear waves which propagate horizontally (Mooney, 1974). The shuttle can be allowed to fall under gravity (a 'down' blow), or raised briskly by means of a rope to provide an 'up' blow. As with surface sources, the combination of 'up' blows and 'down' blows allows the shear waves to be readily identified on the seismic records.

Some surface wave techniques also use a range of hammers to produce Rayleigh waves rich in different frequencies, in order to investigate different depths (spectral analysis of surface waves, SASW). The SASW technique, however, also produces P- and S-waves which can obscure Rayleigh wave detection. Surface waves are the only type of wave measured when using electromagnetic vibrators operating at a single frequency, as used in the continuous surface wave (CSW) technique (Mooney, 1974). Research at the University of Surrey and GDS Instruments Ltd has led the authors to prefer the CSW technique since it was found to give more repeatable results. A disadvantage of this method, however, is that a vibrator must be used which is capable of giving a single frequency input – typically between about 4 Hz and 100 Hz. Electromagnetic vibrators can supply clean signals above about 6 to 10 Hz, but below this more expensive mechanical or hydraulic vibrators are required (Madshus and Westerdahl, 1990).

Traditionally, field geophysical data are acquired using a 12-channel or 24-channel seismograph, but most geotechnical surveys (with the exception of seismic refraction) do not require more than about six channels, so that many engineering researchers (and more recently commercial manufacturers of SASW and CSW equipment (see Fig. 4.69)) have used high-speed data acquisition cards operating within a PC. Seismographs, although more expensive, are not only robust, but also often have other desirable features. Signal enhancement seismographs allow the stacking of data to reduce random noise, which might be considered an advantage. It is preferable, however, to work with clean single traces since stacking inevitably modifies the data. Another feature of modern seismographs is 'digital instantaneous floating point', which allows a large dynamic range (typically in excess of 32 bits) to be optimized for the signal strength received by each channel. Effectively this useful feature avoids the necessity of setting amplifier gain levels before data acquisition is started.

Fig. 4.69 Continuous surface wave system being used on site on a trial road subgrade. The computer-controlled vibrator is in the foreground with a line of six geophones behind (permission of GDS Instruments Ltd)

Geophones vary considerably. For most down-hole, cross-hole and refraction surveying, where the dominant frequency of the incoming signal is typically between 30 and 100 Hz, there are no special requirements, since most commonly available geophones operate satisfactorily down to about 6 Hz. For down-hole and cross-hole work it is normal to use 'three-component' geophones, that is three geophones mounted together and orientated in mutually perpendicular directions. Surface wave techniques require special low-frequency geophones (1–2 Hz natural frequency). These geophones are expensive, however, and also require careful handling.

Choice of survey type

Seismic refraction has been used for many years for mapping and targeting subsurface features such as the soil/rock interface. This method can be used for stiffness investigations. It has the advantage that it is relatively simple to carry out since work can be done from the ground surface with a standard seismograph and geophones using a sledge hammer source as described above. In general, the depths and spacing of stiffness measurements coupled with the requirements for space normally make this technique unsuitable except as a quick means of obtaining preliminary estimates of near-surface stiffness.

In softer ground conditions a down-hole seismic cone survey will be fast and effective; examples of this are given by Butcher and Powell (1996) and Jacobs and Butcher (1996). Some seismic cones contain three-component geophones at only one level. If available, it is far better to use a cone which has two sets of three-component geophones mounted at about 1 m from each other, since traces from the same (rather than consecutive) surface blows can then be obtained and processed to give interval shear wave velocities using cross-correlation techniques (see below) if desired. Cones routinely carry inclinometers, so that any out-of-verticality that might significantly affect the assumed distance between the surface source and the 'down-hole' geophones can be detected.

It has been found that seismic cone testing is extremely valuable even in the determination of stiffness for more routine foundation design, since methods of determining the stiffness of granular soils from penetration tests are notoriously unreliable (e.g. see Clayton *et al.*, 1988, and Fig. 4.73). The seismic cone can also be used in stiff clay (Jacobs and Butcher, 1996), where depths of tens of metres are readily covered. A weakness of the method lies in the possible inability of the cone to penetrate hard or coarse granular strata, and a potential lack of near-surface detail due to the need to locate the hammer some distance away from the top of the cone hole.

When horizontal near-surface variability is sought, the surface wave technique should be considered. With CSW testing the depth of penetration is a function of the stiffness of the ground itself. At a given frequency stiffer ground will be sampled to a greater depth. Given the limited frequency range of most low-cost electromagnetic vibrators, the depth of a CSW survey will be restricted to about 10 m in clay, and 30 m in weak rock. Many stiffness determinations can, however, be made rapidly in the top few metres, and since the method is non-invasive it can be used in all ground conditions, including gravel, rock and contaminated land. The disadvantage of the CSW method is that it requires sophisticated equipment. This is more than compensated for, however, by the speed at which data can be obtained. In addition, computer control can simplify the operator interface and data assessment (e.g. Fig. 4.69).

When the aim of the survey is to investigate greater depths in detail, perhaps where a cone might not penetrate, cross-hole surveying should be used. The major advantage of the cross-hole method is that the depth of investigation can be guaranteed, since the geophysical testing is carried out using three pre-drilled co-linear boreholes, within which

100 mm diameter plastic casing has been grouted. Disadvantages of this method are its cost and the time required to drill and case the holes, although these can be reduced to some extent by re-using one soil sampling hole for geophysics. It is common to use three holes in order to avoid timing errors between source and receiver associated with the seismograph being triggered early or late relative to time at which the energy is released (source triggering errors). Some modern sources, however, use an electronically operated hammer with electronic triggering with the possible advantage of using only two boreholes for the survey. A further disadvantage is that the boreholes should be surveyed for verticality, in order to calculate the travel distance for each depth. This is usually carried out by a specialist contractor, requiring further time and increasing the cost of the survey. Nonetheless, cross-hole surveying has been found to be extremely useful in all ground conditions where holes can be bored.

Surface wave stiffness profiling

Surface wave geophysics in general and CSW methods (i.e. using a vibrator as the seismic source) in particular have the following major attributes.

- The methods are non-invasive, i.e. no penetration or boring is required – an important consideration where penetration or boring is not practical (e.g. Boulder clay) or desirable (e.g. contaminated ground).
- Stiffness–depth profiles can be viewed on-line.
- The computer-controlled vibrator of the CSW system allows frequency magnitudes and increments to be re-selected so that stiffness–depth profiles can be mapped in greater detail (this cannot be done with the SASW system that uses a hammer as the seismic source because frequency content is a characteristic of a particular hammer).

A typical set-up of a commercial CSW system is shown in Fig. 4.69. A computer controls the frequency of a vibrator mounted on a suspension unit. The phase velocity of the Rayleigh wave is measured by an array of vertically polarized sensors called 'geophones'. An on-line plot is displayed of the shear modulus–depth profile where depth is taken as one third of the wavelength (Gazetas, 1982).

The velocities determined over a range of frequencies will form a characteristic relationship between phase velocity and wavelength (the so-called 'dispersion curve') for the ground under investigation. The dispersion curve can be inverted (or interpreted) using a variety of different methods to give a velocity–wavelength relationship from which the stiffness–depth profile can be determined.

The process of converting a field dispersion curve to a Rayleigh wave velocity–depth relationship is known as inversion. There are three principal inversion methods:

- the factored wavelength method
- finite element approaches
- linear models.

The factored wavelength method is the simplest, but least exact, of the methods. It is of practical value because it offers a relatively quick way of processing data on site and so enables preliminary assessment. If using either of the other techniques, then the factored wavelength method can provide a useful initial estimate of the velocity–depth profile to input to the other algorithms. In the factored wavelength method the representative depth is taken to be a fraction of the wavelength λ, i.e. λ/z is assumed to be a constant. A ratio of 2 is commonly, but arbitrarily, used (Jones, 1958; Ballard and McLean, 1975; Abbis, 1981). Gazetas (1982) recommended that $D = \lambda/4$ is used at sites where the stiffness

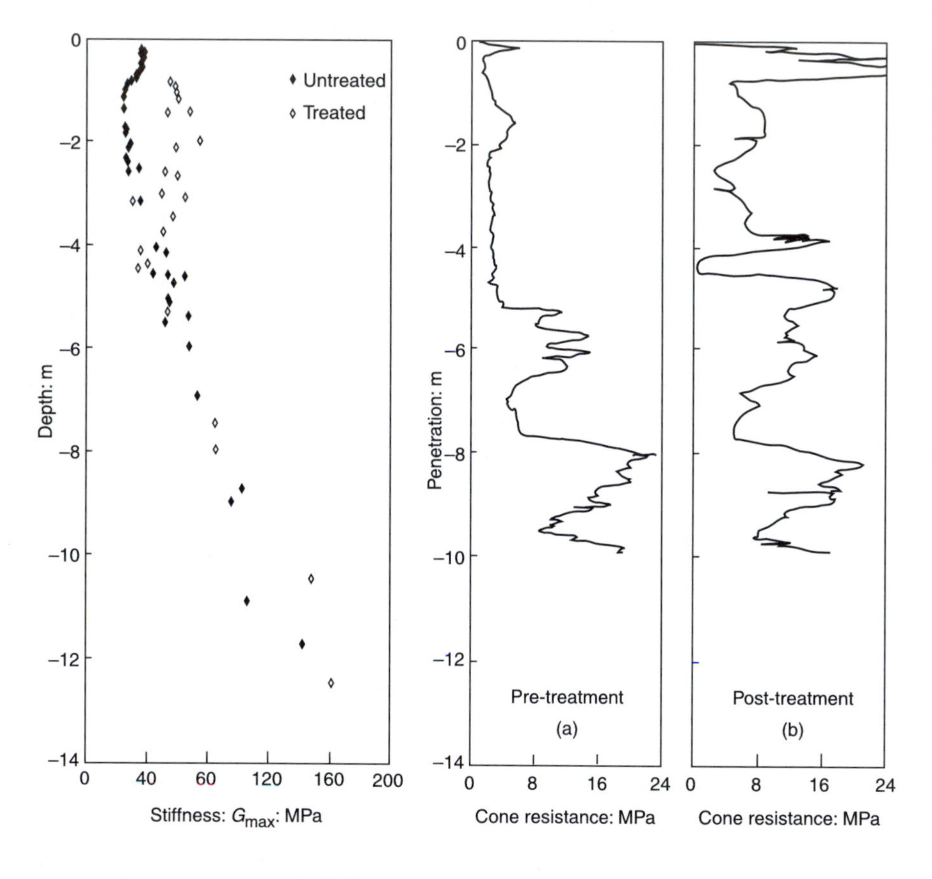

Fig. 4.70 Comparison of 'before' and 'after' CSW stiffness–depth profiles and CPT cone resistance–depth profiles for loose sandy fill between vibro-stone columns for a site in the UK (Sutton and Snelling, 1998)

increases significantly with depth, and that $D = \lambda/2$ is suitable at more homogeneous sites. Gazetas also suggested that taking $D = \lambda/3$ is a reasonable compromise. Grabe and Vrettos (1989), Vrettos and Prange (1990) and Vrettos (1990) also use a $\lambda/3$ rule. Vrettos (1990) gave an analytical solution for surface wave propagation for soils with modulus increasing with depth using the $\lambda/3$ rule (actually $0{\cdot}3\lambda$) for a broad range of parameters. The $\lambda/3$ rule is used for the CSW profiles given in Fig. 4.70. It can be seen that this method is quite adequate for demonstrating ground improvement and correlates reasonably well with CPT results. For several case studies demonstrating the use of the CSW system for measuring ground improvement, see Moxhay *et al.* (2001); two of the case studies are described later in this chapter.

Using the finite element method a synthetic dispersion curve is generated and the stiffness distribution is progressively adjusted until the synthetic dispersion curve matches the curve obtained in the field (Clayton *et al.*, 1995a) as shown in Fig. 4.71. The ground is divided into layers of constant stiffness. For a simple subsurface geometry a two-dimensional, axially symmetric, idealization of surface wave tests can be made. The equations of motion are integrated with respect to time to model the ground motion at the actual geophone locations used in the field. These data are used to determine the synthetic dispersion curve.

Tou *et al.* (2001) compared stiffness profiles obtained by the CSW method with bender element stiffness measurements in the triaxial test (Fig. 4.72). It can be seen that there is reasonable agreement between the bender element measurements and the CSW stiffness profiles. Linear models have been proposed by and Nazarian and Stokoe (1984), Lai and Rix (1998), and Rix (2001).

Advantages and disadvantages of geophysical methods

Like all methods of parameter determination, field geophysical techniques have both advantages and disadvantages. For most engineers, the primary difficulty has been a belief that geophysics measures dynamic stiffness at very small strain levels, and that this stiffness is very different from that required for geotechnical design. It is now realized, however, that this is not the case. *On the contrary, dynamic stiffness can be close to operational static values.* For example, in fractured chalk the ratio between stiffnesses predicted using geophysics and those from large-diameter plate tests is close to unity (Matthews, 1993). In addition, in comparison with other methods such as the standard penetration test, geophysics appears to provide the best way to determine stiffnesses (Fig. 4.66). This is not surprising given the difficulty of determining stiffness from an SPT N value. As shown in Fig. 4.73, the value of elastic

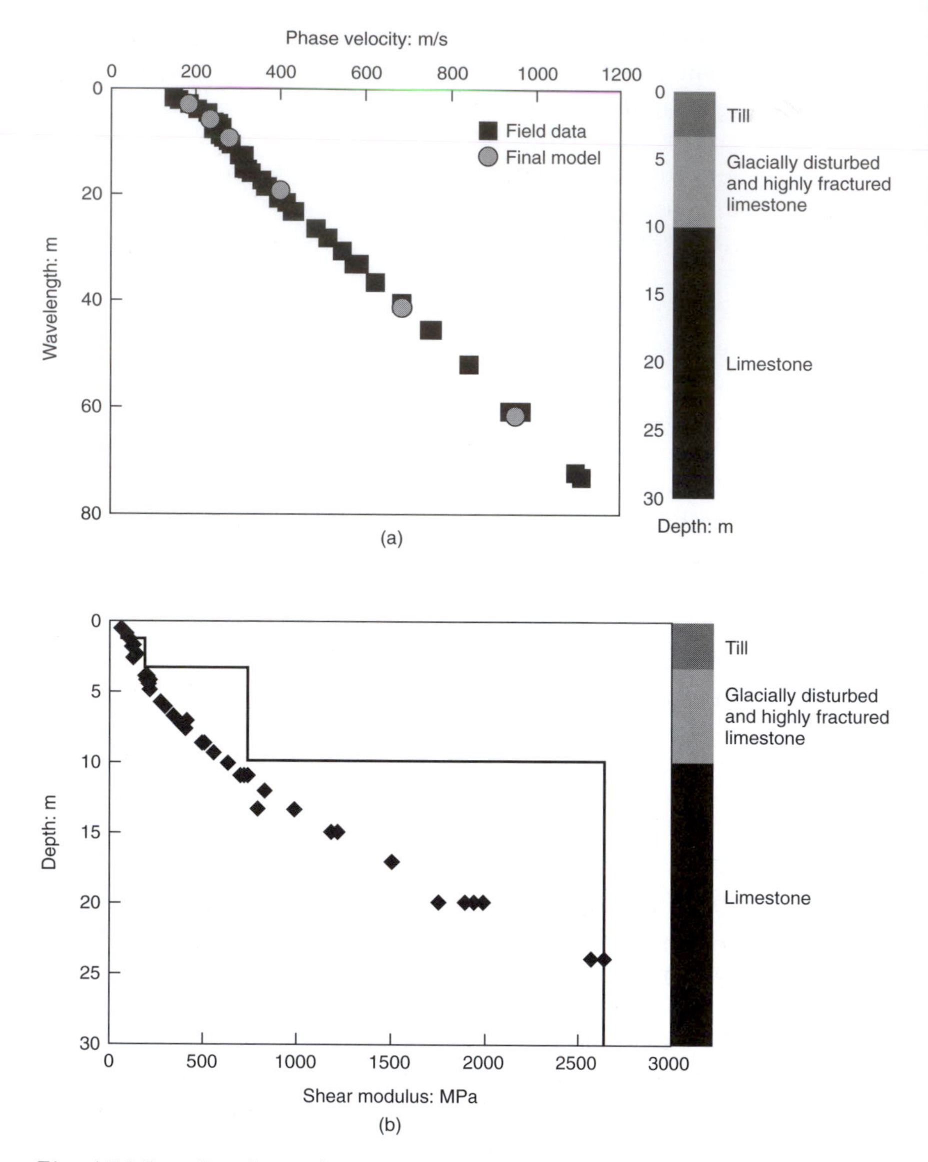

Fig. 4.71 Results of a surface wave survey at Lernacken, Sweden, showing (a) observed and computed curves, (b) stiffness model and $\lambda/3$ interpretation (Clayton et al., *1995a)*

modulus may vary by one or even two orders of magnitude depending on the empirical correlation adopted (see also the section 'Is it time to retire the SPT?' earlier in this chapter).

For clays it has also been found that the maximum stiffness in the highest quality laboratory specimens is close to that determined from field geophysics (Clayton and Heymann, 2001). Table 4.8 earlier in this chapter

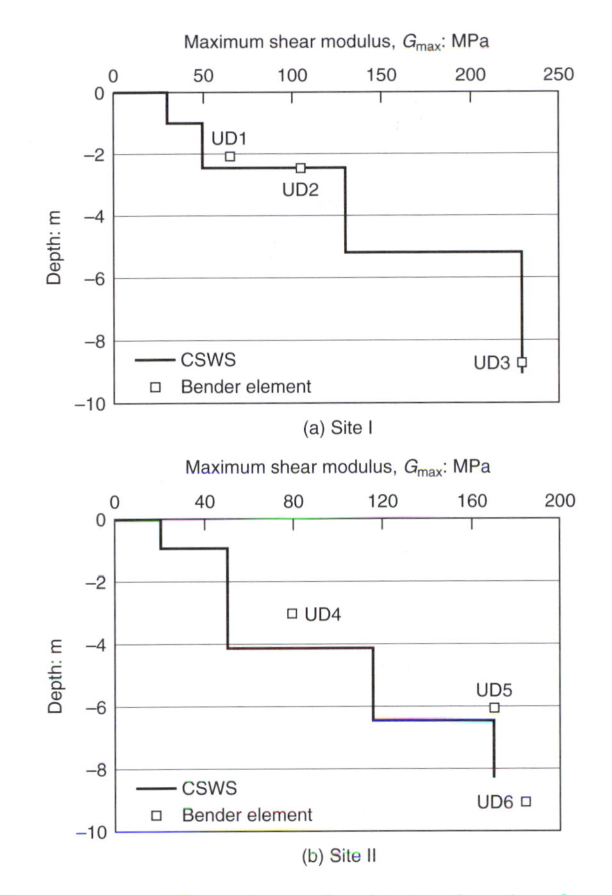

Fig. 4.72 Comparison of G_{max} *determined using bender elements in the triaxial test and by a GDS Instruments Ltd CSW system, after Tou* et al. *(2001)*

shows the stiffness degradation with strain level for soft clay (Bothkennar clay), stiff fissured clay (London Clay) and an intact weak rock (chalk) determined in the triaxial apparatus using local strain instrumentation described by Heymann *et al.* (1997). It is clear from Table 4.8 that stiffness at operational strain levels E_{op} is between 40% and 80% of the maximum stiffness E_0 derived from seismic velocity measurements. Given the rates of stiffness degradation discussed above, it would seem entirely reasonable to estimate stiffness from field geophysical results in combination with some (perhaps conservative) reduction factor to take account of the expected strain level around the proposed construction. Referring to Table 4.8, notional values for factoring E_0 could be:

$E_{op} \approx 0{\cdot}50E_0$ for soft clays

$E_{op} \approx 0{\cdot}85E_0$ for stiff clays and weak rocks.

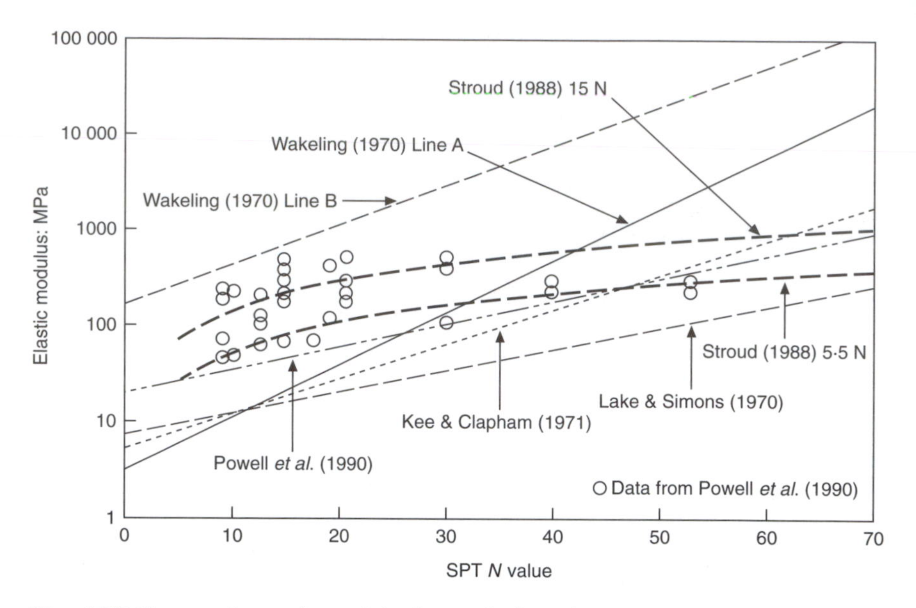

Fig. 4.73 Comparison of empirical correlations between E *and SPT* N *value for chalk (Matthews et al., 1996)*

Values of stiffness estimated in this way can be expected to be far superior to many techniques routinely used today, for example, oedometer testing, external strain triaxial testing, and penetration testing, where poor performance in predicting stiffness has been known for decades (e.g. Burland and Hancock, 1977; Clayton *et al.*, 1988; Izumi *et al.*, 1997). As an example, Fig. 4.74 shows not only stiffnesses back-analysed from ground movements around a number of major structures in the London area, but also the very much lower values from routine oedometer and triaxial tests conducted on samples from the Grand Buildings site. It can be seen that these stiffnesses are about an order of magnitude too low. In contrast, the cross-hole stiffnesses for the London Clay sites are close to those derived from back-analysis.

For sands in Perth, Western Australia, Fahey (2001a) used a self-boring pressuremeter (SBP) to provide a lower bound stiffness, and seismic methods to provide the upper bound stiffness G_0, the maximum shear modulus. Hyperbolic relationships were fitted between these upper and lower bounds to obtain operational modulus. On this basis, Fahey (personal communication) uses $40\% G_0$ for sands in Perth but suggests that for other soils different rules will apply, e.g. in some cases G_0 can be used directly, in other cases, maybe $60\% G_0$.

Geophysics has other advantages. For example, although it may be necessary to install cased holes for sources and receivers, the vast bulk of the ground tested remains at its in situ stress and saturation level,

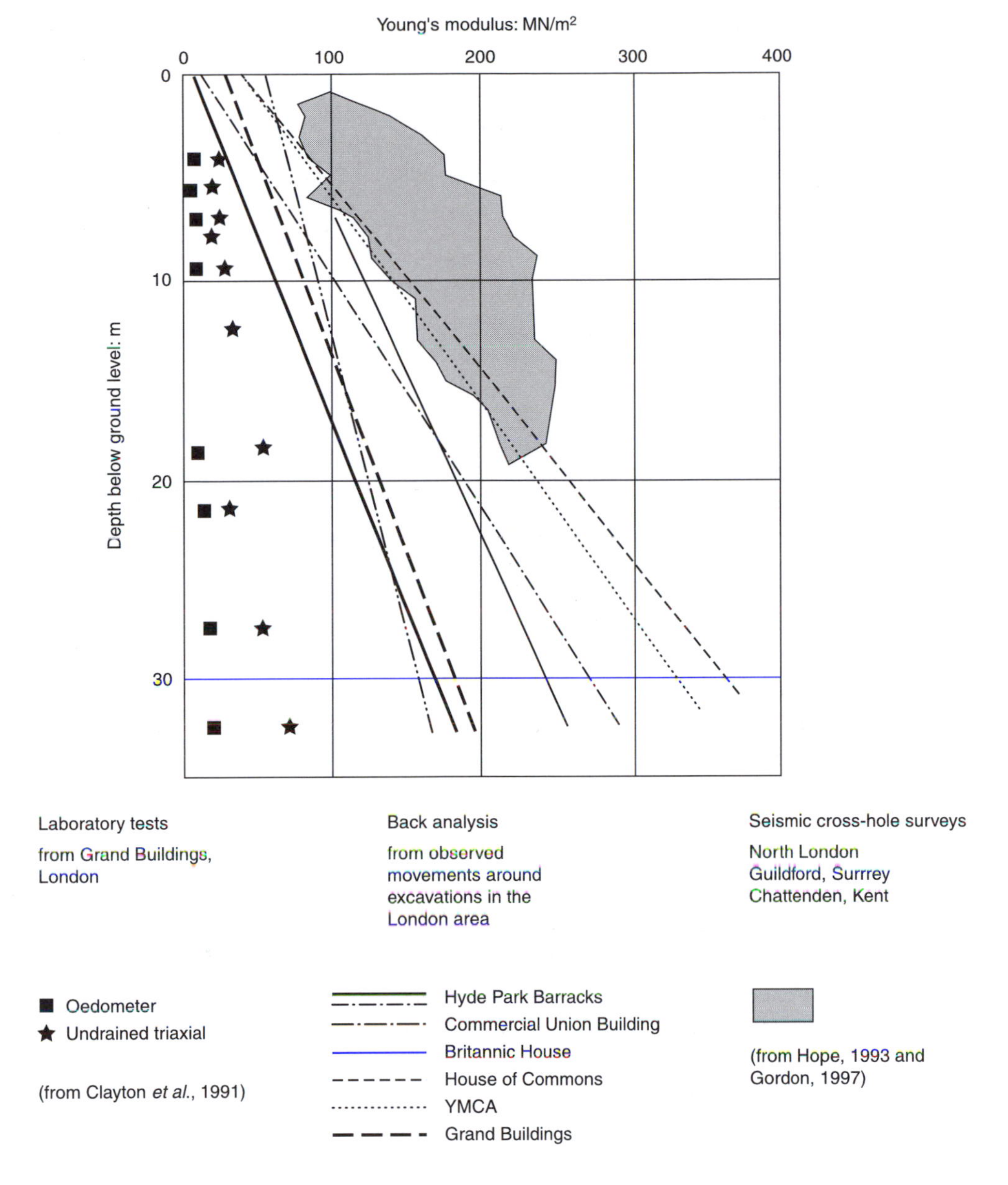

Fig. 4.74 Stiffness profiles for various London Clay sites (Matthews et al., *2000)*

and undisturbed. In contrast, laboratory tests require that specimens be taken. Sample disturbance involves not only mechanical disturbance to the soil structure, but also stress relief. In addition, field geophysical results are representative of a large volume of ground, so that layering and fracturing are taken into account; also, some techniques (for example CSW) are non-invasive. Because boreholes are not required such tests are very fast, the results can be obtained relatively cheaply, and significant contact with contaminated land can be avoided.

Of course, there are disadvantages. The equipment and techniques used in field geophysics are unfamiliar to most engineers. Equipment is still relatively scarce, and because of its high capital cost (a signal enhancement seismograph can cost in excess of £20 000) it is common for it to be hired in on a job-by-job basis. Because of this it may not be possible to obtain the required equipment exactly when site access can be obtained. There are, however, complete PC-based SASW systems now commercially available for about £10 000 and CSW systems for about £40 000.

Case study: quarry rehabilitation, Swanscombe UK

The site was formerly a chalk quarry and had been filled during the 1950s with up to 32 m of Greensand, known locally as Thanet sand (Moxhay *et al.*, 2001). The proposal was to build 173 houses together with associated structures on the site. Thanet sand usually shows high variation in grain size and silt content. Particle size distribution tests indicated that around 75% of the sample fell outside the grading normally accepted as suitable for improvement by mechanical methods (Baumann and Bauer, 1974; Hughes and Withers, 1974). A blanket treatment was proposed, using stone columns to stiffen the upper layers of the sand fill. Thousands of columns were installed in a regular grid pattern to a depth of between 4 and 4·5 m and the ground in between tested by both static and dynamic cone penetrometers. The penetration tests failed to show any consistency in improvement, even in areas with closer column centres, while load tests on the columns themselves and on zoned areas showed excellent results in terms of settlement. CSW testing was introduced and the anticipated improvement to a depth in the region of 5 m could be observed (see Fig. 4.75(a)). Furthermore, additional CSW testing seven months after stone column installation showed that improvement had actually continued and that the stiffness down to around 2 m depth had increased significantly while the site remained untouched (see Fig. 4.75(b)).

Case study: industrial development, Basildon UK

A 4500 m^2 site investigation was undertaken for an industrial development (Moxhay *et al.*, 2001). The site had been filled some years previously with clay and small quantities of hardcore and rubble. The underlying natural material below the fill at between 2·5 and 3·5 m depth was made up of bands of sand and gravel overlying firm to stiff London Clay. Groundwater was not found in the upper soil and fill, but the high

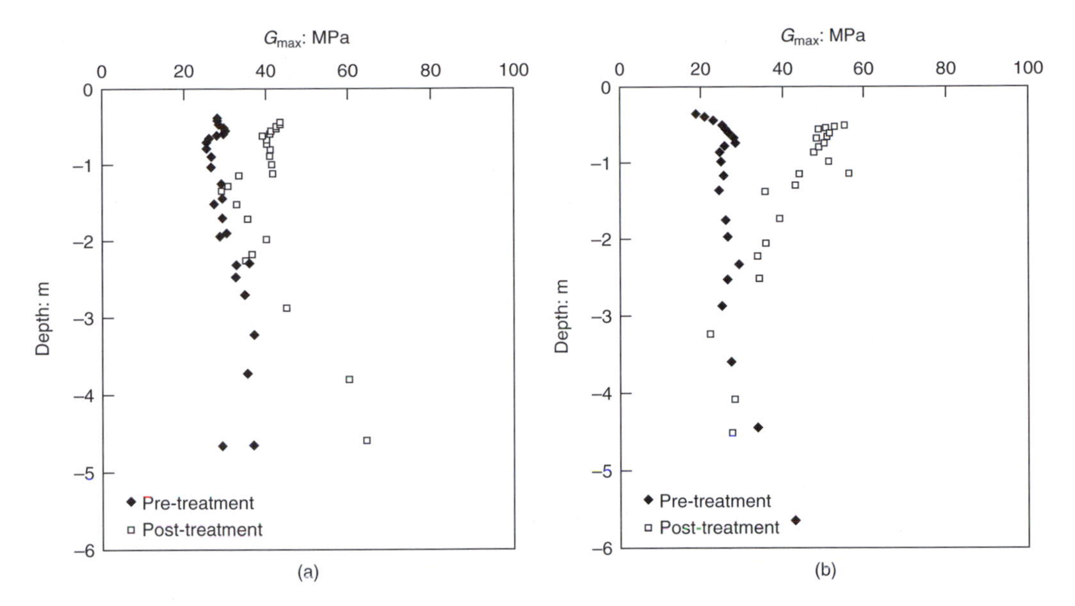

Fig. 4.75 Continuous surface wave (CSW) stiffness–depth profiles for a site in Swanscombe, (a) immediately after treatment and (b) seven months after treatment (Moxhay et al., *2001)*

density of the fill in places meant that in order to install stone columns, the holes would have to be pre-bored, thus adding expense. The highly variable densities meant in situ ground treatment was essential to control future settlements. Dynamic compaction incorporating stone pillars was chosen as the most technically viable and cost effective solution to support both ground-bearing slab loadings up to 75 kPa and main building foundations with loads of 150 kPa. Excavations on site following initial weight dropping indicated the pillars were around 1·75 m deep. A programme of CSW testing prior to treatment helped identify areas of the site with less stiff soil profiles and hence in need of the greatest improvement effort. A second set of tests after treatment showed a significant overall increase in ground stiffness as well as demonstrating that all areas of the site, including those known to be less stiff, were satisfactorily improved to a uniform standard. Typical results from an initially stiffer area of the site are illustrated in Fig. 4.76(a) and from an initially softer area in Fig. 4.76(b). The difference in stiffness between the two pre-treatment profiles and similarity between the two post-treatment ones is notable. It is also evident that the stiffened soil raft that has been created extends to a greater depth than the rock pillars. A cold food processing unit has now been constructed on the site and is in use.

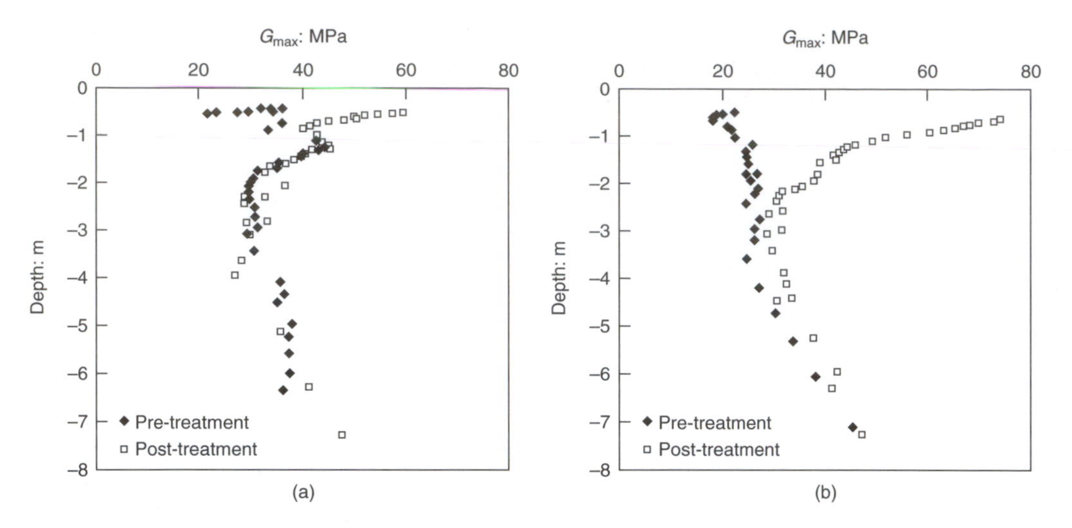

*Fig. 4.76 Continuous surface wave (CSW) stiffness–depth profiles for a site in Basildon pre-treatment and post-treatment by dynamic compaction incorporating stone pillars (a) for an initially stiffer area of the site and (b) for an initially softer area (*Moxhay et al.*, 2001)*

Summary of advantages of seismic methods

- Stiffness measured using geophysics is close to that required for the calculation of displacement around a range of civil engineering structures.

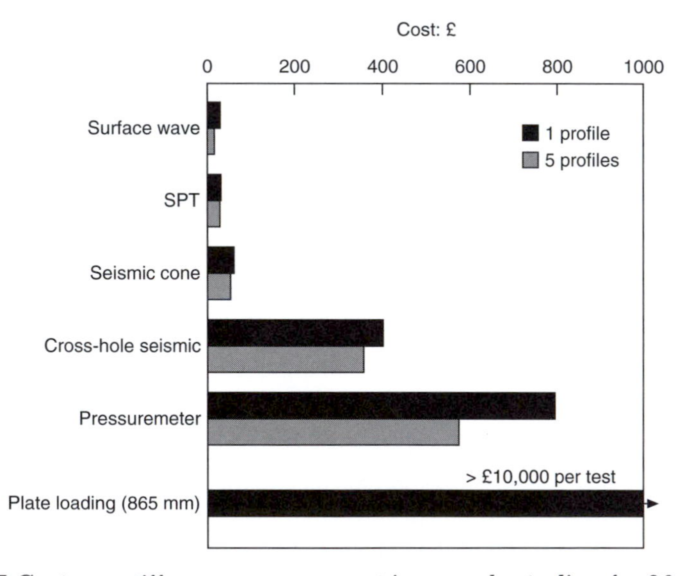

*Fig. 4.77 Cost per stiffness measurement in pounds sterling for 20 m deep profiles in chalk for various methods of stiffness assessment (*Matthews et al.*, 1996)*

- Maximum shear modulus G_0 may be used to benchmark stiffness measurements using other methods.
- Seismic methods may provide the most reliable means of stiffness measurement in geomaterials that are difficult or impossible to sample.
- Field geophysical techniques for stiffness profiling are relatively cheap compared with some other in situ tests (Fig. 4.77).
- Cross-hole surveys are useful for deep investigations.
- A combination of down-hole and cross-hole profiling gives an indication of anisotropy.
- Surface wave surveys are best used for relatively shallow investigations (up to 30 m in soil and 50 m in rock).
- The surface wave method is a quick and cost-efficient means for graphically demonstrating near-surface ground improvement by, say, cement stabilization, compaction, dynamic compaction, vibro-flotation, vibro-stone columns, etc.

APPENDIX 1

Recommended list of units

Recommended list of units, unit abbreviations, quantity symbols and conversion factors for use in soil and rock mechanics

Part 1. SI base units, derived units and multiples.

Quantity and symbol	Units and multiples	Unit abbreviations	Conversion factors for existing units	Remarks
Length (various)	kilometre metre millimetre micrometre	km m mm μm	1 mile = 1·609 km 1 yard = 0·9144 m 1 ft = 0·3048 m 1 in = 25·40 mm	1 micrometre = 1 micron
Area (A)	square kilometre square metre square millimetre	km^2 m^2 mm^2	1 $mile^2$ = 2·590 km^2 1 yd^2 = 0·8361 m^2 1 ft^2 = 0·0929 m^2 1 in^2 = 645·2 mm^2	
Volume (V)	cubic metre cubic centimetre cubic millimetre	m^3 cm^3 mm^3	1 yd^3 = 0·7646 m^3 1 ft^3 = 0·02832 m^3 1 in^3 = 16·39 cm^3 1 UK gallon = 4546 cm^3	To be used for solids and liquids
Mass (m)	megagram (or tonne) kilogram gram	Mg (t) kg g	1 ton = 1·016 Mg 1 lb = 0·4536 kg	Megagram is the SI term
Unit weight (γ)	kilonewton per cubic metre	kN/m^3	100 lb/ft^3 = 15·708 kN/m^3 (62·43 lb/ft^3 pure water = 9·807 kN/m^3 = specific gravity 1·0 approx.)	Unit weight is weight per unit volume
Force (various)	Meganewton kilonewton Newton	MN kN N	1 tonf = 9·964 kN 1 lbf = 4·448 N 1 kgf = 9·807 N	
Pressure (p, u)	Meganewton per square metre Megapascal	MN/m^2 MPa	1 $tonf/in^2$ = 15·44 MN/m^2 (1 MN/m^2 = 1 N/mm^2)	To be used for shear strength, compressive strength, bearing capacity, elastic moduli and laboratory pressures of rock

Quantity and symbol	Units and multiples	Unit abbreviations	Conversion factors for existing units	Remarks
Stress (σ, τ) and Elastic moduli (E, G, K)	kilonewton per square metre kilopascal	kN/m^2 kPa	$1\,lbf/in^2 = 6{\cdot}895\,kN/m^2$ $1\,lbf/ft^2 = 0{\cdot}04788\,kPa$ $1\,tonf/ft^2 = 107{\cdot}3\,kPa$ $1\,bar = 100\,kPa$ $1\,kgf/cm^2 = 98{\cdot}07\,kPa$	Ditto for soils
Coefficient of volume compressibility (m_v) or swelling (m_s)	square metre per meganewton square metre per kilonewton	m^2/MN m^2/kN	$1\,ft^2/tonf = 9{\cdot}324\,m^2/MN$ $= 0{\cdot}009324\,m^2/kN$	
Coefficient of water permeability (k_w)	metre per second	m/s	$1\,cm/s = 0{\cdot}01\,m/s$	This is a velocity, depending on temperature and defined by Darcy's law $V = k_w \dfrac{\delta h}{\delta s}$ V = velocity of flow $\dfrac{\delta h}{\delta s}$ = hydraulic gradient
Absolute permeability (k)	square micrometre	μm^2	$1\,darcy = 0{\cdot}9869\,\mu m^2$	This is an area which quantifies the seepage properties of the ground independently of the fluid concerned or its temperature $V = \dfrac{kpg}{\eta}\dfrac{\delta h}{\delta s}$ p = fluid density g = gravitational acceleration η = dynamic viscosity
Dynamic viscosity (η)	millipascal second (centipoise)	mPa s (cP)	$1\,cP = 1\,mPa\,s$ $(1\,Pa = 1\,N/m^2)$	Dynamic viscosity is defined by Stokes' Law. A pascal is a newton per square metre
Kinematic viscosity (ν)	square millimetre per second (centistoke)	mm^2/s (cSt)	$1\,cSt = 1\,mm^2/s$	$\nu = \eta/\rho$

Quantity and symbol	Units and multiples	Unit abbreviations	Conversion factors for existing units	Remarks
Celsius temperature (t)	degree Celsius	°C	t°F = 5(t − 32)/9°C	The Celsius temperature t is equal to the difference $t = T - T_0$ between two thermodynamic temperatures T and T_0 where $T_0 = 273{\cdot}15$ K (K = kelvin)

Part 2. Other units

Quantity and symbol	Units and multiples	Unit abbreviations	Conversion factors for existing units	Remarks
Plane angle (various)	Degree Minute second (angle)	° ′ ″		To be used for angle of shearing resistance (ϕ) and for slopes
Time (t)	year day hour second (time)	year d h s	1 year = 31·557 × 10^6 s 1 d = 86·40 × 10^3 s 1 h = 3600 s	'a' is the abbreviation for year The second (time) is the SI unit
Coefficient of consolidation (c_v) or swelling (c_s)	square metre per year	m^2/a	1 ft^2/a = 0·0929 m^2/a	

APPENDIX 2

Writing Reports by David Palmer

Introduction

There follows a reproduction of this timeless monograph, written by the late David Palmer and published by Soil Mechanics Ltd in 1957.

As H.Q. Golder commented in the preface:

> *'It is unfortunately only too apparent to anyone who has to read, to correct, and to pass reports for publication that many young engineers, however well qualified technically, are lamentably unable to express themselves in good, clear English. The only way to remedy this defect is to arouse in these engineers an interest in words, in their meaning, in their sound and in their appearance....'*

Certainly, it is our experience as university teachers marking assignments and lab reports that we were correcting the layout, style and English spelling and grammar as much as the soil and rock mechanics principles involved!

David wrote in his introduction:

> *This manual... was written particularly to help engineers, new to writing reports, who are engaged on site investigations...'*

Clearly, he had young engineers in mind, and wrote his 'manual' (he was too modest to call it a 'monograph') with a light touch, e.g.:

> *'Elegance is all very well, but a fairly golden rule whenever... we are thoroughly delighted with the turn of phrase, is 'strike it out'. Dr Cooling advises 'when in doubt, wheel it out'. An example of such elegance... is the engineer's description 'sky-blue clay streaked with claret'. This was wheeled out!'*

Under 'Relevance', David amusingly noted:

> *'... check that what you have said really means something, or sooner or later someone will evaluate the 'Harding coefficient of so-whatness' for your statements.'*[1]

– a worthy principle to set alongside our own 'sanity test'!

[1] Sir Harold Harding was a director of Soil Mechanics Ltd and a past President of the Institution of Civil Engineers.

Of course, this monograph reflects the time it was published – some 45 years ago. Then, the units of measurement were tons, pounds, feet and inches. In addition, the present-day production and transmission of text and images by electronic means was unknown. Nevertheless, the essential principles of site investigation report writing are clearly explained by this classic little book.

We are extremely grateful to Sylvia Palmer, Soil Mechanics Ltd and Environmental Services Group Ltd for permission to reproduce *Writing Reports* by David Palmer.

WRITING REPORTS

BY

D. J. PALMER

M.A., A.M.I.C.E.

LONDON, 1957

SOIL MECHANICS LIMITED

65, OLD CHURCH STREET, S.W.3

PREFACE

This manual was originally written as an internal note to guide members of the Staff of Soil Mechanics Ltd. in the general practice or ' house style ' of the firm.

The rules given in the manual are the result of many years of experience including several thousand reports. The application of the rules, not too rigidly, results in a reasonably consistent style and helps to avoid the omission of important data.

It is unfortunately only too apparent to anyone who has to read, to correct, and to pass reports for publication that many young engineers, however well qualified technically, are lamentably unable to express themselves in good, clear English. The only way to remedy this defect is to arouse in these engineers an interest in words, in their meaning, in their sound and in their appearance—this latter may well improve the standard of spelling too. It is hoped that Mr. Palmer's manual may assist in achieving this end.

So many engineers, having seen the internal note, have begged a confidential copy that it has been decided to publish this manual in the hope that it will prove of use to the profession as a whole and help not only our young engineers, but those of our clients and possibly even of our competitors.

H. Q. GOLDER,
Director.

CONTENTS

	Page
INTRODUCTION	5
THE FRAMEWORK AND CONTENT OF A REPORT	6
Title Page	9
Table of Contents	9
Abstract	9
Introduction	11
Description of Site	11
Geology of Site	11
Description of Borings	11
Field Tests	12
Laboratory Tests	13
Discussion	13
Conclusions	14
References	14
Appendices	14
Tables	14
Figures	14
PREPARING THE DRAFT	17
STYLE	17
CALCULATIONS	18
CHECKING REPORTS	19
FINIS	19
REFERENCES	19
APPENDICES	20
A. House Style	20
B. Typical names of geological formations	23
C. Signs for correcting drafts	24
FIGURES	
Title page	7
Table of Contents	8
Abstract	10
Correction of a Draft	15–16

INTRODUCTION

This manual is written in the firm conviction that an Engineering Report should be a work of literature within its own limited field. The author makes no claim that it is comprehensive. It was written particularly to help engineers, new to writing reports, who are engaged on site investigations, and therefore its scope is necessarily limited. It includes experience gained over a number of years by the staff of Soil Mechanics Limited. In this respect the author wishes to acknowledge with gratitude what he has been taught about the job by present and former colleagues.

No doubt some of the practice described in this note is not to everybody's taste but it will be readily appreciated that conformity with standard rules is the first step to acquiring a clear and a good style. Revision of the rules is necessary from time to time because the content and approach of reports changes with increasing experience, particularly in a subject like Soil Mechanics.

Because of the manual's limited scope it is recommended that it is read in conjunction with 'The Complete Plain Words' by Sir Ernest Gowers (1954)* which is a most readable guide to the use of English in everyday life.

D. J. P.

* See reference at end of text p. 19

THE FRAMEWORK AND CONTENT OF A REPORT

The framework of our engineering reports is as follows and in the order shown:—

1. Title page.
2. Table of Contents.
3. Abstract.
4. Text.
5. Conclusions.
6. References or Bibliography.
7. Appendices.
8. Tables of Results.
9. Drawings and Photographs.

An example of the more detailed structure, suitable for the average report of a site investigation, can be seen illustrated in Fig. 2, a Table of Contents. Naturally the titles of the individual sections depend on the nature of the work carried out but the list given in Fig. 2 will be suitable for many investigations.

The essentials to remember regarding content are that the report should be an account of the whole job from start to finish and that it should contain all the technical facts, good and bad, without reference to any personal administrative difficulties. Attention should be given to detail, but in such a way as to avoid incomprehensible catalogues, e.g. of strata borehole by borehole already better depicted in the borehole logs. The best approach is to make generalizations about the problem and then work to the particular, illustrating exceptions to the generalizations. Facts should be given first, theories afterwards. Drawings and sketches often help where words fail.

It is important to remember that one is not only writing to assist the client but also to assist one's colleagues.

The client (*a*) wants his problem stated simply,
(*b*) wants a simple answer, and
(*c*) wants to be able to pass the information on to a specialist or contractor.

We want to be able to get all technical information relating to the job without consulting the file.

Each report should be written to suit the particular client. If the client is a Consulting Engineer who has considerable knowledge of geotechnical problems, the content of the report will differ considerably from a report written for a Mechanical Engineer or a Solicitor. In each case the client should not be left with the impression either that he is being 'talked down to' or that a bunch of boffins are 'talking over his head'. It must be assumed even with the humblest of clients that there are no geotechnical problems that are so complex that they cannot be explained in simple terms to him. If the client fails to understand the report it is almost certain that the deficiency is ours.

It is also important to examine carefully what the terms of reference are for the job and to 'take good care to include anything that comes within the framework set by these terms and reject anything that does not come within the framework' (Kapp, 1948).

The content of individual sections of the report is now discussed in detail:—

REPORT NO. 2217/1

SITE INVESTIGATION AT FAWLEY REFINERY, HANTS.

EXTENSION TO THE MARINE TERMINAL JETTY.

March, 1956.

SOIL MECHANICS LIMITED,
65, Old Church Street,
L O N D O N, S. W. 3.,
England.

Fig 1. Title Page.

\- 2 -

REPORT NO. 2217/1

SITE INVESTIGATION AT FAWLEY REFINERY, HANTS.

EXTENSION TO THE MARINE TERMINAL JETTY.

CONTENTS

ABSTRACT Page 3

INTRODUCTION " 4

THE SITE AND GEOLOGY " 4

THE BORINGS " 4

FIELD TESTS " 5

LABORATORY TESTS " 5

DISCUSSION " 6

CONCLUSIONS " 8

APPENDICES
- A. The standard penetration test " 9
- B. List of all investigations on jetty site " 10
- C. Previous test piles " 11

TABLES
- I. Index properties " 12
- II. Triaxial compression test results " 13

DRAWINGS
- Borehole logs Fig. 1 to 7
- Section through borings " 8
- Mohr circles " 9
- Site plan " 10
- Key plan " 11

----ooOoo----

Fig. 2. Table of Contents.

TITLE PAGE

An example is illustrated in Fig. 1. This gives the number of the report, the title of the report, the name and address of the firm, and the month and year of issue. The title of the report should in effect be a very brief précis of the report. It should state:—

(*a*) The nature of the work, e.g. site investigation.

(*b*) The purpose of the work, e.g. for a jetty.

(*c*) The situation or location, e.g. Shellhaven, Essex.

The title thus reads:—

'Site Investigation for a Jetty at Shellhaven, Essex.'

The name of the country and region should be added where necessary to avoid obscurity, thus:—

'Site Investigation for a New Runway at Palisadoes Airport, Jamaica, B.W.I.'

TABLE OF CONTENTS

An example is given in Fig. 2. The following points should be noted:—

(*a*) The report number and full title are repeated at the top of the sheet.

(*b*) All titles of sections are given in capitals as in the text.

(*c*) All sub-titles of *Appendices*, *Tables* and *Figures* are in lower case, indented five spaces.

(*d*) Double spacing is observed throughout between main sections, separation of the single spaced sub-sections being obtained adequately by indenting in lower case.

(*e*) The contents will normally be typed last, the typing staff inserting corrected page numbers according to the final typing of the text.

(*f*) All pages of the report, including Appendices and Tables should be numbered consecutively in Arabic numerals starting with the title page as page 1. The page number should be omitted from the title page.

ABSTRACT

An example is given in Fig. 3. This should follow the table of contents. Normally it will be typed on a separate sheet by itself with the number and title of the job again repeated. It should be a précis of the subject matter of the report and therefore should be brief. Its purpose is to tell the reader quickly what the job is about including the findings. As such it covers a wider field than the conclusions. Where the report is very short it is permissible to type the *Contents* and the *Abstract* on one page.

Phrases such as 'This report describes . . .' can be overworked. Go straight to the point, e.g. 'Five borings about 50 ft. deep revealed about 34 ft. of alluvium overlying Keuper Marl'. It is often not necessary to give the location, etc., because the descriptive title appears on the same page.

The layout of the Abstract sheet should be determined by the length of the *Abstract*. For very short ones some attempt should be made to preserve symmetry, if necessary by increasing the width of margin on all four sides. Such adjustments can be left to the final typing.

- 3 -

REPORT NO. 2217/1

SITE INVESTIGATION AT FAWLEY REFINERY, HANTS.

EXTENSION TO THE MARINE TERMINAL JETTY.

A B S T R A C T

Seven borings put down from a floating craft showed the following ground succession: soft alluvial clay, silt and peat (16 ft. to 24 ft. thick), compact river gravel (5 ft. to 15 ft. thick), stiff to hard Barton Clay (22 ft. maximum penetrated). The results of laboratory tests, chiefly on the Barton Clay, are presented in the report, where it is seen that the shear strength confirms the stiff to hard consistency of the clay. The ground conditions are discussed and reference is made to founding the extension of the jetty on monoliths or on piles. The results of two pile driving and loading tests are presented, with comment on the behaviour of the test piles under the proposed working loads.

----oo0oo----

Fig. 3. Abstract.

INTRODUCTION

The main purpose of the introduction is to answer the questions:—

Who ? When ? What ? Why ? Where ?

The order in which these questions are answered is a matter of taste, but the introduction should answer the following points:—

Who? The name of the client, of his consulting engineers, of the contractor on the site, of any other people concerned. Discretion is sometimes necessary on this point; sometimes the names are better left out.

When? Dates when the site work was carried out and other relevant dates.

What? Nature of the investigation, e.g. number of borings, soil tests, loading tests, pumping tests, etc. Include reference to previous reports.

Why? Purpose of the investigation and background of problem, e.g. foundation pressures and settlements.

Where? Location of site, e.g. Karachi Dry Dock, Pakistan. Refer to site plan if necessary.

DESCRIPTION OF THE SITE

Describe in more detail the location of the site with reference to the topography, local roads, railways and rivers, etc. Reference should be made to the site plan and a key plan showing the neighbouring area, drawn to a scale of, say, 2½ in. to 1 mile.

Any other matters of interest about the site and neighbouring area should be noted here.

GEOLOGY OF THE SITE

When brief this can be treated under *Description of the Site*. When more detailed it is best treated separately. An attempt should always be made to relate the soil to generally accepted geological terms. In the United Kingdom a study should be made of the series of pamphlets 'British Regional Geology' published by the Stationery Office in eighteen parts, other regional memoirs, and geological maps. (Maps prepared by the Geological Survey are not completely reissued owing to the destruction of the plates during the war. However, information can often be obtained from original manuscripts, etc., at the Survey.)

A geologist should always be consulted about your job before, during and after the job. All available information about the geology of the site should be obtained before the site work starts. The geology should be investigated thoroughly at the sample description stage of the work.

DESCRIPTION OF THE BORINGS

After stating briefly the method of boring and sampling the first thing to get over to the client is a simple picture of the soil conditions at the site. This is not always possible. The first essential, however, is to plot the boreholes on one sheet of paper, if possible in sectional form and related to a well established datum such as the Ordnance Datum (stated as Liverpool Datum or Newlyn Datum, *not* Ordnance Datum). From such sections a clear picture can be obtained in one's own mind. An endeavour should be made to present the sections as a figure in the report. Interpolations should always be done judiciously and with the guidance of a geologist.

Having seen the soil conditions pictorially it is possible to describe them in the text. Preferably this should be presented as a generalized profile throughout the site. The descriptions need not be as detailed as those presented on the borehole logs, e.g. :—

(*a*) *Topsoil* ... ... 1 to 3 ft. thick.
(*b*) Soft grey organic silty *clay* ... ... 6 to 9 ft. thick.
(*c*) Very soft dark grey *peat* ... ... About 2 ft. thick.
(*d*) Soft grey organic silty *clay* ... ... 10 to 12 ft. thick.
(*e*) Dense sandy *gravel* ... ... 11 to 14 ft. thick.
(*f*) Very stiff red *marl* (found in Borehole 1 only) ... 2 ft. 6 in. thick.
(*g*) Medium to hard red *sandstone* ... ... 13 to 15 ft. penetrated.

Detailed remarks can now be presented with exceptions to the generalized profile. But above all, avoid catalogues of strata borehole by borehole. These are given in the borehole logs.

On sites where it is difficult to relate strata from borehole to borehole, such as extended sites, some difficulty is bound to arise. This can be dealt with sometimes by treating the site area by area, i.e. in groups of boreholes, but in any case there are no golden rules except that the finished product *must* be intelligible, not only to the client but to other people who are obliged to read the report.

Almost invariably include a paragraph on ground water. In certain cases this can form a separate section of the report. Ground water levels are most intelligibly plotted on borehole logs in the form of a table set below the log, e.g. :—

Date	Time	Depth of borehole	Depth of tubes	Depth to water
17 Sept., 1952	10.30 a.m.	4 ft. 6 in.	—	4 ft. 6 in.
,,	12.30 p.m.	7 ft.	8 ft.	4 ft. 4 in.
,,	6.00 p.m.	14 ft.	13 ft. 6 in.	Borehole dry
		etc.		

Having examined the table and sequence of strata it is quite often possible to make definite statements about the ground water table and to make the appropriate symbol at the correct depth on the logs. If the table of water levels is inconclusive this should be stated ; in such cases the matter can often be resolved by the installation of filter tubes. Special attention should be given to variations of ground water due to artesian conditions and with the tide. Record when water is added to help boring or to prevent piping at the base.

FIELD TESTS

These can be reported in the previous section if few, but if substantial they require a section of their own. A description of the method of testing should be given or reference given to description in published works, e.g. say what N is in the Standard Penetration (Raymond) test—the number of blows of a 140 lb. monkey freely falling through thirty inches to drive the standard penetration tool twelve inches into the ground beyond an initial penetration of six inches.

Having described the testing it is then necessary to say what the test results are factually without entering the realms of hypothesis, e.g. the shear strength of clays measured by the vane test, or the relative density of sands and gravels measured by the Raymond tests.

LABORATORY TESTS

The tests carried out should be referred to in the following general order:—

Index Properties—water content, liquid and plastic limits, etc.
Mechanical analyses.
Permeability tests.
Limiting density tests.
Shear strength tests, including triaxial compression test.
Consolidation tests.
Sulphate tests.

If a substantial amount of testing has been carried out each type of test should be given a separate sub-heading, e.g. *Consolidation Tests*. Standard soil testing procedure such as for index property, triaxial and consolidation tests need not be described except where special techniques such as remoulding samples for triaxial testing have been carried out. Where tests are not so well known or standardized, such as limiting density tests, some description of method or reference must be given.

Having referred the reader to the figures and tables in which the results have been presented, the results of the tests should be factually presented in generalized form. Exceptions to general trends should be noted, the whole object being to arrive at figures of soil properties which can be referred to in the *Discussion* when analysing the problems.

When analysing and presenting the results always attend carefully to the following checks:—

(*a*) Cross-check moisture contents on index property and triaxial compression test tables.
(*b*) Cross-check consistency from sample description with shear strength of soil.
(*c*) Cross-check composition from mechanical analysis with sample description.
(*d*) Check pre-treatment loss in mechanical analysis.
(*e*) Compare moisture contents and shear strength.
(*f*) Cross-check table of triaxial compression tests with Mohr circles.
(*g*) Shear strengths should be in round figures and should read the same on tables and on Mohr circles.
(*h*) Cross-check consolidation curves with table of consolidation results.
(*i*) Inspect variations in coefficients of consolidation.

DISCUSSION

First state the problems in sufficient detail with figures for such matters as column loadings. Apply all the information and technical facts to the problems, using well-tried methods of analysis to produce readily understood practical advice. At this stage statements should be qualified where necessary, e.g.:—

'According to the test results on sample No.......'

'The results indicate that . . .'

'The bearing capacity of piles has been calculated, and theoretically is . . .'

However, no opportunity should be lost to give a definite answer when justified by the facts. When the facts do not justify the presentation of definite answers very often the reason can be attributed to parsimony on the part of a client, a matter which has to be treated with considerable tact in the report. It is most important, of course, to ensure that all the necessary work and tests have been done or suggested to the client to ensure that definite answers are given.

CONCLUSIONS

These should be numbered, each complete in itself, brief, developing the findings in logical order and to the point. As a general rule nothing should appear in the *Conclusions* that has not already appeared in the body of the Report.

REFERENCES

References to books, papers, etc., if substantial, should be given on a separate sheet to follow the last page of the report. Normally, however, they should follow on the same page as the signatures, but a clear space should always be allowed between the signatures and the references.

All references to technical papers and books should be given in the manner used in 'Géotechnique' (see Appendix A).

APPENDICES

Appendices are used for any documentary material outside the body of the report and tables of results. Calculations are normally included as Appendices. An appendix may often be used with advantage when, for example, a detailed description of a test is needed, which would interrupt the flow of the report, and thereby the ease with which the main arguments may be understood.

TABLES

Tables of test results should be reproduced on standard tabular sheets, numbered in Roman numerals, and are commonly as follows:—

Table I Index properties.
Table II Undrained triaxial compression tests.
Table III Consolidation tests.

Further numbers can be added for other types of tests.

FIGURES

The list of figures in the average report will usually take the following form and order:—

Borehole logs ... ... Fig. 1 to 5.
Particle size distribution curves ... ... Fig. 6, a, b, c.
Mohr circles ... ... Fig. 7, a, b, c.
Pressure-void ratio relationship ... ... Fig. 8, a, b, c.
Sections through boreholes ... ... Fig. 9.
Site plan ... ... Fig. 10.
Key plan ... ... Fig. 11.

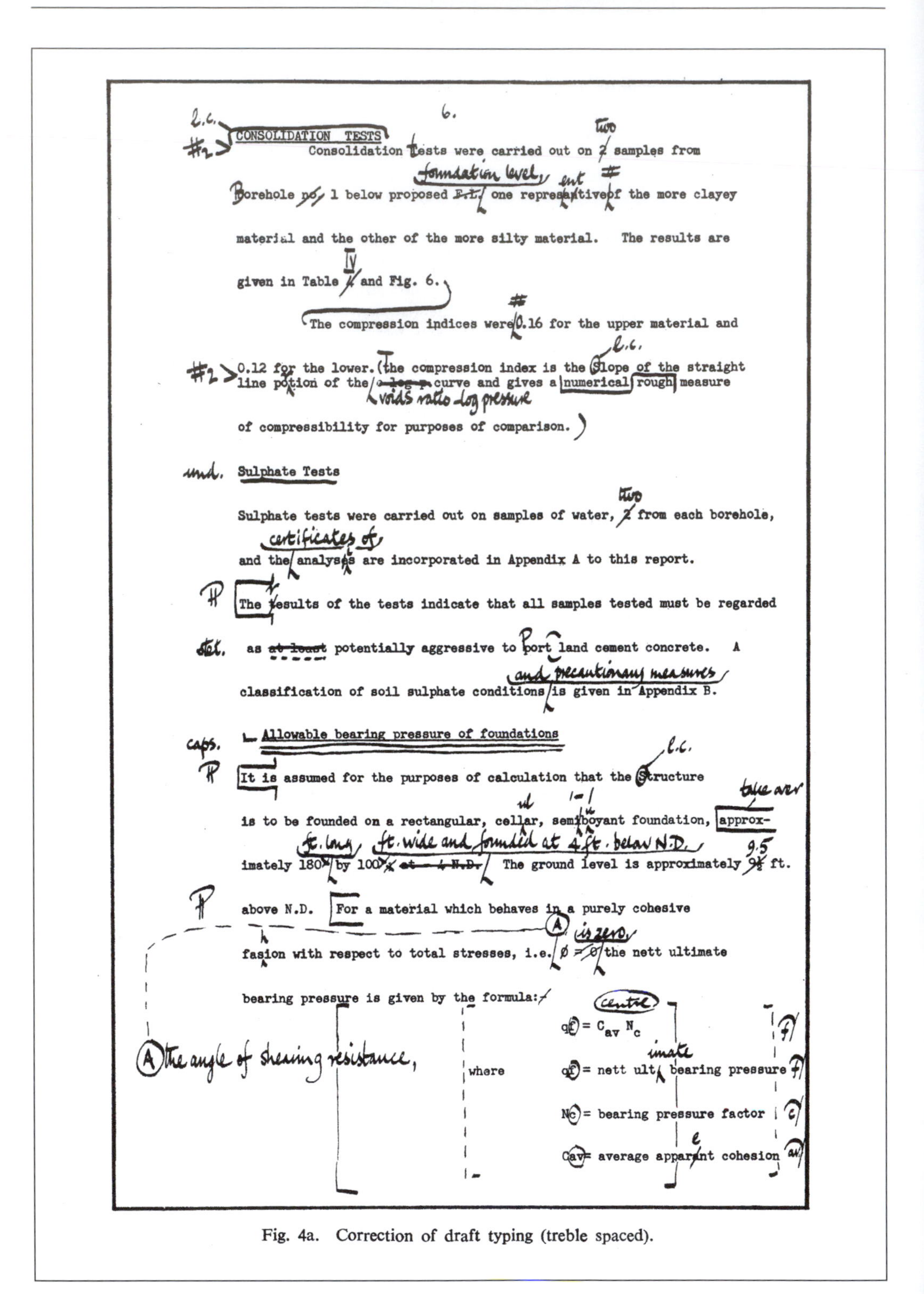

6.

CONSOLIDATION TESTS

Consolidation Tests were carried out on 2 samples from Borehole no. 1 below proposed F.L. one representive of the more clayey material and the other of the more silty material. The results are given in Table 4 and Fig. 6.

The compression indices were 0.16 for the upper material and 0.12 for the lower. (the compression index is the Slope of the straight line potion of the e-log p curve and gives a numerical rough measure of compressibility for purposes of comparison.)

Sulphate Tests

Sulphate tests were carried out on samples of water, 2 from each borehole, and the analyses are incorporated in Appendix A to this report.

The results of the tests indicate that all samples tested must be regarded as at least potentially aggressive to port land cement concrete. A classification of soil sulphate conditions is given in Appendix B.

Allowable bearing pressure of foundations

It is assumed for the purposes of calculation that the Structure is to be founded on a rectangular, cellar, semiboyant foundation, approximately 180' by 100' at -4 N.D. The ground level is approximately 9½ ft. above N.D. For a material which behaves in a purely cohesive fasion with respect to total stresses, i.e. ø = 0 the nett ultimate bearing pressure is given by the formula:

$q_f = C_{av}\ N_c$

where

q_f = nett ult bearing pressure

N_c = bearing pressure factor

C_{av} = average apparant cohesion

Fig. 4a. Correction of draft typing (treble spaced).

- 6 -

Consolidation Tests

Consolidation tests were carried out on two samples from Borehole 1 below proposed foundation level, one representative of the more clayey material and the other of the more silty material. The results are given in Table IV and Fig. 6. The compression indices were 0.16 for the upper material and 0.12 for the lower. (The compression index is the slope of the straight line portion of the voids ratio-log pressure curve and gives a rough numerical measure of compressibility for purposes of comparison.)

Sulphate Tests

Sulphate tests were carried out on samples of water, two from each borehole, and the certificates of analysis are incorporated in Appendix A to this report.

The results of the tests indicate that all samples tested must be regarded as at least potentially aggressive to Portland cement concrete. A classification of soil sulphate conditions and precautionary measures is given in Appendix B.

ALLOWABLE BEARING PRESSURE OF FOUNDATIONS

It is assumed for the purposes of calculation that the structure is to be founded on a rectangular, cellular, semi-buoyant foundation, approximately 180 ft. long by 100 ft. wide, and founded at 4 ft. below N.D. The ground level is approximately 9.5 ft. above N.D.

For a material which behaves in a purely cohesive fashion with respect to total stresses, i.e. the angle of shearing resistance, Ø is zero, the nett ultimate bearing pressure is given by the formula:

$$q_f = c_{av} N_c$$

where q_f = nett ultimate bearing pressure
N_c = bearing pressure factor
c_{av} = average apparent cohesion

The nett ultimate bearing pressure is the increase in pressure, which applied at a given depth will just cause shear failure in the ground.

Fig. 4b. Corrected copy (double spaced).

The following points should be noted:—

(*a*) The figure numbers of borehole logs should be the same as the borehole numbers wherever possible.

(*b*) Figures not given above should be put in some logical order, e.g. vane test results should follow triaxial test results since they both measure shear strength.

(*c*) The site and key plans should always be presented last, preferably adjacent to a blank sheet so that they can be pulled out and referred to whilst reading the report.

(*d*) Don't allow drawings and graphs to contain more detail than is required. Avoid having too many curves on the same figure and be scrupulous in labelling co-ordinate axes with the units that they represent and the scales.

(*e*) Where possible avoid using sheets of more than foolscap size for drawings. They make the report difficult to handle.

PREPARING THE DRAFT

Once the site work, description of samples and plotting of logs are under way, no time should be lost in preparing the draft report. Details can be filled in and draft typing carried out when all information is available.

The House Style formulates rules for layout and typing of reports and the use of abbreviations in them. A well-tried set of rules is given in Appendix A to this manual. Where typed drafts depart from the rules, the appropriate corrections should be made by engineers. In correcting drafts the signs given in B.S. 1219 (1945) should be used. A selection of the most useful and appropriate of these signs is given in Appendix C. An example of the use of these signs is given in Fig. 4.

Before presenting the draft report for checking the engineer must initial the report to certify that the positions of the boreholes have been checked. This is of utmost importance. There should be no shadow of doubt about borehole positions being correct.

The progress of the draft report to final approved text is as follows:—

When the draft has been completed and all errors of style, layout and typography corrected it should be placed in a draft report folder with all other report documents and passed to the Senior Engineer for his comments and after any necessary amendment to the Geologist for checking accuracy of geological data. Finally it is passed to a Director who makes any necessary final amendments. At each stage, the report is signed by the person checking.

When the text of the report is approved it should be passed immediately for final typing. At the same time the remainder of the drawings should be passed to the Drawing Office. (It is most desirable to pass as many drawings as possible to the Drawing Office at an earlier date.)

STYLE

It would be presumptuous to lay down the style of writing. Moreover, plenty of latitude given to individual taste results in holding the attention of those who have to read a large number of reports, and also helps the reader to improve his own style. There are, however, a few important rules of good style that should be noted.

DIRECTNESS

Directness and the use of short, well-understood words is always to be encouraged. For example, don't let *approximately* serve for *about* or *roughly*. *Materialize* is a bad substitute for *happen* and *take place*, and *most* will often do the work of *majority*. Gowers gives many other examples. Long sentences, unless absolutely clear in sense, should be avoided.

MEANINGS OF WORDS

Many words are used without any thought about what they mean. Examples are:—

'The test results are *summarized* in Table III' when Table III contains all the results, and 'the test is *outlined* in . . .' when a very full description is given.

Other examples are the use of *anticipate* for *expect*, *alternately* for *alternatively*, *appreciate* for *understand*, *infer* for *imply*.

ELEGANCE

Elegance is all very well, but a fairly golden rule whenever we think we have written something particularly good and elegant, in fact when we are thoroughly delighted with the turn of phrase, is 'strike it out'. Dr. Cooling advises 'when in doubt, wheel it out'. An example of such elegance worthy of note is the engineer's description 'sky-blue clay streaked with claret'. This was wheeled out !

RELEVANCE

Test each statement for relevance to the terms of reference. This is difficult and sometimes you must be ruthless in deleting well-loved matter which is not relevant. At the same time check each statement for double meanings and check that what you have said really means something, or sooner or later someone will evaluate the 'Harding coefficient of so--whatness' for your statements.

MATHEMATICAL SYMBOLS

Mathematical Symbols should not be used as abbreviations in the text of the report.

CALCULATIONS

The following general rules should be noted in manuscript preparation of calculations for reports:—

(*a*) All calculations should be set out on foolscap, preferably on special calculation sheets provided in pads by the firm. In any case, they should have a one-inch margin, a title, location number, date, and should be signed.

(*b*) Calculations should be set out tidily with neat sketches and the writing and numbers should be clearly legible. They should in fact be set out in such a manner that at any time a photostat may be made of them so that they can be sent to a client.

(*c*) So far as is possible, the standard nomenclature of Soil Mechanics symbols given in 'Géotechnique' vol. II, 1, pp. 84–86, should be followed in all calculations, supplemented by other symbols shown in internal technical notes. Choose your notation with the limitations of the typewriter in mind.

(*d*) All figures quoted in reports should be supported by a calculation sheet, no matter how short. All steps and assumptions in calculations should be clearly noted.

(*e*) Each calculation sheet should be numbered consecutively, and for extensive calculations an index should be provided.
(*f*) All superseded calculations should be marked clearly as such and crossed out or, where possible, destroyed.
(*g*) When the calculations are completed they should be bound in a calculation folder which should be clearly labelled. This should be done even if there is only one sheet of calculations.
(*h*) When reproducing calculations for reports a mixture of typescript and handwriting in equations often produces a patchy appearance. For this reason it is preferable to write in the equations by hand. Reproduction of calculations in manuscript is often most satisfactory.

Calculations should be included in a report as an Appendix wherever possible, even if only in skeleton form.

CHECKING REPORTS

Thorough checking is necessary at all stages of the report. Mistakes made by typists in the final reproduction of the text of reports are often caused by bad 'copy'. If your approved draft has been extensively altered in manuscript, you will find it time-saving to have it typed again for the final 'copy'.

When the final typing is complete, check it with the help of another person. It is preferable that you should read the final text, your assistant reading the 'copy'. From time to time it is as well to make a deliberate mistake in reading to find out if your assistant is awake.

Checking of the drawings must also be rigorous. In doing this you should remember that a polished tracing from the Drawing Office may often blind you to the fact that some of the captions make absolute nonsense. Errors which are common are spelling mistakes and the absence of scales and North points.

The importance of checking final typescripts and tracings before the stencils are run off, the prints made and the report sent for binding, cannot be over-emphasized. Apart from the waste of money in pulling a bound report apart at the last moment, you will get nothing but black looks from the staff associated in getting the report prepared, and complaints from the client because he has not received his report on time. It is no use claiming contributory negligence in matters where you alone are responsible.

FINIS

In conclusion it should be stated that the author has felt increasing diffidence whilst preparing this manual for publication because he is very much aware of his own shortcomings in knowledge of the English language, and also that the ground has been well covered by eminent men. He will be grateful for any comments and suggestions, and if there are any unwitting errors he begs leave to quote Boswell, who reported that a lady once asked Johnson 'how he came to define *pastern*, the *knee* of a horse; instead of making an elaborate defence, as she expected, he at once answered, "Ignorance, Madam, pure ignorance" '.

REFERENCES

1. BRITISH STANDARD 1219 : 1945. 'Printers' and Authors' Proof Corrections.' British Standards Institution.
2. GEOLOGICAL SURVEY. 'British Regional Geology.' H.M.S.O. (in 18 parts).
3. GOWERS, SIR ERNEST, 1954. 'The Complete Plain Words.' H.M.S.O.
4. KAPP, R. O., 1948. 'The Presentation of Technical Information.' Constable.
5. SIMON, O., 1954. 'Introduction to Typography.' Pelican Books.

SOIL MECHANICS LIMITED—Works Instruction No. 89

APPENDIX A

HOUSE STYLE OF REPORTS

To ensure uniformity rules are needed for the layout and typing of reports and for the use of abbreviations in them. This instruction incorporates the rules laid down in Works Instruction No. 70, December 1952, together with revisions and additions that have become necessary with experience. It is mainly concerned with abbreviations and the typographical layout, some of which can be seen exemplified in it. It should be stressed again that strict conformity with the House Style, though it may not satisfy everybody's taste, ensures tidiness and improves the appearance of reports. A glance at a few old reports issued before 1952 will demonstrate this.

The rules are these:—

TYPOGRAPHY AND LAYOUT

1. *Headings and paragraphs.* The following typing style should be observed:—
   - (*a*) The title and section headings should be in capitals, underlined.
   - (*b*) Sub-section headings should be in lower case, underlined, with the following matter on a new line.
   - (*c*) Paragraph headings should be in lower case, underlined, indented and with the following matter on the same line, [as shown in this instruction].*
   - (*d*) Extravagant indentations for paragraph openings should be avoided; five spaces should be used. Lists of notes, strata, etc., should be indented five spaces, to distinguish them from the main text. To amplify, the reference letter or number to the note should be indented five spaces. The note itself should be indented ten spaces. Notes should be single-spaced [as on this page].

2. *Division of words* at the end of a line should be avoided wherever possible.

3. *Quotations* should be given as single marks, thus ('). Quotations within quotations should be given as double marks ("). Long extracts should be indented five spaces and single-spaced, and quotation marks need not be used. Punctuation marks belonging to the quotation should be placed inside the quotation marks.

4. *Margins.* A margin of nine line-spaces (1½ in.) should be left between the top of the page and the text, and a margin of seven line-spaces (1¼ in.) between the bottom of the pages and the bottom of the text. These spacings should be the same on every page. If footnotes are inserted then the 1¼-in. margin should be left at the bottom of the page. A 1½-in. margin should be allowed on the left-hand side of the page, to allow for binding, and a clear ½-in. margin allowed on the right-hand side. Page numbering should be at the top of the page [and at the spacing indicated on this page].

5. *Footnotes* should be avoided where possible and never used for references. Where it is necessary to use footnotes they should be spaced from the text and separated from it by a ruled line.

---

* The matter within square brackets, strictly speaking, applied only in the original typed version of this instruction.

6. *References.* References to books, papers, etc., if substantial will be given on a separate sheet to follow the last page of the report. Normally, however, they should follow on the same page as the Director's signature but a clear space should always be allowed between the signature and the references.

All references to technical papers and books are to be given in the manner used in 'Géotechnique'. In the text the reference will be given as the author's name followed by the year, both in brackets, e.g. (Skempton, 1948). At the end of the report a list of references will be given in alphabetical order of authors' names, followed by authors' initials, the year of the reference (and in date order where the same author appears more than once), the title, the name of the journal or book, the volume number, month and page number. Volume will be written 'vol.' and page 'p.' e.g.:—

SKEMPTON, A. W., 1948. 'A study of the geotechnical properties of some post-glacial clays.' Géotechnique, vol. I, 1, (June), pp. 7–22.

ABBREVIATIONS AND STYLES

7. *Titles.* The title of a report should be in effect a very brief précis of the report. It should state:—

(*a*) The nature of the work, e.g. Site Investigation.

(*b*) The purpose of the work, e.g. for a jetty.

(*c*) The situation or location, e.g. Shellhaven, Essex, the title thus reading:— 'Site Investigation for a Jetty at Shellhaven, Essex'.

The name of the country and the region should be added where necessary to avoid obscurity, thus:—

'Site Investigation for a New Runway at Palisadoes Airport, Jamaica, B.W.I.'

8. *Simple abbreviations including Areas and Volumes.* Abbreviations in reports will always be in the singular (i.e. the letter 's' will not be added) and will be followed by a full stop. Thus, for example, inches and feet will be written 'in.' and 'ft.' The only exception to this is when the abbreviation is followed by an oblique stroke when the stop is omitted, e.g. lb/cu. ft. Abbreviations should not be made for short words such as 'tons', e.g. write 2 tons/sq. ft., ½ ton/sq. ft.

Areas and volumes will be written as 'sq.' and 'cu.', thus 'pounds per square foot' will be lb/sq. ft. Indices will only be used in mathematical equations.

9. *Use of numbers.* Generally in descriptive matter numbers under 100 should be in words, but figures should be used when the matter consists of a sequence of stated quantities, numbers, ages, etc. Spell out indefinite numbers, e.g. 'has been done a thousand times'. Insert commas with four or more figures, e.g. 4,367. To represent an approximate date use the fewest figures possible, 1931–2, not 1931–32. Dates should be written in full in descriptive matter, i.e. 'on the 20th of January, 1954'. The word 'number' will either be written in full or abbreviated to 'No.' The latter form should be avoided in the text wherever possible, and used only for lists of equipment, etc. To amplify, avoid writing '4 No. boreholes were put down', but write 'four boreholes were put down'.

10. *Dimensions.* When giving the dimensions of a sampling tube for instance, two variants are possible and either may be used:—

(*a*) 4 in. dia. by 6 in. long, or

(*b*) four inches diameter by six inches in length.

11. *Capitals.* Capital letters will be used for proper nouns, including the names of geological formations. Descriptions of soils and rocks which are not geological formations should be written without capital letters. The use of capital letters in other cases is to be discouraged. Examples are given in Appendix B.

12. *Figures* will be referred to in the text as 'Fig.' Table and Appendix will be written out in full, i.e. Table II, Appendix B. Appendices and Tables will be referred to in Roman numerals or capital letters, figures in Arabic numerals.

13. *Boreholes* will be referred to in the text of a report as 'Borehole 2'.

14. *Datum levels* will be referred to as Newlyn Datum or Liverpool Datum, abbreviated to N.D. or L.D. The first time these terms are used in a report they will be written out in full, followed by the appropriate abbreviation in brackets. The term 'Ordnance Datum' will be avoided wherever possible. In the text, levels will be referred to by the words 'above datum' or 'below datum', e.g. 20 ft. above N.D. In tables and drawings levels will be given as + or −.

15. *Points of the Compass.* The Cardinal points will be written in full in lower case, e.g. north, south. Semi-cardinal points will also be written in full, e.g. north-east, south-west.

16. *The sequence of strata* in a borehole is best described by tabulating by letters from the top stratum. The strata should not be numbered since the normal geological method of calling the oldest layer 1 and the next oldest 2 and so on, is confusing to engineers and is not the most logical method for foundation problems.

17. *Previous reports* by Soil Mechanics Limited will be referred to in reports as S.M. Report No. 1030/2.

August, 1955.

APPENDIX B

TYPICAL NAMES OF GEOLOGICAL FORMATIONS

(To be written with capital initial letters)

Carboniferous Limestone
Millstone Grit
Coal Measures
Magnesian Limestone
Bunter Sandstone
Keuper Marl
Oxford Clay
Lower Greensand
Gault
Chalk
Reading Beds
London Clay
Norwich Crag
Chalky Boulder Clay
Boyn Hill Terrace
River Alluvium

TYPICAL GEOLOGICAL TERMS WHICH ARE NOT NAMES OF FORMATIONS

(To be written with small initial letters)

sandy limestone
red marls
boulder clay
shale

APPENDIX C: CORRECTION OF DRAFT REPORTS

The following signs from B.S. 1219 (1945) should be used in the correction of typed drafts. Other signs given in B.S. 1219 may be used where appropriate. (Some of them are obviously inappropriate, being related to the technique of printing, e.g. No. 14 'Wrong fount'.)

No. in B.S. 1219	Marginal mark	Meaning	Corresponding mark in text
2	δ/	Delete (take out)	/
5	stet	Leave as printed	---- under letters or words to remain
6	caps	Change to capital letters	≡ under letters or words to be altered
9	l.c.	Change to lower case	Encircle letters to be altered
12*	und.	Underline word or words	___ under words affected
17		Substituted letters or signs under which this is placed to be 'superior	Encircle letters or signs to be altered
19		Substituted letters or signs over which this is placed to be 'inferior'	do.
23		Close up—delete space between letters	linking words or letters
24	#	Insert space	⅄
25	#>.	Space between lines or paragraphs (indicate amount of space)	
28	trs.	Transpose	between letters or words numbered when necessary
30*	5⅄	Indent five spaces	
32*		Move to the left	(This sign indicates how far)
33*		Move to the right	do.
34		Move lines to the right	at the left side of group to be moved
35*		Move lines to the left	at left side of group to be moved
36	[ ]	Move portion of matter so that it comes within the position indicated	[ ] at limits of required position
37	take over	Take letter or word from end of one line to beginning of next	
38	take back	Take letter or word from beginning of one line to end of preceding line	
44	n.p. or ¶	Begin a new paragraph	before first word of new paragraph
45	run on	No fresh paragraph here	between paragraphs
47*	⅄	Insert appended matter	Ⓐ or for matter written with text
48	⅄	Insert matter indicated in margin	⅄
63	|-|	Insert hyphen	⅄

NOTES

(*a*) Signs marked * have been adapted to suit the special circumstances of typed drafts.

(*b*) Full stops, etc., should be deleted with the sign × to avoid confusion using the B.S. sign / with an oblique stroke such as in 'lb/sq. ft.' The sign ⊙ for deleting full stops, etc., should not be used.

(*c*) B.S. 1219 recommend making all corrections marginal. This is not often necessary with typed reports because plenty of space is given within the text for corrections.

(*d*) Where an alteration is to be made that cannot easily be explained by improvised signs, a clear instruction for the typists is to be given in the margin.

References and bibliography

Aas, G. (1967). Vane tests for investigation of anisotropy of undrained shear strength of clays. *Proc. Geotech. Conf., Oslo*, **1**, 3–8.

Abbis, C.P. (1979). A comparison of the stiffness of chalk at Mundford from a seismic survey and large scale tank test. *Géotechnique*, **29**, 461–468.

Abbiss, C.P. (1981). Shear wave measurements of the elasticity of the ground. *Géotechnique*, **31**(1), 91–104.

Aboshi, H., Yoshikumi, H. and Maruyama, S. (1970). Constant loading rate consolidation test. *Soils Found.*, **10**(1), 43–56.

Acar Y.B., Tumay, M.T. and Chan, A. (1982). Interpretation of the dissipation of penetration pore pressures. *Proc. Int. Symp. on Numerical Models in Geomechanics, Zurich*. Polkema, Rotterdam.

Addo, K. and Robertson, P.K. (1992). Shear wave velocity using Rayleigh surface waves. *Can. Geotech. J.*, **29**(4), 558–568.

Andersen, A. and Kolstad, P. (1979). The NGI 54 mm sampler for undisturbed sampling of clays and representative sampling of coarser materials. *Proc. Int. Symp. on Soil Sampling*, Singapore, 13–21.

Apted, J.P. (1977). *Effects of weathering on some geotechnical aspects of London clay*. PhD Thesis. Imperial College, London.

Arthur, J.R.F. and Menzies, B.K. (1972). Inherent anisotropy in a sand. *Géotechnique*, **22**(1), 115–128.

Arthur, J.R.F., Chua, K.S. and Dunstan, T. (1977). Induced anisotropy in a sand. *Géotechnique*, **27**(1), 13–30.

ASTM D1194-72 (Re-approved 1987). *Standard test method for bearing capacity of soil for static load and spread footings*. American Society for Testing Materials, Philadelphia, USA.

Auld, B. (1977). Cross-hole and down-hole V_s by mechanical impulse. *J. Geotech. Eng. Div., ASCE*, **103**(GT12), 1381–1398.

Bailey, J.J. and Reitz, H.M. (1970). Building damage from expansive steel slag backfill. *J. Soil Mech. Found. Div. Am. Soc. Civ. Eng.*, **96**, 1810–1813.

Baligh, M.M. (1985). The strain path method. *J. Geotech. Eng. Div, ASCE*, **111**(GT9), 1108–1136.

Baligh, M.M., Azzouz, A.S. and Chin, C.T. (1987). Disturbance due to ideal tube sampling. *J. Geotech. Eng. Div., ASCE*, **113**(GT7), 739–757.

Ballard, R.F. and McLean, F.G. (1975) Seismic field methods for in-situ moduli. *Proc. Conf. on Insitu Measurement of Soil Properties*. Speciality Conference of the Geotechnical Engineering Division ASCE., Raleigh, North Carolina, **1**, 121–150.

Ballard, R.J. (1976). Method for crosshole seismic testing. *J. Geotech. Eng. Div., ASCE*, **102**(GT12), 1261–1273.

Baria, R., Jackson, P.D. and Green, A.S.P. (1986). Application of cross-hole seismic measurements in site investigation surveys. *Geophysics*, **51**, 914–929.

Barton, D.C. (1929). The seismic method of mapping geologic structure. *Geophys. Prospect. Trans. Am. Inst. Mining Metall. Eng.*, **81**, 572–624.

Bates, C.R. (1989). Dynamic soil property measurements during triaxial testing. *Géotechnique*, **39**(4), 721–726.

Baumann,V. and Bauer, G.E.A. (1974). The performance of foundations on various soils stabilized by the vibro compaction method. *Can. Geotech. J.*, **11**, 509–529.

Begemann, H.K.S. (1965). The friction jacket cone as an aid in determining the soil profile. *Proc. 6th Int. Conf. Soil Mech. and Found. Eng., Montreal*, Vol 1, pp. 17–20.

Begemann, H.K.S. (1969). The Dutch static penetration test with the adhesion jacket cone. *Lab. Grondmech. Delft, Meded*, **12**(4), 69–100; **13**(1), 1–86.

Bellotti, R., Ghionna, V., Jamiolkowski, M., Robertson, P.K. and Peterson, R. (1989). Interpretation of moduli from self-boring pressuremeter tests in sand. *Géotechnique*, **39**(2), 269–292.

Bishop, A.W. and Eldin, G. (1950). Undrained triaxial tests on saturated sands and their significance in the general theory of shear strength. *Géotechnique*, **2**, 13–32.

Bishop, A.W. and Henkel, D.J. (1962). *The Measurement of Soil Properties in the Triaxial Test*. Second edition, Edward Arnold, London.

Bishop, A.W. and Wesley, L.D. (1975). A hydraulic triaxial apparatus for controlled stress path testing. *Géotechnique*, **25**(4), 657–670.

Bishop, A.W., Green, G.E., Garga, V.K., Andresen, A. and Brown, S.D. (1971). A new ring shear apparatus and its application to the measurement of residual strength. *Géotechnique*, **21**, 273–328.

Bishop, A.W., Green, G.E. and Skinner, A.E. (1973). Strength and deformation measurements on soils. *Proc. Eighth Int. Conf. Soil Mech. Found. Eng., Moscow*, Vol. 1, pp. 57–64.

Bjerrum, L. (1972). Embankments on soft ground. *ASCE Proc. Conf. on Performance of Earth and Earth Supported Structures, Purdue*, **2**, 1–54.

Bjerrum, L. (1973). Problems of Soil Mechanics and Construction on Soft Clays. *Proc. 8th Int. Conf. Soil Mech and Found. Eng.*, **3**, 111–159.

Bjerrum, L. and Simons, N. E. (1960). Comparison of shear strength characteristics of normally consolidated clays. *Proc. Res. Conf. on Shear Strength of Cohesive Soils*, Boulder, Colorado, ASCE, 711–724.

Bjerrum, L. and Eggestad, A. (1963). Interpretation of loading tests on sand. *Proc. Eur. Conf. Soil Mech. and Found. Eng., Wiesbaden*, Vol. 1, pp. 199–204.

Bjerrum, L. and Landva, A. (1966). Direct simple shear tests on Norwegian quick clay. *Géotechnique*, **16**(1), 1–20.

Blockley, D.I. and Godfrey, P.S. (2000). *Doing it Differently: Systems for Rethinking Construction*. Thomas Telford, London.

Blyth, F.G.H. and de Freitas, M.H. (1984). *A Geology for Engineers*. Seventh edition. Edward Arnold, London.

Bolton M.D. and Whittle R.W. (1999). A non-linear elastic/perfectly plastic analysis for plane strain undrained expansion tests. *Géotechnique*, **49**(1), 133–141.

Brandon, P. and Betts, M. (1995). *Integrated Construction Information*. E and FN Spon, London.

Brierley, G.S. (1998). Subsurface investigations and geotechnical report preparation. In *Subsurface Conditions. Risk Management for Design and Construction Management Professionals*, edited by D.J. Hatem. Wiley, New York, ch. 3, ISBN 0 471 15607 8.

British Standards Institution. (1990). *British Standard Methods of Test for Soils for Civil Engineering Purposes, Part 2, Classification tests.* BS 1377: 1990. BSI, London.

British Standards Institution. (1992). *Specification for Environmental Management Systems.* BS 7750: 1992. BSI, London.

British Standards Institution (1999). *Code of Practice for Site Investigations.* BS 5930: 1999. BSI, London.

Bromhead, E.N. (1979). *The Stability of Slopes.* E and FN Spon, London.

Broms, B.B. (1980). Soil sampling in Europe: state of the art. *J. Geotech. Eng. Div., ASCE*, **106**(GT1), 5–98.

Broms, B.B. and Casbarian, A.O. (1965). Effect of rotation of the principal stress axes and the intermediate principal stress on the shear strength. *Proc. Sixth Int. Conf. Soil Mech. Found. Eng., Montreal*, Vol 1, pp. 179–183.

Building Research Establishment (1980). Low-rise buildings on shrinkable clay soils: Part 1. *BRE Digest*, **240**, HMSO, London.

Building Research Establishment (1981). Assessment of damage in low-rise buildings, with particular reference to progressive foundation movement. *BRE Digest*, **251**, HMSO, London.

Building Research Establishment (1996). Desiccation in clay soils. *BRE Digest*, **412**, CRC, London.

Building Research Establishment (1999). Low-rise building foundations: the influence of trees in clay soils. *BRE Digest*, **298**, CRC, London.

Burland, J.B. (1970). Discussion Session A, *Proc. Conf. On Insitu Investigations in Soils and Rocks*, British Geotechnical Society, London, 1969, pp. 61–62.

Burland, J.B. and Hancock, R.J.R. (1977), Underground car park at the House of Commons, London: geotechnical aspects. *Struct. Eng.*, **55**(2), 87–100.

Burland, J.B. and Lord, J.A. (1970). The load deformation behaviour of Middle Chalk at Mundford, Norfolk: a comparison between full-scale performance and in-situ and laboratory measurements. *Proc. Conf. on In situ Investigations in Soils and Rocks.* British Geotechnical Society, London. pp. 3–15.

Butcher, A.P. and Powell, J.J.M. (1996). Practical considerations for field geophysical techiques used to assess ground stiffness. *Advances in Site Investigation Practice*, edited by C. Craig. Thomas Telford, London, pp. 701–714.

Butcher, A.P. and Tam, W.S.A. (1994). An example of the use of Rayleigh waves to detect the depth of a shallow landfill. *Modern Geophysics in Engineering Geology. Proc. 30th Conf. Engineering Group of the Geological Society*. Liege, in press.

Casagrande, A. (1936). The determination of preconsolidation load and its practical significance. *Proc. 1st Int. Conf. Soil Mech and Found. Eng., Harvard*, Vol 3, pp. 60–64.

Casagrande, A. (1941). Triaxial shear research (US Corps Eng.). *Harvard Progress Report No. 3.*

Casagrande, A. and Carrillo, N. (1944). Shear failure of anisotropic materials. *Proc. Boston Soc. Civ. Eng.*, **31**, 74–87.

Chandler, R.J., Crilly, M.S. and Montgomery-Smith, G. (1992a). A low cost method of assessing clay desiccation for low-rise buildings. *Proc. Inst. Civ. Eng.*, **92** 82–89.

Chandler, R.J., Harwood, A.H. and Skinner, P.J. (1992b). Sample disturbance in London Clay. *Géotechnique*, **42**(4), 577–585.

Cheney, J.E. (1988). 25 years' heave of a building constructed on clay after tree removal. *Ground Eng.*, **21**(5), 13–27.

CIRIA (1989). *The Engineering Implications of Rising Groundwater in the Deep Aquifer Beneath London*. Special Publication **69**. CIRIA, London.

CIRIA (1994). Foundations in chalk. *CIRIA Project Report 11*. CIRIA, London.

CIRIA (2001) *RiskCom: Software tool for managing and communicating risks*. CIRIA, London.

Clayton, C.R.I. (1993). *The Standard Penetration Test (SPT): Methods and Use*. CIRIA Funder Report CP/7. CIRIA, London.

Clayton, C.R.I. (1998). Talking point. *Ground Eng.*, **31**(4), 15.

Clayton, C.R.I. (2001). Managing geotechnical risk: time for a change? *Proc. ICE, Geotech. Eng.*, Paper 149, Pages 3–11.

Clayton, C.R.I. and Heymann, G. (2001). Stiffness of geomaterials at very small strains. *Géotechnique*, **51**(3), 245–255.

Clayton, C.R.I. and Siddique, A. (1999). Tube sampling disturbance – forgotten truths and new perspectives. *Proc. Inst. Civ. Eng. Geotech. Eng.*, **137**, July, 127–135.

Clayton, C.R.I., Simons, N.E. and Matthews, M.C. (1982). *Site Investigation*, Granada, London.

Clayton, C.R.I., Simons, N.E. and Instone, S.J. (1988). Research on dynamic penetration testing in sands. *Proceedings of the 1st International Conference on Penetration Testing, ISOPT 1, Florida*, 1988, pp. 415–422.

Clayton, C.R.I., Hight, D.W. and Hopper, R.J., (1992). Progressive destructuring of Bothkennar Clay: implications for sampling and reconsolidation procedures. *Géotechnique*, **42**(2), 219–239.

Clayton, C.R.I., Gordon, M.A. and Matthews, M.C. (1994). Measurements of stiffness of soils and weak rocks using small strain laboratory testing and geophysics. *Proc. Int. Symp. on Pre-failure Deformation Characteristics of Geomaterials*, Vol. 1. Balkema, Rotterdam, pp. 229–234.

Clayton, C.R.I., Matthews, M.C., Gunn, M.J., Foged, N. and Gordon, M.A. (1995a). Reinterpretation of surface wave test for the Øresund crossing. *Proc. 11th European Conf. on Soil Mech. and Found. Eng., Copenhagen*, Vol 1. Danish Geotechnical Society, Copenhagen, pp. 141–147.

Clayton, C.R.I., Matthews, M.C. and Simons, N.E. (1995b). *Site Investigation*. Second edition. Blackwell Science, Oxford.

Clayton, C.R.I., Siddique, A. and Hopper, R.J. (1998). Effects of sampler design on tube sampling disturbance – numerical and analytical investigations. *Géotechnique*, **48**(6), 847–867.

Cole, K.W. and Burland, J.B. (1972). Observations of Retaining Wall Movements associated with a Large Excavation. *Proc. 5th Euro. Conf. Soil Mech and Found. Eng.*, **1**: 445–453.

Conlon, R.J. (1989). On being a geotechnical engineer. In *The Art and Science of Geotechnical Engineering at the Dawn of the Twenty-first Century*, edited by E.J. Cording *et al.* Prentice Hall, New Jersey, pp. 1–11.

Cooper, D.F. and Chapman, C.B. (1987). *Risk Analysis for Large Projects: Models, Methods and Cases*. Wiley, Chichester.

Copperthwaite, W.C. (1902). The Greenwich Footway Tunnel. *Proc. Instn. Civ. Eng.*, **150**, Session 1901–1902.

Crawford, C.B. and Burn, K.N. (1969). Building damage from expansive steel slag backfill. *J. Soil Mech. Found. Div. Am. Soc. Civ. Eng.*, **95**, 1325–1334.

Crilly, M.S. and Chandler, R.J. (1993). *A Method of Determining the State of Desiccation in Clay Soils*. BRE Inf. Paper No. 4/93, BRE, London.

Cruickshank, J. (1993). Limehouse Link Supplement. *New Civil Engineer*, May 1993.

Cuccovillo, T. and Coop, M.R. (1997). Yielding and pre-failure deformation of structured sands. *Géotechnique*, **47**(3), 491–508.

Cuellar, V., Monte, J.L. and Valerio, J. (1995). Characterization of waste landfills using geophysical methods. *Proc. 11th European Conf. on Soil Mech. and Found. Eng., Copenhagen*, Vol 2. Danish Geotechnical Society, Copenhagen, pp. 33–38.

D'Appolonia, D.J., D'Appolonia, E. and Brisette, R.F. (1970). Discussion on settlement of spread footings on sand. *J. Soil Mech. Found. Eng. Div., ASCE*, **96**(SM2), 754–761.

Dascal, O., Tournier, J.P. Tavenas, F. and La Rochelle, P. (1972). Failure of a test embankment on sensitive clay, *Proc. ASCE Spec. Conf. on Performance of Earth and Earth-Supported Structures, Purdue University*, Vol 1.1, pp. 129–158.

Davila, R.S., Sego, D.C. and Robertson, P.K. (1992). Undisturbed sampling of sandy soils by freezing. *Proc. 45th Can. Geo. Conf.*, Toronto, Paper 13A-1-13A-10.

De Beer, E. (1967). Clay strength characteristics of the Boom Clay, *Proc. Geotech. Conf.*, Oslo, 1967, **1**, 83–88.

de Ruiter, J. (1971). Electric penetrometer for site investigations. *J. Soil Mech. Found. Eng. Div., ASCE*, **97**(SM2), 457–472.

Decourt, L. (1990). The Standard Penetration Test: state-of-the-art report. *Proc 12th Int. Conf. Soil Mech. and Found. Eng., Rio de Janeiro*. Balkema, Rotterdam.

Department of the Environment (1991). *Review of Mining Instability in Great Britain Regional Report South East England Including Greater London*. Vol. 1, DOE, London.

Department of the Environment, Transport and the Regions. *PIT Project on Managing Geotechnical Risk*. DETR, unpublished report.

Department of the Environment, Transport and the Regions. (1998) *Rethinking Construction. The report of the Construction Task Force*. DETR, London, ISBN 1 65 1 12 094 7.

Donald, I.B., Sloan, S.W. and Chiu, H.K. (1980). Theoretical analysis of rock socketed piles. *Proc. Int. Conf. On Structural Foundations on Rock, Sydney*, Vol 1, edited by P.J.N. Pells. Balkema, Rotterdam, pp. 303–316.

Dougherty, M.T. and Barsotti, N.J. (1972). Structural damage and potentially expansive sulfide minerals. *Bull. Ass. Eng. Geol.*, **9**(2), 105–125.

Driscoll, R. (1984). The effects of clay soil volume changes on low-rise buildings. In *Ground Movements and their Effects on Structures*, edited by P.B. Attewell and R.K. Taylor. Surrey University Press, Blackie, Glasgow, 303–320.

Drnevich, V.P. and Massarsch, K.R. (1980). Sample disturbance and stress–strain. *J. Geotech. Eng. Div., ASCE,* **105**(GT9), 1001–1016.

Dumbleton, M.J. (1983). *Air photographs for investigating natural changes, past use and present condition of engineering sites,* TRRL Laboratory Report 1085, Transport Research Laboratory, Crowthorne, Berks.

Dunnicliff, J. and Green, G.E. (1988). *Geotechnical Instrumentation for Monitoring Field Performance*. John Wiley, New York.

Edmunds, C.N., Green, C.P. and Higginbottom, J.E. (1987). Subsidence Hazard Prediction for Limestone Terrains, as applied to the English Cretaceous Chalk. *Planning and Engineering Geology*. Geological Society Engineering Geology Special Publication No. 4, pp. 283–293.

Edwards R. (1969). AMEC Civil Engineering Ltd. Personal communication.

Eide, O. (1968). *Geotechnical problems with soft Bangkok clay on the Nakhon Sawan highway project*. NGI Pub. No 78, Norwegian Geotechnical Institute, Oslo.

Eide, O. and Holmberg, S. (1972). Test fills to failure on the soft Bangkok clay, *Proc. ASCE Spec. Conf. Performance of Earth and Earth-Supported Structures*, Purdue University, 1972, **1.1**, 159–180.

Engineering Council (1993). *Guidelines on Risk Issues.* Engineering Council, London.

Fahey, M. (2001a) Measuring soil stiffness for settlement prediction. *Proc. 15th Int. Conf. on Soil Mech. and Geotech. Eng., Istanbul, August 2001*. To be published, Balkema, Rotterdam.

Fahey, M. (2001b). Soil stiffness values for foundation settlement analysis. *Prefailure Deformation Characteristics of Geomaterials*, edited by Jamiolkowski, Lancellotta and Lo Presti. Swets and Zeitlinger, Lisse, ISBN 90 5809 075 2.

Fasiska, E., Wagenblast, N. and Dougherty, M.T. (1974). The oxidation mechanisms of sulphide minerals. *Bull. Ass. Eng. Geol.*, **11**, 75–82.

Flaate, K. and Preber, T. (1974). Stability of road embankments on soft clay, *Can. Geotech. J.*, **11**, 72–89.

Flanagan R. and Norman, G. (1993). *Risk Management and Construction.* Blackwell Science, London, ISBN 0 632 02816 5.

Fletcher, G.F.A. (1965). Standard penetration test: its uses and abuses. *J. Soil Mech. Found. Eng. Div., ASCE,* **91**(SM4), 67–75.

Fookes, P.G. (1997). Geology for engineers: the geological model, prediction and performance. The First Glossop Lecture. *Q. J. Eng. Geol.*, **30**, 293–424.

Fookes, P.G., Baynes, F. and Hutchinson, J.N. (2000). Total geological history: a model approach to the anticipation, observation and understanding site conditions. *Proc. Int. Conf. on Geotech. and Geol. Engrg, GeoEng 2000*, Vol 1, Technomic, Pennsylvania, pp. 370–421.

Fookes, P.G., Baynes, F. and Hutchinson, J.N. (2001). Total geological history: a model approach to understanding site conditions. *Ground Eng.*, **34**(3), 22–23.

Gabriel K. (2001). What's on the agenda? *Ground Eng.*, **34**(7), 22–23.

Garga, V. A. (1970). *Residual shear strength under large strains and the effect of sample size on the consolidation of fissured clay*. PhD Thesis, University of London.

Gazetas, G. (1982). Vibrational characteristics of soil deposits with variable velocity. *Int. J. Numeric. Analyt. Methods Geomech.*, **6**, 1–20.

Georgiannou, V.N. and Hight, D.W. (1994). The effects of centre-line tube sampling strains on the undrained behaviour of two stiff overconsolidated clays. *ASTM Geotech. Test. J.*, **17**(4), 475–487.

Georgiannou, V.N., Rampello, S., Kim, Y. and Sato, T. (1991). Static and dynamic measurement of undrained stiffness of natural over consolidated clays. *Proceedings of the 10th European Conference on Soil Mechanics and Foundation Engineering*, **1**, 91–96.

Gibson, R.E. (1967). Some results concerning displacements and stresses in a non-homogeneous elastic half-space. *Géotechnique*, **17**(1), 58–67.

Gibson, R.E. and Anderson, W.F. (1961). Insitu measurement of soil properties with the pressuremeter. *Civ. Eng. Public Works Rev.*, **56**(658), May, pp. 615–618.

Gilbert, W.S. (1879). *The Pirates of Penzance.* The Savoy Operas, London.

Gillot, J.E., Penner, E. and Eden, W.J. (1974). Microstructure of Billings Shale and biochemical alteration of products, Ottawa, Canada. *Can. Geotech. J.*, **11**, 482–489.

Giroud, J.P. (1972). *Tables Pour le Calcul des Fondations.* Dunod, Paris.

Glossop, R. (1968). The rise of geotechnology and its influence on engineering practice. Eighth Rankine Lecture. *Géotechnique*, **18**, 105–150.

Glossop, R. and Skempton, A.W. (1945). Particle size in silts and sands. *J. Inst. Civ. Eng.*, **25**, 81–105.

Godfrey P.S. (1996). *Control of Risk – a Guide to the Systematic Management of Risk from Construction.* CIRIA Special Publication 125. CIRIA, London, ISBN 0 86017 441 7.

Golder, H.Q. and Palmer, D.J. (1955). Investigation of a bank failure at Scrapsgate, Isle of Sheppey, Kent. *Géotechnique*, **5**, 55–73.

Gordon, M.A., Clayton, C.R.I., Thomas, T.C. and Matthews, M.C. (1995). The selection and interpretation of seismic geophysical methods for site investigation. *Proc. Inst. Civ. Engrs Conf. on Advances in Site Investigation Practice.* Thomas Telford, London.

Grabe, J. and Vrettos, C. (1989). Dispersion measurements to estimate compaction effect on granular soils. *Soil Dynamics and Liquefaction*, edited by A.S. Cakmak and I. Herrera. Computational Mechanics Publ., Southampton, pp. 143–148.

Graham, J. and Au, V.C.S. (1985). Effects of freeze–thaw and softening on a natural clay at low stresses. *Can. Geotech. J.*, **22**(1), 69–78.

Grainger, P., McCann, D.M. and Gallois, R.W. (1973). The application of the seismic refraction technique to the study of the fracturing of the Middle Chalk at Mundford, Norfolk. *Géotechnique*, **23**(2), 219–232.

Green, G.E. and Bishop, A.W. (1969). A note on the drained strength of sand under generalised strain conditions. *Géotechnique*, **19**(1), 144–149.

Grey, S. (1995). *Practical Risk Assessment for Project Management.* Wiley, Chichester.

Guy, S. (1992). *English Local Studies Handbook.* University of Exeter Press, Exeter.

Haberfield, C.M. and Johnston, I.W. (1990). The interpretation of pressuremeter tests in weak rock – theoretical analysis. *Proc. 3rd International Symposium on Pressuremeters*, Oxford, pp. 169–178.

Haefeli, R. (1951). Investigations and measurements of the shear strengths of saturated cohesive soils. *Géotechnique*, **2**, 186–208.

Haimes Y.Y. (1998). *Risk Modelling, Assessment, and Management.* Wiley, New York, ISBN 0 471 24005 2.

Haldane, A.D., Carter, E.K. and Barton, G.M. (1970). The relationship of pyrite oxidation in rock–fill to highly acid water at Carin Dam, A.C.T., Australia. *Proc. Int. Conf. Eng. Geol.* **2**, 1113–1124.

Hall, J.W., Cruickshank, I.C. and Godfrey, P.S. (2001). Software-supported risk management for the construction industry. *Proc. ICE, Civ. Instn Engrs.*, Paper 12272, 42–48.

Hambly, E.C. (1969). *Plane strain behaviour of soft clay.* PhD Thesis. University of Cambridge, UK.

Harrison, I.R. (1991). A pushed thinwall sampling system for stiff clays. *Ground Eng.*, April, 30–34.

Hartikainen, J.K. (1981). Ground Improvement of the Grain Silo at Hodeidah. *Proc. 10th Int. Conf. Soil Mech and Fdn Engrg*, **1**: 137–144.

Hatem D.J. (1998). Introduction. In *Subsurface Conditions. Risk Management for Design and Construction Management Professionals*, edited by D.J. Hatem. Wiley, New York.

Haupt, R.S. and Olson, J.P. (1972). Case history–embankment failure on soft varied silt. *Proc. ASCE Spec. Conf. Performance of Earth and Earth-Supported Structures*, Purdue University, 1972, **1**, 29–64.

Hertwig, A. (1931). Die Dynamische Bodenuntersuchung. *Der Bauingenieur*, **25**, 457–461; **26**, 476–480.

Heukolom, W. and Foster, C.R. (1962). Dynamic testing of pavements. *Trans. Am. Soc. Civ. Eng.*, **127**(1), 425–445.

Heymann, G. (1998). *The stiffness of soils and weak rocks at very small strains.* PhD Thesis. Department of Civil Engineering, University of Surrey.

Heymann, G., Clayton, C.R.I. and Reed, G.T. (1997). Laser interferometry to evaluate the performance of local displacement transducers. *Géotechnique*, **47**(3), 399–405.

Hight, D.W. (1993). A review of sampling effects in clays and sands. *Proc. Int. Conf. on Offshore Site Investigation and Foundation Behaviour*, Society for Underwater Technology, 115–146.

Hight, D.W. (1996). Moderator's Report to Session 3: Drilling, Boring, Sampling and Description. *Proc. Int. Conf. on Advances in Site Investigation Practice.* Thomas Telford, London, pp. 337–360.

Hight, D.W. (2000). Sampling methods: evaluation of disturbance and new practical techniques for high quality sampling in soils. Keynote Lecture, *Proc. 7th Nat. Cong. of the Portuguese Geotech. Soc.*, Porto.

Hight, D.W. (2002). Soil characterisation: the importance of structure, anisotropy and natural variability. 38th Rankine Lecture. To be published in *Géotechnique*.

Hight, D.W. and Higgins, K.G. (1994). An approach to the prediction of ground movements in engineering practice: background and application. *Proc. Int. Symp. on Pre-failure Deformation Characteristics of Geomaterials*, Vol 3. Balkema, Rotterdam.

Hight, D.W. and Jardine, R.J. (1993). Small-strain stiffness and strength characteristics of hard London tertiary clays. In *Geotechnical Engineering of Hard Soils – Soft Rocks*, edited by A. Anagnostopoulos *et al.* Balkema, Rotterdam, pp. 533–552.

Hight, D.W., Hamza, M.M. and El Sayed, A.S. (2000). Engineering characterisation of the Nile Delta clays. *Proc. Conf. IS Yokohama 2000. Coastal Geotechnical Engineering in Practice.* Published in Volume 2, 2002.

Highways Agency (1996). *Value for Money Manual.* HMSO, London.

Hillier, R.P. (1992). *The plate test on clay: a finite element study*. PhD Thesis. University of Surrey.

Hobbs, N.B. (1975). Factors affecting the prediction of settlement of structures on rock: with particular reference to the Chalk and Trias. *Proc. Conf. on Settlement of Structures, Cambridge*. Pentech Press, London, pp. 579–610.

Hodges, W.G.H. (1976). Insitu tests in chalk. A report on plate bearing tests, dynamic and static soundings in chalk, carried out at Whitchurch and Otterbourne, Hampshire. *Ground Eng.*, April, pp. 51–54.

Hoek, E. and Bray, J.W. (1981). *Rock Slope Engineering*. 3rd edition. E and FN Spon, London.

Holtz, W.G. (1947). The use of triaxial shear tests on earth materials. *Proc. ASTM*, **47**, 1067–1076.

Holtz, W.G. and Gibbs, J.J. (1957). Engineering properties of expansive clays. *Trans ASCE*, **121**, 641–663.

Howland, A.W. (1991). The engineering geology of the London Docklands. *Proc. Inst. Civ. Eng.*, Part 1, **90**, 1153–1178.

Hudson, J.A. (1992). *Rock Engineering Systems – Theory and Practice.* Ellis Horwood, Chichester.

Hughes, J.M.O. and Withers, N.J. (1974). Reinforcing of soft soils with stone columns. *Ground Eng.* **7**(3), 42–44; 47–49.

Hughes, J.M.O., Wroth, C.P. and Windle, D. (1977). Pressuremeter tests in sands. *Géotechnique*, **4**, 455–477.

Hutchinson, J.N. (1988). General Report, Morphological and geotechnical parameters of landslides in relation to geology and hydrogeology. *Proc. 5th Int. Symp. on Landslides, Lausanne*, Vol 1. Balkema, Rotterdam, pp. 3–35.

Hvorslev, M.J. (1949). *Subsurface exploration and sampling of soils for civil engineering purposes.* Waterways Experiment Station, Vicksburg, Mississippi.

ICE (1991). *Inadequate Site Investigation.* Thomas Telford, London, pp. 1–26.

Institute of Field Archaeologists (1994). *Standards and Guidance for Archaeological Desk-based Assessments*, The Institute of Field Archaeologists, Reading, UK.

Institution of Civil Engineers and Federation of Civil Engineering Contractors. (1991) *General Conditions and Forms of Tender Agreement and Bond for Use in Connection with Works of Civil Engineering Construction.* lst edn 1945, 4th edn, 5th edn 1973, 6th edn 1991.

Institution of Civil Engineers and the Faculty and Institute of Actuaries. (1998). *RAMP: Risk Analysis and Management for Projects.* Thomas Telford, London.

Izumi K., Ogihara M. and Kameya H. (1997). Displacements of bridge foundations on sedimentary soft rock: a case study on small–strain stiffness. *Géotechnique*, **47**(3), 619–632.

Jacobs, P.A. and Butcher, A.P. (1996). The development of the seismic cone penetration test and its use in geotechnical engineering. *Advances in Site Investigation Practice*, edited by C. Craig. Thomas Telford, London, pp. 396–406.

Janbu, N., Bjerrum, L. and Kjaernsli, B. (1956). Veiledning ved Løsning av Fundamenteringsoppgaver. *NGI Publication No. 16.*

Janbu, N. (1963). Soil compressibility as determined by oedometer and triaxial tests. *Eur. Conf. Soil Mech. and Fdn Engng*, Wiesbaden, **1**, 19–25.

Janbu, N., Tokheim, O. and Senneset, K. (1981). Consolidation tests with continuous loading. *Proc. 10th Int. Conf. Soil Mech. and Found. Eng., Stockholm*, Vol 4. Balkema, Rotterdam, pp. 645–654.

Jardine, R.J., Symes, M.J. and Burland, J.B. (1984). The measurement of soil stiffness in the triaxial apparatus. *Géotechnique*, **34**(3), 323–340.

Jardine, R.J., Potts, D.M., Fourie, A.B. and Burland, J.B. (1986) Studies of the influence of non-linear stress–strain characteristics in soil-structure interaction. *Géotechnique*, **36**(3), 377–396.

Jefferies, M.G. (1988). Determination of horizontal geostatic stress in clay with self-bored pressuremeter. *Can. Geotech. J.* **25**(3), 559–573.

Jennings, J.E. and Knight, K. (1957). The prediction of total heave from the double oedometer test. *Trans. South. African Inst. Civ. Eng.*, **7**, 285–291.

Joint Contracts Tribunal (1980). *JCT8O. Standard Form of Building Contract.* JCT, 1980. Revised as JCT98.

Jones, G.A. (1974). *Method of estimation of settlements of fills over alluvial deposits from the results of field tests.* R5/6/74. NITRR. CSIR, South Africa.

Jones, G.A. (1975). Deep sounding – its value as a general investigation technique with particular reference to friction ratios and their accurate determination. *Proc. 6th Reg. Conf. Africa Soil Mech. Found. Eng., Durban*, Vol. 1, pp. 167–175.

Jones, G.A. and Rust, E. (1982). Piezometer penetration testing. *Proc. 2nd Eur. Symp. On Penetration Testing, Amsterdam.* Balkema, Rotterdam, pp. 607–614.

Jones, R.B. (1958) In-situ measurement of the dynamic properties of soil by vibration methods. *Géotechnique*, **8**(1), 1–21.

Jones, R.B. (1962). Surface wave technique for measuring the elastic properties and thickness of roads: theoretical development. *J. Appl. Phys.*, **13**, 21–29.

Jovičić, V. and Coop, M.R. (1997). Stiffness of coarse-grained soils at small strains. *Géotechnique*, **47**(3), 545–562.

Jovičić, V., Coop, M.R. and Simiæ, M. (1996). Objective criteria for determining G_{max} from bender element tests. *Géotechnique*, **46**(2), 357–362.

Kee, R. (1974). *The behaviour and design of foundations in chalk.* MPhil Thesis. Department of Civil Engineering and Construction, Hatfield Polytechnic, UK.

Kee, R. and Clapham, H.G. (1971) An empirical method of foundation design in Chalk. *Civ. Eng. Public Works Rev.*

Kennie, T.J.M. and Matthews, M.C. (1985). *Remote Sensing in Civil Engineering*, Surrey University Press, Glasgow.

Kim, Y.S., Ochi, K. and Tatsuoka, F. (1990). Strength and deformation properties of mudstones in triaxial compression. *Proceedings of the 8th Japanese Symposium on Rock Mechanics*, 357–362.

Kjaernsli, B. and Simons, N.E. (1962). Stability investigations of the North Bank of the Drammen River. *Géotechnique*, **12**, 147–167.

Kjellman, W. (1951). Testing the shear strength of clay in Sweden. *Géotechnique*, **2**, 225–232.

La Gatta, D.P. (1970). Residual strength of clays and clay shales by rotation shear tests, *Harvard Soil Mechanics Series, No. 86*, Cambridge, Mass.

La Rochelle, P., Trak, B., Tavenas, F. and Roy, M. (1974). Failure of a test embankment on a sensitive Champlain Clay deposit. *Can. Geotech. J.*, **11**, 142–164.

La Rochelle, P., Sarrailh, J., Tavenas, F., Roy, M. and Leroueil, S. (1981). Causes of sampling disturbance and design of a new sampler for sensitive soils. *Can. Geotech. J.*, **18**(1), 52–66.

La Rochelle, P., Leroueil, S. and Tavenas, F. (1986). A technique for long-term storage of clay samples. *Can. Geotech. J.*, **23**(4), 602–605.

Ladd, C.C. (1972). Test embankment on sensitive clay, *Proc. ASCE Spec. Conf. on Performance of Earth and Earth-Supported Structures*, Purdue University, 1972, **1.1**, 101–128.

Ladd, C.C. and Lambe, T.W. (1963). The strength of undisturbed clay determined from undrained tests. *Symp. on Laboratory Shear Testing of Soils*, ASTM, STP 361, 342–371.

Lade, P.V. and Duncan, J.M. (1973). Cubical triaxial tests on cohesionless soil. *J. Soil Mech. Found. Eng. Div., ASCE*, **99**(SM10), 793–812.

Lai, C.G. and Rix, G.J. (1998). *Simultaneous inversion of Rayleigh phase velocity and attenuation for near-surface site characterization*. Georgia Institute of Technology, Atlanta, USA.

Lake, L.M. and Simons, N.E. (1970). Investigations into the engineering properties of chalk at Welford Theale, Berkshire. *Proc. Conf. on In situ Investigations in Soils and Rocks*. British Geotechnical Society, London, pp. 23–29.

Lake, L.M. and Simons, N.E. (1975). Some observations of a four-storey building founded in chalk at Basingstoke, Hampshire. *Proc. Conf. on Settlement of Structures, Cambridge*, Pentech Press, London, pp. 283–291.

Latham M. (1994). *Constructing the Team*. Final report of the government/ industry review of procurement and contractual arrangements in the construction industry. HMSO, London.

Lefebvre, G. and Poulin, C. (1979). A new method of sampling in sensitive clay. *Can. Geotech. J.*, **16**(1), 226–233.

Leroueil, S. (2001). Some fundamental aspects of soft clay behaviour and practical implications. *Proc. 3rd Int. Conf. on Soft Soil Engineering, Hong Kong*, edited by Lee *et al.* Balkema, Rotterdam, pp. 37–53.

Leroueil, S. and Vaughan, P.R. (1990). The general and congruent effects of structure in natural soils and weak rocks. *Géotechnique*, **40**(3), 467–488.

Leroueil, S., Tardif, J., Roy, M., Konrad, J.-M. and La Rochelle, P. (1991). Effects of frost on the mechanical behaviour of Champlain Sea clays. *Can. Geotech. J.*, **28**(5), 690–697.

Levy, J.F. and Morton, K. (1975). Loading tests and settlement observations on granular soils. *Proc. Conf. on Settlement of Structures*, BGS Cambridge, Pentech Press, London, pp. 43–52.

Littlejohn, G.S., Cole, K.W. and Mellors, T.W. (1994). Without site investigation ground is a hazard. *Proc. Inst. Civ. Eng.*, Paper 10379, May, pp. 72–78; *Civ. Eng.*, **102**.

Lo, K.Y. (1965). Stability of slopes in anisotropic soils. *J. Soil Mech. Fed. Eng. Div., ASCE*, **91**(4), 85–106.

Lo, K.Y. and Stermac, A.G. (1965). Failure of an embankment founded on varved clay. *Can. Geotech. J.*, **2**(3), 243–253.

Lord, J.A. (1970). Discussion: Session A, *Proc. Conf. On Insitu Investigations in Soils and Rocks*, British Geotechnical Society, London, 1969, pp. 40–42.

Lowe, J. III., Jonas, E. and Obrician, V. (1969). Controlled gradient consolidation test. *J. Soil Mech. Found. Eng. Div., ASCE*, **95**(SM4), 77–97.

Lucas, H.C. and Robinson, V.K. (1995). Modelling of rising groundwater levels in the Chalk Aquifer of the London Basin. *Q. J. Eng. Geol.*, **28**, Supplement 1.

Lunne, T., Berre, T. and Strandvik, S. (1997a). Sample disturbance effects in soft low plastic Norwegian clay. *Symp. on Recent Developments in Soil and Pavement Mechanics, Rio de Janeiro*, 81–102.

Lunne, T., Robertson, P.K. and Powell, J.J.M. (1997b). *Cone Penetration Testing in Geotechnical Practice*. E and FN Spon, London.

Lunne, T., Berre, T., Strandvik, S., Anderson, K.H. and Tjelta, T.I. (2001). *Deep water sample disturbance due to stress relief*. Norwegian Geotechnical Institute, Oslo.

Madhloom, A.A.W.A. (1973). *The undrained shear strength of a soft silty clay from King's Lynn, Norfolk*. MPhil Thesis. University of Surrey.

Madshus, C. and Westerdahl, H. (1990). Surface wave measurements for construction control and maintenance planning of roads and airfields. *3rd International Conference on Bearing Capacity of Roads and Airfields, Trondheim*, pp. 233–242.

Managing Health and Safety. Five Steps to Success. HSE Books, Sudbury.

Manassero, M. (1989). Stress–strain relationships from drained self boring pressuremeter tests in sand. *Géotechnique*, **39**(2).

Marsland, A. (1971). The use of insitu tests in a study of the effects of fissures on the properties of stiff fissured clays. *Proc. 1st Australia–New Zealand Conf. On Geomechanics, Melbourne*, Vol 1. pp. 180–189.

Marsland, A. and Eason, B.J. (1973). Measurements of the displacements in the ground below loaded plates in deep boreholes. *British Geotech. Soc. Symp. on Field Instrumentation in Geot. Eng.*, **1**, pp. 304–317.

Matthews, M.C. (1993). *Mass compressibility of fractured chalk*. PhD Thesis. Department of Civil Engineering, University of Surrey.

Matthews, M.C. and Clayton, C.R.I (1993). Influence of intact porosity on the engineering properties of a weak rock. *Geotechnical Engineering of Hard Soils – Soft Rock*, edited by A. Anagnostopoulos *et al.*, Vol. 1. Balkema, Rotterdam, pp. 693–702.

Matthews, M.C., Hope, V.S. and Clayton, C.R.I. (1996) The use of surface waves in the determination of ground stiffness profiles. *Proc. Inst. Civ. Eng. Geotech. Eng.*, **119**, April, 84–95.

Matthews, M.C., Clayton, C.R.I., and Own, Y. (2000a). The use of field geophysical techniques to determine geotechnical stiffness parameters. *Proc. Inst. Civ. Eng. Geotech. Eng.* January, 31–42.

Matthews, M.C., Clayton, C.R.I. and Rigby-Jones, J. (2000b). Locating dissolution features in the Chalk. *Q. J. Eng. Geol. Hydrogeol.*, **33**(2), 125–140.

Mayne, P.W. (2001). Ground property characterisation by insitu tests. *Proc. 15th Int. Conf. on Soil Mech. and Geotech. Eng., Istanbul, August 2001.* To be published: Balkema, Rotterdam.

McCann, D.M., Jackson, P.D. and Green, A.S.P. (1986). Application of cross-hole seismic measurements in site investigation surveys. *Geophysics*, **51**, 914–929.

McDowell, P.W. 1989. Ground subsidence associated with doline formation in chalk areas of Southern England. In *Engineering and Environmental Impacts of Sinkholes and Karst*, edited by J.A. Beck. Balkema, Rotterdam, pp. 129–134.

McDowell, P.W. and Poulsom, A.J. 1996. Ground subsidence related to dissolution of chalk in southern England. *Ground Eng.*, **29**(2), 29–33.

Meigh, A.C. (1987). *Cone Penetration Testing*. Butterworths, London, p. 141.

Ménard, L. and Broise, Y. (1975). Theoretical and practical aspects of dynamic consolidation. *Géotechnique*, **25**(1), 3–18.

Menzies, B.K. (1975). A device for measuring volume change. *Geotechnique*, **25**(1), 133–134.

Menzies, B.K. (1976a). An approximate correction for the influence of strength anisotropy on conventional shear vane measurements used to predict field bearing capacity, *Géotechnique*, **26**(4), 631–634.

Menzies, B.K. (1976b). Strength, stability and similitude. *Ground Eng.*, **9**(5), 32–36.

Menzies, B.K. (1988). A computer controlled hydraulic triaxial testing system. *Advanced Triaxial Testing of Soil and Rock*. ASTM STP 977, 82–94.

Menzies, B.K. (1997) Applying modern measures. *Ground Eng.*, July.

Menzies, B.K. (2000) Near-surface site characterisation by ground stiffness profiling using surface wave geophysics. *Instrumentation in Geotechnical Engineering. H.C. Verma Commemorative Volume*, edited by K.R. Saxena and V.M. Sharma. Oxford & IBH Publishing Co. Pvt. Ltd., New Delhi, Calcutta, pp. 43–71.

Menzies, B.K. and Hooker, P. (1992). PC and local microprocessor controlled geotechnical testing systems. *Proc. Int. Conf. Geotechnics and Computers*. Paris.

Menzies, B.K. and Mailey, L.K. (1976). Some measurements of strength anisotropy in soft clays using diamond-shaped shear vanes, *Géotechnique*, **26**, 535–538.

Menzies, B.K. and Matthews, M.C. (1996) *The Continuous Surface Wave System: A Modern Technique for Site Investigation*. Special Lecture: Indian Geotech. Conf., Madras.

Menzies, B.K. and Merrifield, C.M. (1980). Measurements of shear stress distribution on the edges of a shear vane blade. *Geotechnique*, **30**(3), 314–318.

Menzies, B.K. and Simons, N.E. (1978). Stability of embankments on soft ground. *Developments in Soil Mechanics 1*. Applied Science Publishers Ltd, London, Chapter 11, pp. 393–436.

Menzies, B.K. and Sutton, H. (1980). A control system for programming stress paths in the triaxial cell. *Ground Eng.*, **13**(1), 22–23.

Menzies, B.K. and Sutton, H. (1981). The effects of simulated sampling and construction stress paths on the stress–strain properties of a sand. *Rivista Italiana di Geotecnica*, **16**(1), 55–62.

Menzies, B.K., Sutton, H. and Davies, R.E. (1977). A new system for automatically simulating K_0 consolidation and K_0 swelling in the conventional triaxial cell. *Geotechnique*, **27**(4), 593–596.

Millot, G. (1970). *Geology of Clays*, translated by W.R. Farrand and H. Paquet. Springer-Verlag, New York.

Mitchell, R.J. (1973). An apparatus for plane strain and true triaxial testing of undisturbed soil samples. *Can. Geotech. J.*, **10**(3), 520–527.

Mooney, H.M. (1974). Seismic shear waves in engineering. *J. Geotech. Eng., ASCE*, **100**(GT8), 905–923.

Mott MacDonald and Soil Mechanics Ltd (1994). *Study of the Efficiency of Site Investigation Practices.* Transport Research Laboratory, Crowthorne, Berks, 1994, Project Report 60, ISSN 0968 4093.

Mottershead, D.N. 1976. The quaternary history of the Portsmouth region. *Portsmouth Geographical Essays.* Department of Geography, University of Portsmouth.

Moum, J. and Rosenqvist, I.Th. (1959). Sulfate attack on concrete in the Olso region. *J. Am. Concrete Inst. Proc.*, **56**, 257–264.

Moxhay, A.L., Tinsley, R.D. and Sutton, J.A. (2001). Monitoring of soil stiffness during ground improvement using seismic surface waves. *Ground Eng.*, January, 34–37.

Muir Wood, D. (1990). Strain dependent soil moduli and pressuremeter tests. *Géotechnique*, **40**, 509–512.

Mukabi, J.N., Tatsuoka, F. and Hirose, K. (1991). Effect of strain rate on small strain stiffness of kaolin. *Proceedings of the 26th Japanese National Conference on Soil Mechanics and Foundation Engineering, Nagano, Japan.* pp. 659–662.

Mukabi, J.N., Tatsuoka, F., Kohata, Y, Tsuchida, T. and Akino, N. (1994). Small strain stiffness of Pleistocene clays in triaxial compression. *Proc. Int. Symp. on Pre-failure Deformation Characteristics of Geomaterials*, Vol 1. Balkema, Rotterdam, pp. 189–196.

Nazarian, S. and Stokoe, K.H. (1984). Insitu shear wave velocities from spectral analysis of surface waves. *Proc. 8th World Conf. on Earthquake Engineering*, **3**, 31–38.

Nicholson, D., Tse, C.-M. and Penny, C. (1999). *The Observational Method in Ground Engineering: Principles and Applications.* CIRIA Report R185. CIRIA, London, ISBN 0 86017 497 2.

Nienhuis, H. and Price, D.G. (1990). The scale effect with regard to deformability of calcarenite. *Proc. 24th Annual Conf. of the Eng. Group of the Geol. Soc. Field Testing in Engineering Geology*, edited by F.G. Bell *et al.* Eng. Geol. Special Publication No. 6, Geol. Soc., London, pp. 205–215.

Nixon, P.J. (1978). Floor heave in buildings due to the use of pyretic shales as fill material. *Chem. Ind.*, 4 March, 160–164.

Pahl, G. and Beitz, W. (1996). *Engineering Design. A Systematic Approach.* Springer Verlag, London, ISBN 3 540 19917 9.

Palmer, A.C. (1972). Undrained plane-strain expansion of a cylindrical cavity in clay: a simple interpretation of the pressuremeter test. *Géotechnique*, **22**(3), 451–457.

Park, C.S. (1993). *Deformation and strength characteristics of a variety of sands by plane strain compression tests.* Doctor of Engineering Thesis. University of Tokyo.

Parry, R.H.G. (1972). Stability analysis for low embankments on soft clays. *Stress–Strain Behaviour of Soils.* G.T. Foulis, Henley-on-Thames, pp. 643–668.

Parry, R.H.G. and McLeod, J.H. (1967). Investigation of slip failure in flood levee at Launceston, Tasmania. *Proc. 5th Australia–NZ Conf. Soil Mech. Found. Eng., Auckland*, pp. 249–300.

Peacock, W.S. and Whyte, I.L. (1986). Site investigation practice. *Proc. Inst. Civ. Eng. Munic. Eng.*, **5**, 235–245.

Peck, R.B. (1948). History of building foundations in Chicago. *Univ. of Ill. Eng. Exp. Sta. Bull.*, **373**, 64 pp.

Peck, R.B. (1962). Art and science in subsurface engineering. *Géotechnique*, **12**, 60–68.

Pells, P.J.N. (1983). Plate loading tests on soil and rock. *Proc. Extension course on insitu testing for geotechnical investigations*, Sydney, *May–June*. Balkema, Rotterdam, pp. 73–86.

Pells, P.J.N. and Turner, R.M. (1979). Elastic solutions for the design and analysis of rock socketed piles. *Can. Geotech. J.*, **16**, 481–487.

Penner, E., Eden, W.J. and Gillott, J.E. (1973). Floor heave due to biochemical weathering of shale. *Proc. 8th Int. Conf. Soil Mech. Found. Eng., Moscow*, Vol 2. pp. 151–158.

Perry, J. and West, G. (1996). *Sources of Information for Site Investigations in Britain*. (Revision of TRL Report LR 403). Transport Research Laboratory Report 192. Transport Research Laboratory, Crowthorne, UK.

Peterson, R., Iverson, N.L. and Rivard, P.J. (1957). Studies of several dam failures on clay foundations. *Proc. 4th Int. Conf. Soil Mech. and Found. Eng., London*, Vol 2. pp. 348–352.

Phien-Wej, N. and Chavalitjiraphan, S. (1991). Properties of coastal Bangkok clay and application to cut slope analysis. *Int. Symp. Geo-Coast '91, Yokohama*, Vol 1. pp. 69–74.

Pilot, G. (1972). Study of five embankment failures on soft soils, *Proc. ASCE Spec. Conf. Performance of Earth and Earth-Supported Structures*, Purdue University, 1972, **1**(1), 81–100.

Porovic, E. and Jardine, R.J. (1994). Some observations on the static and dynamic shear stiffness of Ham River sand. *Proc. Int. Symp. on Pre-failure Deformation Characteristics of Geomaterials*, Vol 1. Balkema, Rotterdam, pp. 25–30.

Poulos, H.G. and Davies, F.H. (1974). *Elastic Solutions for Soil and Rock Mechanics*. John Wiley, New York.

Powell, J.J.M., Marsland, A., Longworth, T.I. and Butcher, A.P. (1990). Engineering properties of Middle Chalk encountered in investigations for roads near Luton, Bedfordshire. *Int. Chalk Symposium, Brighton*. Thomas Telford, London, pp. 327–341.

Pugh R.S., Parnell, P.G. and Parkes, R.D. (1995). A rapid and reliable on-site method of assessing desiccation in clay soils. *Proc. Inst. Civ. Eng., Geotech. Eng.*, Paper 10595, 25–30.

Quigley, R.M. and Vogan, R.W. (1970). Black shale heaving at Ottawa, Canada. *Can. Geotech. J.*, **7**, 106–112.

Ray, R.P. and Morris, K.B. (1995). Automated laboratory testing for soil/water characteristic curves. *Unsaturated Soils*, edited by Alonso and Delage. Balkema, Rotterdam, pp. 547–552.

Raymond, G.P. (1967). The bearing capacity of large footings and embankments on clays. *Géotechnique*, **17**, 1–10.

Reeves, G. (1996). *The Geologists' Directory*. Geological Society, London.

Rendulic, L. (1937). Ein Grundgesetz der Tonmechanik und sein experimenteller Beweis. *Bauingenieur*, **18**, 459–467.

Richart, F.E., Hall, J.R. and Woods, R.D. (1970). *Vibrations of Soils and Foundations*. Prentice-Hall, Englewood Cliffs, New Jersey.

Ricketts, G.A., Smith, J. and Skipp, B.O. (1996). Confidence in seismic characterisation of the ground. In *Advances in Site Investigation Practice*, edited by C. Craig. Thomas Telford, London, pp. 673–686.

Ridley, A.M. and Burland, J.B. (1993). A new instrument for the measurement of soil moisture suction. *Géotechnique*, **43**(2), 321–324.

Rigby-Jones, J., Clayton, C.R.I. and Matthews, M.C. 1993. Dissolution features in the chalk: from hazard to risk. In *Risk and Reliability in Ground Engineering*, edited by B.O. Skipp. Thomas Telford, London, pp. 87–99.

Rix, G.J. (2001). Review of recent developments in insitu seismic testing. *Prefailure Deformation Characteristics of Geomaterials*, edited by Jamiolkowski, Lancellotta and Lo Presti. Swets and Zeitlinger, Lisse, ISBN 90 5809 075 2, Vol 2, pp. 1333–1337.

Rolfsen, E.N. and Simons, N.E. (1971). *Measurements of Soil Conditions, Deformations and Porewater Pressures Under a Runway Embankment overlying Soft Clay, with Vertical Sand Drains, at Fornebu Airport, Oslo.* Norwegian Geotechnical Institute Technical Report, No 10. NGI, Oslo.

Roscoe, K.H. (1953). An apparatus for the application of simple shear to soil samples. *Proc. Third Int. Conf. Soil Mech. Found. Eng., Zurich*, Vol 2. pp. 186–191.

Roscoe, K.H. (1970). The influence of strains in soil mechanics. *Géotechnique*, **20**(2), 129–170.

Rowe, P.W. (1972). The relevance of soil fabric to site investigation practice. 12th Rankine Lecture. *Géotechnique*, **22**(2), 195–300.

Rowe, P.W. and Barden, L. (1966). A new consolidation cell. *Géotechnique*, **16**(2), 162–170.

Roy, M. (1975). *Predicted and observed performance of motorway embankments on soft alluvial clay in Somerset.* MPhil Thesis. University of Surrey.

Rust E. and Jones G.A. (1990). *Prediction of performance of embankments on soft alluvial deposits using the piezometer probe (CPTU).* RDAC Research Project No. 89/14, South African Road Board.

Schjetne, K. (1971). The measurement of pore pressure during sampling. *Proc. Speciality Session on Quality in Soil Sampling, 4th Asian Regional Conf. Int. Soc. Soil Mech. and Found. Eng.*, Bangkok, p. 12016.

Scott, R.F. and Ko, H.Y. (1969). Stress-deformation and strength characteristics. *Proc. Seventh Int. Conf. Soil Mech. Found. Eng., Montreal*, Vol 1. pp. 359–363.

Seed H. and Booker J. (1976). *Stabilisation of potentially liquefiable sand deposits using gravel drain system.* Earthquake Engineering Research Centre, Berkeley. Report No EERC 76/10.

Sembelli, P. and Ramirez, A. (1969). Measurement of residual strength of clays with a rotation shear machine. *Proc. 7th Int. Conf. Soil Mech. Found. Eng., Mexico*, Vol 3. Sociedad Mexicana de Suelos, pp. 528–529.

Serota, S. (1966). Discussion. *Proc. ICE*, **35**(9), 522.

Shamburger, J.H., Patrick, D.M. and Lutton, R.J. (1975). Design and construction of compacted shale embankments. Vol. 1 Survey of problem areas and current practices. *Report No. FHWA–RD–75–61 Prepared for Federal Highway Administration, Washington, D.C.* (N.T.I.S. Springfield Virginia).

Shibuya, S., Hwang, S.C. and Mitachi, T. (1997). Elastic shear modulus of soft clays from shear wave velocity measurement. *Géotechnique*, **47**(3), 593–601.

Shilston, D., Parsons, A., Harrison, E. and Lee, K. (1998). Giants' shoulders: the cost effective use of geotechnical desk studies in civil and structural engineering. *Proc. Seminar on The Value of Geotechnics in Construction* (Nov. 1998), Construction Research Communications Ltd, London, pp. 25–36.

Shirlaw, J.N., Dazhi, W., Ganeshan, V. and Hoe, C.S. (1999). A compensation grouting trial in Singapore marine clay. *Int. Symp. on Geotechnical Aspects of Underground Construction in Soft Ground, Tokyo*, pp. 149–154.

Simon, P., Hillson, D. and Newland, K. (1997). *PRAM: Project Risk Analysis and Management Guide*. Association for Project Management, Norwich.

Simons, N.E. (1967) Contribution to the Discussion. *Geotechnical Conference, Oslo*, Vol 2. pp. 159–160.

Simons, N.E. (1971). The stress path method of settlement analysis applied to London clay. *Stress–strain Behaviour of Soils*, edited by R.H.G. Parry. *Proc. Roscoe Mem. Symp.* G.T. Foulis, Henley-on-Thames, pp. 241–252.

Simons, N.E. and Menzies, B.K. (1974). A note on the principle of effective stress. *Geotechnique*, **24**(2), 259–261.

Simons, N.E. and Menzies, B.K. (1978). The long-term stability of cuttings and natural clay slopes. *Developments in Soil Mechanics 1.* Applied Science Publishers Ltd, London, Chapter 10, pp. 347–391.

Simons, N.E. and Menzies, B.K. (2000). *A Short Course in Foundation Engineering.* Thomas Telford, London.

Simons, N.E., Menzies, B.K. and Matthews, M.C. (2001). *A Short Course in Soil and Rock Slope Engineering.* Thomas Telford, London.

Simpson, B. and Driscoll, R. (1998). *Eurocode 7: A Commentary*. DETR-BRE-ARUP.

Site Investigation In Construction (1993). **4** *Guidelines for the safe investigation by drilling of landfills and contaminated land.* Thomas Telford, London.

Site Investigation Steering Group (1993). Site Investigation in Construction, Part 1. *Without site investigation: ground is a hazard.* Thomas Telford, London.

Skempton, A.W. (1955). *Soil mechanics and its place in the university*. Inaugural Lecture as Professor of Soil Mechanics, Imperial College.

Skempton, A.W. (1964). Long-term stability of clay slopes. *Géotechnique*, **14**, 77–101.

Skempton, A.W. (1977). Slope Stability of Cuttings in Brown London Clay. Special Lecture. *Proc. 9th Int. Conf. Soil Mech. and Found. Eng.*, Vol 3. pp. 261–270.

Skempton, A.W. and Bjerrum, L. (1957). A contribution to the settlement analysis of foundations on clay. *Géotechnique*, **7**(4), 168–178.

Skempton, A.W. and La Rochelle, P. (1965). The Bradwell Slip: a short term failure in London clay. *Géotechnique*, **15**, 221–242.

Skempton, A.W. and Petley, D.J. (1967). The strength along structural discontinuities in stiff clays. *Proc. Geotech. Conf., Oslo*, Vol 2. pp. 29–46.

Smallwood, A.R.H., Morley, R.S., Hardingham, A.D., Ditchfield, O. and Castleman, J. (1997). Quantitative risk assessment of landslides – case histories from Hong Kong. *Engineering Geology and the Environment*, edited by P.G. Marinos *et al.* Balkema, Rotterdam, pp. 1055–1060.

Smith, P.R. (1992). *The behaviour of natural high compressibility clay with special reference to construction on soft ground.* PhD Thesis. Imperial College, London.

Smith, R.E. and Wahls, H.E. (1969). Consolidation under constant rates of strain. *J. Soil Mech. Found. Div., ASCE*, **95**, SM2, 519–539.

Souto, A., Hartikainin, J. and Özüdoğru, K. (1994). Measurement of dynamic parameters of road pavement materials by the bender element and resonant column tests. *Géotechnique*, **44**(3), 519–526.

Spanovich, M. and Fewell, R. (1970). Building damage from expansive steel backfill. *J. Soil Mech. Found. Div., ASCE*, **96**, 1808–1810.

Stamatopoulos, A.C. and Kotzias, P.C. (1965). Construction and performance of an embankment in the sea on soft clay. *Proc. 6th Int. Conf. Soil Mech. and Found. Eng., Montreal*, Vol 2. University of Toronto Press, pp. 566–571.

Stokoe, K.H. and Woods, R.D. (1972). In situ shear wave velocity by cross–hole method. *J. Soil Mech. Found. Eng. Div., ASCE*, **98**(SM5), 443–460.

Stroud, M.A. (1988). The standard penetration test: its application and interpretation. *Proc. ICE Conf. On Penetration Testing in the UK, University of Birmingham*. Thomas Telford, London.

Sutherland, H.B. (1975). Granular materials. Review paper, Session 1. *Proc. Conf. On Settlement of Structures, BGS Cambridge*, Pentech Press, London, pp. 473–499.

Sutton, J.A. and Snelling, K. (1998). Assessment of ground improvement using the continuous surface wave method. *Proc. 4th Conf. Environ. and Eng. Geoph. Soc., Barcelona*. pp. 485–488.

Tanaka, H. and Tanaka, M. (1999). Key factors governing sample quality. To be published.

Tatsuoka, F. and Kohata, Y. (1994). Stiffness of hard soils and soft rocks in engineering applications. *Proc. Int. Symp. on Pre-failure Deformation Characteristics of Geomaterials*, Vol 3. Balkema, Rotterdam.

Tatsuoka, F. and Shibuya, S. (1992). *Deformation Characteristics of Soils and Rocks from Field and Laboratory Tests.* Report of the Institute of Industrial Science, University of Tokyo, **37**(1), 1–136.

Tavenas, S., Leroueil, S. and Roy, M. (1982). The piezocone test in clays: use and limitations. *Proc. 2nd Eur. Symp. on Penetration Testing, Amsterdam*. Balkema, Rotterdam, pp. 889–894.

Tavenas, F. and Leroueil, S. (1987). Laboratory and insitu stress-strain-time behaviour of soft clays: a state-of-the-art. *International Symposium on Geotechnical Engineering of Soft Soils, Mexico City*, Vol. 2, pp. 1–46.

Taylor, D.W. (1944). Triaxial shear research (US Corps Eng.), *MIT Progress Report No. 10.*

Taylor, R.K. and Cripps, J.C. (1984) Mineralogical controls on volume change. In *Ground Movements and Their Effects on Structures*, edited by P.B. Attewell and R.K. Taylor. Surrey University Press, Blackie, Glasgow, pp. 268–302.

Telford, W.M., Geldart, L.P. and Sheriff, R.E. (1990). *Applied Geophysics.* Second Edition. Cambridge University Press, Cambridge.

Terzaghi, K. (1923). Die berechnung der Durchlassigkeitsziffer des Tones aus dem Verlauf der hydrodynamischen Spannungserscheinungen. *Sitzung berichte (Abt. IIa) Akademit der Wissenschaften*, Vienna, Part 20, 32 (3/4), 125–138.

Terzaghi, K. (1932). Tragfahigkeit der Flachgründungen. *Prelim. Publ. 1st Int. Cong. Int. Ass. Brudge Struct. Eng.*, pp. 659–672.

Terzaghi, K. (1936). Relation between soil mechanics and foundation engineering. *Presidential Address, Proceedings of the 1st International Conference on Soil Mechanics and Foundation Engineering, Harvard*, Vol. 3. pp. 13–18.

Terzaghi, K. (1943). *Theoretical Soil Mechanics*. Wiley, New York.

Terzaghi, K. and Peck, R.B. (1948). *Soil Mechanics in Engineering Practice.* Wiley, New York.

Terzaghi, K. and Peck, R.B. (1967). *Soil Mechanics in Engineering Practice*. Second edition. Wiley, New York.

The Concise Oxford Dictionary of Current English. Ninth edition. Oxford University Press, Oxford.

Tiedemann, B. (1937). Über die Schubfestigkeit bindiger Boden. *Bautechnik*, **15**, 433–435.

Timoshenko, S. (1934). *Theory of Elasticity*. McGraw-Hill, New York, p. 193.

Tokimatsu, K., Kuwayama, S., Tamura, S. and Miyadera, Y. (1991). V_s determination from steady state Rayleigh wave method. *Soils Found.*, **31**(2), 153–163.

Tomlinson, M.J. (1980). *Foundation Design and Construction*. Pitman, London.

Torstensson, B. (1977). Time dependent effects in the field vane test. *Proc. Int. Symp. on Soft Clay*. Asian Inst. Tech., Bangkok, pp. 387–397.

Tou, J.H., Leong, E.C., Rahardjo, H. and Cheong, H.K. (2001). Determination of G_{max} of residual soils using bender elements. *Proc. 14th Southeast Asian Geotech Conf., Hong Kong*, Vol 1, pp. 581–586.

Tyrell, A.P., Lake L.M. and Parsons, A.W. (1983). *An Investigation of the Extra Costs Arising on Highway Contracts*. Transport and Road Research Laboratory, Crowthorne, Berks, Supplementary Report SR814.

Uff, J.F. and Clayton, C.R.I. (1986). *Recommendations for the procurement of ground investigation*, CIRIA Special Publication 45.

University of Surrey. (1998). *Managing Geotechnical Risk: Improving Productivity in UK Building and Construction, Task 3 Report Investigation of the use of IT and Software for Geotechnical Risk Management*. Report prepared by the University of Surrey on behalf of the Institution of Civil Engineers, London.

van der Merwe, D.H. (1964). The prediction of heave from the plasticity index and the percentage of clay fraction. *Civ. Eng. South Africa*, **6**(6), 103–107.

Varaksin, S. (1981). Recent developments in soil improvement techniques and their practical applications. *Sols Soils*.

Vaughan, P.R., Chandler, R.J., Apted, J.P., Maguire, W.M. and Sandroni, S.S. (1993). Sample disturbance with particular reference to its effect on stiff clays. *Proc. Wroth Memorial Symp. Oxford*. pp. 685–708.

Vermeiden, J. (1948). Improved sounding apparatus as developed in Holland since 1936. *Proc. 2nd Int. Conf. on Soil Mech. and Found. Eng., Rotterdam*, Vol 1. pp. 280–287.

Viggiani, G. and Atkinson, J.H. (1995). Interpretation of bender element tests. *Géotechnique*, **45**(1), 149–154.

Von Karman, Th. (1911). Festigkeitversuche unter allseitiger Druck. *Verein Deutscher Ingenieure Verlag*, **55**, 1749–1757.

Vrettos, C. (1990a). Dispersive SH-surface waves in soil deposits of variable shear modulus. *Soil Dynam. Earthq. Eng.* **9**, 255–264

Vrettos, C. (1990b). In-plane vibrations of soil deposits with variable shear modulus: I. Surface waves. *Int. J. Num. Anal. Methods Geomech.*, **14**, 209–222.

Vrettos, C. (1991). Time-harmonic Boussinesq problem for a continuously nonhomogeneous soil. *Earthq. Eng. Struct. Dyn.*, **20**, 961–977.

Vrettos, C. and Prange, B. (1990). Evaluation of insitu effective shear modulus from dispersion measurements. *J. Geotech. Eng., ASCE*, **116**, 1581–1585.

Wakeling, T.R.M. (1970). A comparison of the results of standard site investigation methods against the results of a detailed geotechnical investigation in Middle Chalk at Mundford, Norfolk. *Proc. Conf. on In situ Investigations in Soils and Rocks*, British Geotechnical Society, London, pp. 17–22.

Waltham A.C. (1994). *Foundations of Engineering Geology*. Blackie Academic and Professional, London.

Ward, W.H., Burland, J.B. and Gallois, R.M. (1968). Geotechnical assessment of a site at Mundford, Norfolk for a Proton Accelerator. *Géotechnique*, **18**(4), 399–431.

Weeks, A.G. (1969). The stability of natural slopes in South-East England as affected by periglacial activity. *Q. J. Eng. Geol.*, **2**(1), 49–61.

Weeks, A.G. (1970). *The stability of the Lower Greensand Escarpment in Kent*. PhD Thesis. University of Surrey.

West, J.M. (1976). The role of ground improvement in foundation engineering. *Proc. Ground Treatment by Deep Compaction*. ICE, London, pp. 71–78.

Wheeler P. (1999). Scattering predictions – Imperial College predictions – competition shows pile design remains a big uncertainty. *New Civil Engineer*, December, p. 34.

Whittle, R.W (1999). Using non-linear elasticity to obtain the engineering properties of clay. *Ground Eng.*, **32**(5), 30–34.

Wilkes, P.F. (1972). An induced failure at a trial embankment at King's Lynn, Norfolk, England, *Proc. ASCE Spec. Conf. Performance of Earth and Earth–Supported Structures*, Purdue University, Vol 1. pp. 29–63.

Willams, A.A.B. and Donaldson, G.W. (1980). Building on expansive soils in South Africa: 1973–1980. *Proc. 4th Int. Conf. Expansive Soils, Denver. ASCE*, pp. 834–844.

Wissa, A.E.Z., Christian, J.T., Davis, E.H. and Helberg, S. (1971). Consolidation at constant rate of strain. *J. Soil Mech. Found. Eng. Div., ASCE*, **97**(SM10), 1393–1413.

Withers, N.J., Howie, J., Hughes, J.M.O. and Robertson, P.K. (1989). Performance and analysis of cone pressuremeter tests in sands, *Géotechnique*, **39**(3), 433–454.

Wrench, B.P. (1984). Plate tests for the measurements of modulus and bearing capacity of gravels. *Civ. Eng. South Africa*, September, 429–437.

Wroth, C.P. (1969). Some recent developments of the simple shear apparatus at Cambridge. *Proc. Seventh Int. Conf. Soil Mech. and Found. Eng., Mexico*, Vol 3, Discussion: Speciality Session on New Laboratory Methods of Investigating Soil Behaviour. pp. 526–527.

Yoshimi, Y., Hatanaka, M. and Oh-Oka, H. (1978). Undisturbed sampling of saturated sands by freezing. *Soils Found.*, **11**(3), 59–73.

Yoshimi, Y., Tokimatsu, K. and Ohara, J. (1994). In situ liquefaction resistance of clean sands over a wide density range. *Géotechnique*, **44**(3), 479–494.

Index

Note: Figures and Tables are indicated by *italic page numbers*

access (for ground investigation), observations during walk-over survey, 120
acidic ground, hazards due to, 124
activity chart, for expansive clays, *54*
adverse event, monetary consequences, 142
aerial photography, 105–116
  advantages as data source, 105–106
  amount of data presented, 105
  availability, 106
  and classification of ground characteristics, 114
  cost, 106
  detail in image, 105
  emulsion and filter combinations, 107–108
  geometry of image, 107
  image interpretation procedure, 115–116
  image medium, 108–109
  interpretation of, 111–116
  for mining desk studies, 85
  oblique, *102*, 107, 109, 113
  patterns, 113
  photographic parameters affecting interpretation, 106–110
  principles of image interpretation, 111–115
  scale of image, 108
  season of photography, 109–110
  shadows, *102*, 112–113
  shape of features, 113–114
  and site history, 95, *96*, 105
  size of objects, 114
  and slope instability, 48, 100, *101*, *102*, *103*, 104, 119
  as source of data/information, *74*, *76–79*, 85, 92, 95, *96*
  sources, *98*, 106, 116
  spectral sensitivity, 105–106
  texture in image, *104*, 105, 113, 123
  time of day of photography, 109
  tone of image, 111–112
  topographical maps superimposed on, 88, *89*
  vantage point advantage, 105
  vertical, *101*, *104*, 107, 109
alluvium deposits, 80
Amherst test site, seismic piezocone results, *252*
anhydrite, hydration of, 134, 137
archeological surveys, photography for, 109
architectural records, *99*
archives, as sources of information, *98–99*, 121
Ariake clay, sampling of, 175–176
asbestos, as hazard, 125–126
Association of Geotechnical and Geoenvironmental Specialists (AGS), 3
Atherfield Clay, *83*, *103*, 104
Atterberg Limits, expansive potential of clay assessed from, 137
Autoplumb, example(s) of use, 23

Bang Bo test excavation, Thailand, 191
Basildon, Essex
  aerial photographs, 95, *96*
  industrial development, 288–289, *290*
bearing capacity theory, 63, 66
Begemann cone penetration test, 248, 249
bender elements
  example(s) of use, *188*, *285*
  in triaxial test, 199, 203
bibliography, 321–340

Biddulph Moor, Staffordshire, historical maps, 92–95
Bishop ring shear apparatus, *216*, 217
Bjerrum's (shear vane) correction factor, 28, 221
  correlation with plasticity index, *223*
black shales, 137, 138
body waves (seismic waves), 267
  primary/compressional (P) waves, 267, 269
  secondary/shear (S) waves, 267
boreholes
  depth, 9–10
  for foundations, 8–10
  layout, 8–11
  for motorways, 8
  for slopes, 11
  spacing, 8
Bothkennar clay
  sampling of, 178, *185*
  undrained shear strength profiles, *189*
boulder clay, 81, 85, *137*
Bowden, Andrew, quoted, 1–2
Bracklesham Beds, 80, *83*
Bradwell, Essex
  landslip, 41–42
  settlement calculations for circular foundations, *167*
Brent Knoll, Somerset, trial embankment, 226–233
Brierley, G. S., quoted, 151
Bristol, geological map, 87, *88*
British Geological Survey (BGS)
  borehole records, *71*, 85
  field slips, 84
  maps and memoirs, *71*, 81n[1–2]
British Standards Institution, *Code of Practice for Site Investigations*, 154–155
brittleness index, 220
Bromhead, E. N., on peak strength and its measurement, 217, 219
Bromhead ring shear apparatus, 217, *218*
I. K.Brunel Archive, *99*

Cambridge self-boring pressuremeter (Camkometer), 166, *254*, 257, *258*
carbon dioxide, as hazard, 126
carbon monoxide, as hazard, 126
carcinogens, 125–126
Casagrande construction (for determining pre-consolidation pressure), *210*, 211
Casagrande oedometer test, 204, 209–213
  continuous consolidation test equipment, 213
  and pre-consolidation pressure, 210–211
  Skempton–Bjerrum correction, 211–213
  specimen size, 204
Casagrande standpipes, example(s) of use, 22
chalk, settlement predictions for, 243, 274, *275*
clays
  expansive clays, *54*, 136–137
  geotechnical investigations on, 16–28
  shrinkage potential listed, *137*
  see also soft clays; stiff clays
Clayton, C. R. I., quoted, 140–141
Coal Authority Mine Records Office, 85
coal mining activities, 84, 87
coal seams, 87
collapse subsidence, 127–128
collapsing soils, 56–57, *58*
compressible soils, settlement due to, 136
compression index, data in case studies, *31*
conceptual design, 5–6, 14–67, 153
  examples in case studies, 20–28, 29–40, 41–43, 50–52, 61–63
  factors to be considered, 14–16
concrete
  reactive aggregates in, 124–125
  sulphate attack on, 124
cone penetration test (CPT), 166, 247–252
  electric cone testing, 248, 249–250
  example(s) of use, 24, 27, 36, 62, *65*
  historical background, 247–249
  piezocone used, 250–251
  subtraction cone used, *248*, 250
cone pressuremeter, 255–257
Conlon, R. J., quoted, 153
consolidation, 165
consolidation coefficient, 208
consolidation tests, 173, 204, 209–213
  Casagrande oedometer, 204, 209–213
  hydraulic consolidation cell, 204
  triaxial apparatus, 204
consolidation theory, 205–208

Construction Industry Research and Information Association (CIRIA)
  definition of risk, 142
  risk management software, 158–164
construction materials, reactions with, 123–125
construction precedent, as source of information, 6, *72*, 73, *74*, 97, *99*
continuous consolidation testing, 213
continuous surface wave (CSW) method, 169, 271, 281–283
  advantages and disadvantages, *273*, 278, 280, 281
  compared with other field and laboratory methods, 271, *274*, *282*, *285*
  costs, 288
  equipment, 278, *279*, 281
  example(s) of use, *285*, 288–289, *289*, *290*
contoured maps, in site investigations, 48
conversion factors, *292–294*
Coombe Rock, 80
Cornbrash, 80, 87
cost benefit, aerial photograph compared with digital map, 106
cost overruns
  factors affecting, 147–148
  and site investigation costs, 150–151
cost risks, influence over life of project, *146*
costs
  seismic equipment, 288
  stiffness measurement, *290*
Crews Hill, landslip, *45*
crop marks, 110
cross-hole seismic methods, 166, *169*
  advantages and disadvantages, *272*, 280–281
  compared with other field and laboratory methods, 271, *274*
  cost per measurement, *290*
  equipment used, 278
  example(s) of use, *274*, *287*, 291
cross-hole (seismic) tomography, *272*
cross-sections, in site investigations of slope failures, 47–48
CSW *see* continuous surface wave (CSW) method
cut-off trench, failure of, 42–43
cuttings
  long-term stability, 44–46
  short-term stability, 40–43
  site investigation of, 8
  in stiff fissured clays, 40–46
  triaxial tests, 194–195
cyclic triaxial testing, 197–198
cynicism, need for, 7

data sources, for desk study, *71–72*, *74*, *76–79*
debris flows, 131
Delft cone penetration test, 247–248, 249
desiccation
  assessment of, 55
  factors affecting, 54
design interactions, 153–154
design specification, 152–153
desk study, 13, 68–105
  checklist, *76–79*
  example of information obtained, *14*
  objectives, 68
  overview on how study is done, 73–75
  reason for, 68–69
  sources of information, *71–72*, *74*, *76–79*
  what to look for in, 75, 80–105
detailing, geotechnical design during, 154
development concept, geotechnical experts involved, 155
dial penetrometer, desiccation assessment using, 55
Digital Elevation Model (DEM), 91
digital maps, 89
Digital Terrain Model (DTM), 90
dilatometer, 166, 255, *256*
dispersion curve (surface wave geophysics), 281
  factored wavelength method for inversion, 282–283, *284*
  finite element method for inversion, 283
  inversion/interpretation of, 281–283
  linear models for inversion, 283
dissolution features, 123, 127–128, 135
  observation during walk-over survey, 120
  risks associated with, 135
Docklands Light Railway (DLR)
  construction precedent, 72, 73
  geology, 70, *71*, 72
  hydrogeology, *71*, 72

Docklands Light Railway (DLR) (*continued*)
Lewisham Extension, 69–73
site condition/history, *71*, 72–73
double oedometer test, 56
downthrow, 87
draft report, 312
correction of, *310*, *319*
drainage patterns, sources of information, *77–78*, 115
drained strength, 165
drains and weep holes, landslips caused by blockage, 49
Drammen, Norway, landslip, 48, 50–52
drift deposits, 80–81
Dutch cone penetration test, 247–248
example(s) of use, 24, 27
dynamic triaxial test(s), 197–198
strain control, 197–198
stress control, 198

earthquake frequency spectral range, 197
East Port Said, Egypt, quay wall, 35–40
elastic deformation moduli, 166
determination by plate loading test, 238–240
determination by triaxial test, 195, 199
elastic waves
definitions and terminology, 267–268
see also seismic waves
ELE fixed piston sampler, 175, *176*
electric cone testing, 248, 249–250
electro-osmosis, 59, *60*
embankments
factor(s) of safety for, 194, *222*
site investigation of, 8
on soft clay, 28–40, *222*, 226–233
triaxial tests, 193–194
engineering and architectural records, *99*
English Heritage, records held by, *98*, *99*
environment and planning, sources of information, *74*, *98*
erosion, observation during walk-over survey, 120
evaporite minerals, 134–135
expansive clays, 136–137
classification of, *54*, *137*
exposures (of rock or soil), observations during walk-over survey, 118
fabric inspection (for laboratory samples), 184–185
factor of safety
cuttings/excavations, 40, 195
embankments, 194
foundations, 40
failure mechanisms, search for all possible mechanisms, 7
falls, 129, 132
false colour infrared photography, 107–108
field geophysical stiffness measurement methods, 274–281
choice of survey type, 279–281
equipment, 276–279
field tests, 166, 220–291
cone penetration test, 166, 247–252
cost-per-measurement comparison, *290*
plate loading tests, 233–244
pressuremeters, 166, 252–263
report on, 307
seismic methods, 166–167, 264–291
shear vane tests, 220–233
standard penetration test, 166, 244–247
see also cone penetration test; plate loading tests; pressuremeter tests; seismic methods; shear vane tests; standard penetration test
filter paper method of suction determination, 55
finite element analysis, during excavation, 22–23, 23
fixed piston samplers, 175
example(s) of use, 30, *176*, *185*
floating foundation, 19–20
flooding, assessment of likelihood, *78*, 120
flow slides, 131
flows, 129, 131
Fontwell, West Sussex, collapse subsidence, 127–128, *129*
Fookes' Model, 12–13
footings
on sands, 65–66
triaxial tests, 195–196
Fornebu airport, Oslo, 29–35
borehole log, *31*
field instrumentation, 33–34, *33*, *34*
field pumping tests, 32
soil profile, 30, *31*
fossil river channels, 111–112

foundations
factor of safety, 40
possibilities for multistorey buildings, *19*, 25
on sands, 65–67
site investigations for, 9–10
on stiff fissured clays, 16–28
frost heave, 139

Gabriel, Keith, quoted, 3
Gault, 80, *83*, *137*
Gault clay, pressuremeter test results, *254*
geographic information system (GIS)
photographic images incorporated into, 109
topographical data incorporated into, 89, 91
geological formations, typical names, 318
geological maps
example(s) of use, 85–87, *88*
field notes/slips, 84
for mining desk study, 84
sheet memoirs for, 81n[2], 84
slope instability on, 97, 100, *103*
as sources of information, *74*, 75, *76–79*, 80–82, *83*
strata names in, 76–77
superficial deposits, 80–81
geological model
development of, 12–13
progressive building up of, *14–18*, 75
geology
assessment at desk-study stage, 75
sources of information, *71*, 72, *74*, 75, *76–77*, 80–85
understanding, 7
Geonor piezometers, 52
Geonor torque head (for shear vane test), *230*
geophones, 279, 281
geophysical methods
advantages, 283–287, 290–291
choice of survey type, 279–281
compared with laboratory tests, *274*, 286, *287*
costs, 288
disadvantages, 288
equipment, 276–279
example of information obtained, *16*
see also seismic methods
geotechnical design
accuracy of predictions, 149, *150*
factors affecting, 146–147
improving the design process, 152–154
new methods of working, 151–164
traditional approach, 148–151
geotechnical engineering, requirements for successful practice, 6–8
geotechnical hazards, 127–139
geotechnical risk, managing, 139–164
geotechnical risk system, implementing, 156
geotechnical site investigation
meaning of term, 3
more-effective, 154–156
see also site investigation
Glossop, R., quoted, 4
Golder, H.Q., 295
grading, of soils, 165
Grange Hill, landslip, *45*
granular soils
collapse subsidence in, 127–128, *129*
determination of permeability, 60
prediction of pile bearing capacity, 67
in seismically active areas, 123
Greenwich foot tunnel, *72*, 73
Greenwich railway station, 72
ground conditions
additional costs of unexpected conditions, 147
information about, 79, 97, 100–105
observations during construction, 155
observations during walk-over survey, 118
and risk, 146–148, 149
unforeseen, 155–156
variability, 147, *148*
ground investigation
example of information obtained, *17–18*
meaning of term, 3
phases, 155
speed, 155
ground movements, 134–135
damage caused by, 134
see also heave; settlement; subsidence
ground stiffness
degradation for various geomaterials, *266*
determination using seismic methods, 168–169, 266–274, 281–283
at operational strain levels, 285

ground vibrations, landslips caused by, 49
groundwater
  hazards due to, 124, 127–128
  sources of information, 77–*78*, 120
groundwater level, in London Basin, 72
groundwater lowering, 58–61
  beneficial effects, 59
  factors affecting choice of method, 59
  techniques, 59, *60*
Guildford
  geological map, 82, *83*
  landslip at University of Surrey site, *100*, 101–102, 130
gypsum, dissolution features in, 136

Hadley Wood, landslip, *45*
hazard analysis, 161
hazards
  combination of factors indicating, 122–123
  as health and safety risk, 125–126
  reactions with construction materials, 123–125
  and risk, 142–143
  types, 122
  see also geotechnical hazards
Hazen's formula, 60
Head deposits, 80, 85, 87
heave, 134
  in black shales, 138
  factors causing, 134–135, 136
  prediction of, 166
  see also frost heave
heavy equipment, effects on soil microstructure, 191
historical maps, as sources of information, *74*, 77–*79*, 84, 91–92, 92–95, *98*
historical records, *71*, 72–73, *74*, 92–97, *98–99*
Hodeidah, Yemen, grain silo, 61–63
hollow cylinder apparatus, 166, 173
House of Commons car park, London, 20–24
  borehole log, *21*
  boreholes location plan, *20*
  instrumentation location plan, *24*
  soil profile and properties, *22*
hummocky ground
  aerial photography, 104, *104*, 113
  observation during walk-over survey, *104*, 119
hydraulic fill, sand as, 65
hydrogen sulphide, as hazard, 126
hydrogeology, sources of information, *71*, *72*, *74*
Hythe Beds, 80, *83*, *103*, 104

in situ consolidation characteristics, determination using piezocone, 251
in situ lateral stress, in pressuremeter test, 261
inclinometers, example(s) of use, 23
industrial archaeology, 95, 97
industrial by-products, 137–138
industrial sites, contaminants on, 92, *93*
influence factor, 195, 239
information sources, for desk study, *71–72*, *74*, *76–79*
Institution of Civil Engineers, archives, *99*
intact clays
  landslips in, 50–52
  natural slopes in, 47
International Reference Test Procedures (IRTPs), 245

Japanese standard piston sampler, 175, *176*
Jubilee Line extension, *70*, 73

Kellaways Beds, 85, 87
Keuper Marl, 87
Kilburn, London, pressuremeter test in London Clay, 260–263

laboratory tests, 166, 170–220
  axial tests, 171
  biaxial tests, 172
  compared with geophysical methods, *274*, 286, *287*
  comparison of shear wave velocity measurements with field tests, 186–188
  consolidation tests, 173, 204–213
  cylindrical shear tests, 171
  measurement of initial effective stress in samples, 185
  measurement of strains during reconsolidation, 185–186
  prismoidal shear tests, 172
  report on, 308
  ring shear test, 173, 213–220
  sampling effects, 174–192

simple shear tests, 171–172
soil specimen preparation for, *174*
torsional shear test, 172
triaxial tests, 173, 192–204
types, 170, 171–173
see also oedometer test; ring shear test; triaxial test
land cover, sources of information, 77, 88–91, 114
land use, and aerial photography, 114–115
landforms, and aerial photography, 114
landslips
Bradwell, Essex, 41–42
Brent Knoll trial embankment, 226–228
Drammen, Norway, 48, 50–52
identification of, 132–134
investigation of, 11, 47–50, 97, 100–105
in London Clay, 41–43, *45*, 101
observations during walk-over survey, 119, 132, *133*
possible causes, 49–50
Stag Hill, Guildford, *100*, 101–102
see also slope failures
Laval sampling system, 175
example(s) of use, *176*, *185*
Lernacken, Sweden, surface wave (geophysics) survey, *284*
Light Detection And Ranging (LIDAR) systems, 91
Limehouse Link project, *70*, *72*, 73
lineations, and aerial photography, 115
liquefaction, 63–65, 123
liquid limit, 166
data in case studies, *25*, *232*
liquidity index, 166
data in case studies, 25, *38*
determination of, 211
typical values, 25, 211
local sources of information, *74*, 120–121
London Clay, 80
Basildon, Essex, 95
Bradwell, Essex, 41–42
desiccation assessment of, 55, *56*
first-time slides, *45*
in Guildford area, 82, *83*
House of Commons car park, *21*
in situ permeability, 22
landslip possibilities, 130
measurement of stiffness, *168*
sampling of, 179, *183*
self-boring pressuremeter test in, 260–263
shrinkage potential, *137*
Stag Hill, Guildford, 101, 130
stiffness–depth profiles, *274*, 286, *287*
undrained shear strength, *43*, *181*, *183*
Wraysbury, 42–43
long-term stability, 165
stiff fissured clays, 44–46
Love waves, 268

made ground, observation during walk-over survey, 119–120
magnet extensometers, example(s) of use, 23
main ground investigation, 13
example of information obtained, *17*
matric suction/saturation curve, 200
determination of, 200
Mayne, P.W., 246–247
Mazier rotary coring system, example(s) of use, 37
Ménard pressuremeter, 255
Mercia Mudstone, 87
methane, as hazard, 126
Millennium Map Project, 106
Milton Keynes, geological map, 85–87
mining activity, 83, 84
observations during walk-over survey, 120
subsidence caused by, 136
mining desk study, sources of information for, 84–85
mining instability, 84
mining records, *71*, 85
Mohr–Coulomb failure criterion
for soil loading, 194
for soil unloading, 195
motorway
site investigation for, 8
trial embankment for, 226–233
mudslides, 130

National Audit Office building, London, 69
natural slopes, in clays, 46–47
Necom House, Lagos, Nigeria, 24–28
Nile Delta Clays, 35, 39
North Sea oil pipeline, design, 153–154
Norwegian Geotechnical Institute (NGI), fixed piston sampler, 30, 35, 175, *176*

Norwegian quick clays, 131

oblique aerial photography, *102*, 107, 109, 113
oedometer tests, 173, 204, 209–213
compared with back analysis and seismic surveys, 22–23, *287*
double oedometer test, 56
example(s) of use, 25–26, *26*
on soft clays, 28, 32
on stiff clays, 25–26, *26*
see also Casagrande oedometer test
Opelika test site, seismic piezocone results, *253*
Ordnance Survey products, 89, 90, 92, 116
Osterberg piston samplers, example(s) of use, 36
over-consolidation of clay, determining, 25–26, 210–211
over-consolidation ratio (OCR), in Nile Delta clays, 38–39, *39*
Oxford Clay, 85, 87

Paddington Station, London, historical records, 95, 97, *98–99*
Palmer, D.J., *Writing Reports* monograph, 297–319
parameter determination
by field tests, 166, 220–291
by laboratory tests, 166, 170–220
peak strength, 213
and residual strength, 215
permeability, of clays, 22, 165
Perth, Western Australia, sands, 286
pessimism, need for, 7
photographs, in site investigations, 48
piezocones, 250–251
example(s) of use, 36, 65, 251
limitations, 29
piezometers, example(s) of use, 22–23, 23, *34*, *48*, 51–52, *52*
pile bearing capacity, prediction of, in granular soils, 67
pile capacity, observed vs predicted, *150*
planning, sources of information, *74*, *98*
plastic limit, 166
data in case studies, *25*, *232*
plasticity index, 166
correlation with empirical shear vane correction factor, 221, *223*
correlation with factor of safety, *222*
data in case studies, *38*, *232*
expansiveness of clays related to, *54*
volume change potential related to, *137*
plate loading tests, 233–244
areas of application, 233
constant-rate-of-penetration (CRP) test, 236
cost per measurement, *290*
depth reduction factors, 239, *240*
diameter of plate, 237–238
down-hole tests, 243–244
elastic moduli determined by, 238–240
equipment, *234–235*, *237*
maintained-incremental-load (ML) test, 236
method, 233–238
results, 238–243
settlement prediction using, 241–243
under-plate settlement measurement system, 241, *242*
pneumatic piezometers, example(s) of use, 22–23, 23
Poisson's ratio, 166
and shear modulus, 199
pore water pressure
distributions, 19, 53
measurement of, 29, 34, 250–251
potential hazards, need to look for, 7
Potters Bar, landslips, *45*
pre-bored pressuremeter testing, *254*, 255
precedents, need for knowledge of, 6
pre-consolidation pressure, 166, *210*
Casagrande construction for determining, *210*, 211
determination of, 210–211
pressuremeter test, 166, 252–263
advantages, 259–260
cost per measurement, *290*
disadvantages, 260
example(s) of use, 260–264
insertion systems, 253–258
interpretation of results, 253, *254*, 256
pressuremeters
calibration of, 259
construction, 258–259
pre-bored, *254*, 255
pushed-in, *254*, 255–257
self-boring, *254*, 257–258
procurement, 11–12

progressive failure, 224–225
proof-correction signs, *319*
public utilities, as sources of information, 121
pushed-in pressuremeter testing, *254*, 255–257
pyrites, oxidation of, 138

quarrying activity, 120, 123

radon, as hazard, 126
rainfall, landslips caused by, 49
Rayleigh (R) waves, 167, 268, 269–270
reactive aggregates, 124–125
Reading Beds, 80, 82, *83*, *137*
references listed
  in report, 309
  for this book, 321–340
report
  abbreviations in, 316
  abstract, 304, *305*
  appendices, 309
  calculations, 313
  checking, 314
  conclusions, 309
  description of borings, 306–307
  description of site, 306
  discussion, 308–309
  draft, *310*, 312, 319
  field tests, 307
  figures, 309, 312
  framework and contents of, 301, 303–304
  introduction, 306
  laboratory tests, 308
  purpose of writing, 299
  references listed, 309
  style, 312–314, 315–316
  table of contents, *303*, 304
  tables, 309
  title page, *302*, 304
  typography and layout, 315
report writing, 295–319
residual factor, 219–220, 225
residual risk analysis, 163, *164*
residual strength, 213–216
  definition, 219
  determination of, 216
  meaning of term, 214
  and peak strength, 215
resonant column apparatus, 166, 168
  example(s) of use, *188*
reversing shear box tests, residual strength determined by, 216
ring shear test, 166, 213–220
  apparatus, 216–219
  Bishop's apparatus, *216*, 217
  Broomhead's apparatus, 217, *218*
  definitions, 219–220
risk
  to contractor's investment, 145–146
  effects, 143
  and hazard, 142–143
  meaning of term, 142
  measurement of, 142
  and opportunities, 143
  ownership of, 144
  spreading and transfer of, 145
  true cost of, 144
risk assessment, 161–162
risk control procedures
  effects, 145
  need to update, 144
risk control strategies, software-supported development of, 162–163
risk identification, 161
risk management
  appropriate levels within business context, 160, *161*
  cost–benefit considerations, *146*, 147
  project success affected by, 146
  systematic approach, 141–142, 143, 144, 146
risk management process
  information repositories, 157–158, *157*
  main steps, *157*
  software-supported, 158–164
risk rating, 142, 162, *163*
risk register, 156–157
  software-supported, 159–164
rockfalls, 132
rocks, plate loading test, 233
Royal Commission on the Historical Monuments of England (RCHME)
  aerial photographs collection, 116
  records held by, *98*

sample quality
  assessment of, 184–188
  dangers in improving, 188–190

sampling effects, 174–192
  and disturbance by ground works, 190–192
  in sand(s), 57–58, 182, 184, 186–188
  in soft clays, 174–179
  in stiff clays, 179–182
sampling tubes
  comparison, 175–176, 176–179, *183*
  thick-walled, 174–175, 179, *183*
  thin-walled, 36–37, 175, 176, 177–178, 179, *183*
sand(s)
  foundations on, 65–67
  freezing prior to sampling, 184
  as hydraulic fill, 65
  potential geotechnical problems, *58–65*
  response to geotechnical processes, 61–63
  sampling of, 57–58, 182, 184, 186–188
SASW *see* spectral analysis of surface waves (SASW) method
season, effects on aerial photography, 109–110
secant modulus, in pressuremeter test, 262–263
seismic cone penetration test, 166–167, 169, 251–252, 270–271
  advantages and disadvantages, 270–271, *272*, 280
  compared with other field and laboratory methods, 271, *274*, *290*
  cost per measurement, *290*
seismic methods, 166–167, 264–291
  advantages, 290–291
  choice of survey type, 279–281
  compared with other field and laboratory methods, 271, *274*, *282*, *285*
  continuous surface wave (CSW) method, 169, 271, *273*, 278, *279*, 281–283, *285*, 288–289, *290*
  costs, 288, *290*
  cross-hole methods, 166, *169*, *272*, 280–281, *287*, *290*
  cross-hole tomography, *272*
  down-hole methods, 166, *169*, *272*, 280
  equipment used, 277–278
  ground stiffness measured using, 168–169, 266–274, 281–283
  reflection method, *169*, *273*
  refraction method, *169*, *272*, 279
  spectral analysis of surface waves (SASW) method, 169, 271, *273*, 278, 281, 288
  stiffness–depth profiles determined using, 270–271, *272–273*, 274, 281–283
  sub-surface methods, *169*, *272*
  surface methods, *169*, 271, *272–273*, 280, 281–283
  up-hole shear methods, 166, *169*, *272*
seismic piezocone testing, 252
  typical results, *252*, *253*
seismic waves
  body waves, 267
  surface waves, 268
self-boring pressuremeter (SBP), *254*, 257–258
  construction, 258–259
  depth of penetration, 258
  example(s) of use, 260–263, 286
  types of material suitable for, 257
settlement, 134, 136
  compared with subsidence, 135
  observed vs predicted, 149, *150*
  prediction of, 29, 67, 166, 195, 207–208, 241–243
settlement analysis
  Skempton–Bjerrum correction factor, 211–213
  and triaxial test, 195–196
settlement observations
  data in case study, *27*
  field instrumentation for embankments, *34*
Sevenoaks, Kent, solifluction lobes, 102–105, 131
shear modulus, 166
  determination from shear wave velocity measurements, 199, 269
shear strain
  at limit of elastic behaviour, *265*
  levels in ground around structures, 264
  variation of stiffness with, *168*
shear strength, 165
  see also undrained shear strength
shear vane tests, 220–233
  anisotropy correction factor, 223, 225–226, 232

at Brent Knoll trial embankment, 228, 229–231
Bjerrum's correction factor, 221, 226, 232
compared with laboratory tests, 231
correlations with embankment stability, 221–223, 225–233
diamond-shaped vanes used, 229, *230*
example(s) of use, 30, 35, *48*, 51
shear wave velocity, 252, 269
factors affecting, 186
shear wave velocity measurement
comparison of field and laboratory tests, 186–188
field methods, 166–167, 169, 252, 271
laboratory method, 199
Shelby (sampling) tube, 176
example(s) of use, 36, *176*, *181*, *183*
Sherbrooke down-hole block sampler, 175, *176*
short-term stability, 165
stiff fissured clays, 40–43
shrinking of clay, factors affecting, 53, *137*
similitude, 7
Singapore marine clay
compensation grouting in, 191
sampling of, 174–175
sinkholes, 123, 127–128
site history, sources of information, *71*, 72–73, *74*, *78–79*, 95, 97, *98–99*, 105
site investigation
BS Code of Practice, 154–155
Clayton's comments, 140–141
information and activity flow chart, *5*
meaning of term, 3
objectives, 4
performance indicators, 3–4
for slope failures, 47–50
stages, 4–*5*
success factors, 3
site investigation planning, 4–6
borehole layout and spacing, 8–11
case study examples, 20–28, 29–40, 41–43, 50–52, 61–63
conceptual design and case studies, 14–67
development of geological model, 12–13
and procurement, 11–12
Skempton, A.W.
Tokyo Special Lecture (1977), 44, 46
on triaxial tests, 192
Skempton–Bjerrum correction factor, 211–213
skip test, 244
slags, 137–138
slides, 129, 130–131
Slindon Sand formation, 128
slope angles, measurement during walk-over survey, 118
slope failures
factors triggering, 133–134
in intact clays, 47
site investigations for, 47–50
in stiff fissured clays, 40–43, 46–47
slope instability, 129–135
falls, 129
flows, 129, 131
investigation of, 6, 11, 97, 100–105, 123
observations during walk-over survey, 119, 132, *133*
rockfalls, 132
slides, 129, 130–131
see also landslips
small-strain shear modulus, 186, 252
measurement of, 169
snowfall, aerial photography after, 110
soft clays
construction on, 28–40
effects of strains at periphery of samples, 179, *180*
landslips in, 50–52
sampling effects in, 174–179
stability of embankments on, 28–40, *222*, 226–233
water content distribution in tube samples, *180*
software-supported risk management tool, 158–164
outputs, 164
step 1: setting up risk management process, 159–161
step 2: identifying the risks, 161, *162*
step 3: assessing the risks, 161–162, *163*
step 4: setting up the response/action plan, 162–163, *164*
step 5: responding using the action plan, 164
soil, particle size distribution, 165
soil marks (in aerial photography), 111, *112*
soil mechanics procedures, familiarity with, 6–7

soil moduli, stress path dependency, 167
soil profiling, piezocone used in, 251
soil stiffness, measurement of, 168, 169
soil suction measurements, 55
soil type, and cone tests, *65*, 251
soil–water characteristic curve, 200
solifluction, 131
solifluction lobes, 102–105, 131
spectral analysis of surface waves (SASW) method, 169, 271, 278
  advantages and disadvantages, *273*, 281
  costs, 288
spectral sensitivity, aerial photography, 105–106
stability, *see also* long-term stability; short-term stability
Stag Hill, Guildford, landslip, *100*, 101–102, 130
standard penetration test (SPT), 166, 244–247
  abandonment suggested, 246–247
  compared with geophysical methods, *275*
  cost per measurement, *290*
  example(s) of use, 36, *62*
  factors affecting accuracy, 246, 283, *286*
  historical perspective, 244–246
  International Reference Test Procedure, 245
  national standards, 245–246
  settlement prediction using, 274, *275*
  stiffness determined using, 270, *275*, 283, *286*
standpipes, example(s) of use, 22, 23
stationary piston sampling tubes, example(s) of use, 50
stiff clays
  Bradwell landslip, 41–42
  cuttings in, 40–46
  effect of sampling types on failure envelope, 181, *183*
  effects of strains at periphery of samples, 180, *182*
  foundations on, 16–28
  natural slopes, 46–47
  sampling effects in, 179–182
  Wraysbury landslip, 42–43
stiffness–depth profiles, determination by seismic methods, 270–271, *272–273*, 274, 281–283, *289*, *290*
strain path method, applied to tube sampling, 176–179
strain softening, 214
  and progressive failure, 224, 225
stress path method, 167
  and triaxial test, 196–197
structures, observation during walk-over survey, 119
subdued topography, 113
subsidence, 134, 135–136
  compared with settlement, 135
Sudbury Hill, landslips, *45*
superficial deposits, 80–81
supplementary ground investigation, 13
  example of information obtained, *18*
surface water, sources of information, *77–78*, 120
surface wave (seismic) methods, 271
  advantages and disadvantages, *273*, 278, 280, 281
  continuous surface wave (CSW) method, 169, 271, *273*, 278, *279*, 281–283, *285*, 288–289, *290*
  costs, 288, *290*
  dispersion curve, 281
  equipment used, 278–279, *279*
  spectral analysis of surface waves (SASW) method, 169, 271, *273*, 278, 281, 288
stiffness profiling using, 281–283
surface waves (seismic waves), 268
  Love waves, 268
  Rayleigh (R) waves, 167, 268
Swanscombe, quarry rehabilitation, 288, *289*
swelling of clays, 165
  and consolidation test, 209–210
  factors affecting, 53, 95, 122
swelling soils, 136–138
systematic (engineering) design, 152, *153*
systematic risk management, 141–142, 143, 146
  value added by, 156–157

terrace deposits, 81, 85
Terzaghi, K.
  consolidation theory, 205–208
  quoted, 148–149
topographical maps
  products available in UK, 89–90

as source of information, *74*, *77–79*, *86*, 88–91
superimposed on aerial photography, 88, *89*
topography, sources of information, *74*, *77*, 88–91, *98*
toxic gases, 126
triaxial test(s), 166, 173, 192–204
apparatus described, 192, 201–203
application to settlement, 195–196
application to soil loading, 193–194
application to soil unloading, 42–43, 194–195
application to stress path method, 196–197
bender elements in, 199, 203, *285*
compared with back analysis and seismic surveys, *287*
cyclic, 197–198
drawbacks, 192
dynamic, 197–198
historical perspective, 192
interpretation, 203
modern adaptations, 203–204
residual strength determined by, 216
screw pump for, *201*, 202–203
strain transducers on test specimen, *202*, 203
stresses, 193
tube sampling
in soft clays, 174–179
strain path method applied to, 176–179

U100 sampler, 176–177
example(s) of use, 43, *181*, *183*
penetration test using, 244
undrained shear strength
data in case studies, *22*, *31*, *38*, *43*, *44*, *51*
determination of, 50, 51, 220, 223, 261–262
London Clay, *43*, *44*, *181*, *183*
Singapore marine clay, *175*
undrained strength, 165
units and multiples, *292–294*
University of Surrey site, landslip at, *100*, 101–102, 130
Upper Chalk, *83*, 128
measurement of stiffness, *168*

value management, 144
vane measurements *see* shear vane tests
vegetation
effects on clays, 49, 53–56, 95, 122
and false colour infrared photography, 107–108
observation during walk-over survey, 118–119
vertical aerial photography, *101*, *104*, 107, 109
Victoria railway station, London, 69

walk-over survey, 13, 117–121
equipment needed, 117–118
example of information obtained, *15*
features to be looked for, 49, 118–120, 132, *133*
and local sources of information, 120–121
objectives, 117
overview, 117
Weald Clay, *83*, *103*, 104, *137*
weathering, 138–139
well-point pumping, 59, *60*
West Acton, landslip, *45*
Willen Fault, 87
witness reports, 48–49
Wraysbury
cut-off trench, 42–43
strength–effective stress relationships for blue London Clay, *215*
stress ratio–displacement relationships for blue London Clay, *217*

yield point, 166, *210*
Young's modulus, 166
and shear modulus, 199